Lecture Notes in Mathematics

Edited by A. Dold and B. Eckmann

1303

L. Accardi W. von Waldenfels (Eds.)

Quantum Probability and Applications III

Proceedings of a Conference held in Oberwolfach, FRG, January 25–31, 1987

Springer-Verlag
Berlin Heidelberg New York London Paris Tokyo

Editors

Luigi Accardi
Dipartimento di Matematica, Università di Roma II
Via Orazio Raimondo, 00173, Roma, Italy

Wilhelm von Waldenfels
Institut für Angewandte Mathematik, Universität Heidelberg
Im Neuenheimer Feld 294, 6900 Heidelberg, Federal Republic of Germany

Mathematics Subject Classification (1980): 43A35, 46LXX, 46M20, 47DXX, 60FXX, 60GXX, 60HXX, 60JXX, 80A05, 81C20, 81K05, 81L05, 81M05, 82A15, 83C40, 83C75

ISBN 3-540-18919-X Springer-Verlag Berlin Heidelberg New York
ISBN 0-387-18919-X Springer-Verlag New York Berlin Heidelberg

Printing and binding: Druckhaus Beltz, Hemsbach/Bergstr.
2146/3140-543210

Introduction

This volume contains the proceedings of the first Quantum Probability meeting held in Oberwolfach which is the fourth of a series begun with the 1982 meeting of Mondragone and continued in Heidelberg ('84) and in Leuven ('85). The main topics discussed during the meeting were: quantum stochastic calculus, mathematical models of quantum noise and their applications to quantum optics, the quantum Feynman-Kac formula, quantum probability and models of quantum statistical mechanics, the notion of conditioning in quantum probability and related problems (dilations, quantum Markov processes), quantum central limit theorems.

We are grateful to the Mathematisches Forschungsinstitut Oberwolfach and to its director Prof. M. Barner for giving us the unique opportunity of scientific collaboration and mutual exchange.

We would like to thank also the speakers and all the participants for their contributions to the vivid and sometimes heated discussions.

L. Accardi
W. v. Waldenfels

CONTENTS

A NOTE ON MEYER' S NOTE

Luigi Accardi
Dipartimento di Matematica
Universita' di Roma II

(1.) NOTATIONS AND STATEMENT OF THE PROBLEM

Let us denote
- $\Gamma(L^2(\mathbf{R}_+)$ the Boson Fock space over the one-particle space $L^2(\mathbf{R}_+)$
- $\mathcal{E} = \{\psi(f) : f \in L^2(\mathbf{R}_+)\}$ the set of exponential vectors in $\Gamma(L^2(\mathbf{R}_+)$.
- $\Phi = \psi(0)$ the vacuum state in $\Gamma(L^2(\mathbf{R}_+)$.
- $\Gamma(\chi_{[0,t]})$ the orthogonal projector defined by

$$\Gamma(\chi_{[0,t]})\psi(f) = \psi(\chi_{[0,t]}f)$$

- $\Phi_{t]} := \Gamma(\chi_{[0,t]})\Phi$; $\Phi_{[t} := \Gamma(\chi_{[t,\infty)})\Phi$
- W(f) $f \in L^2(\mathbf{R}_+)$ the Weyl operator characterized by the property

$$W(f)\psi(g) = e^{-\frac{\|f\|^2}{2} - <f,g>}\psi(f+h)$$

- A , A^+ , N the annihilation, creation and number (or gauge or conservation) fields defined, on $\mathcal{E}$ by the relations :

$$A(f)\psi(g) = < f, g > \psi(g)$$

$$A^+(f)\psi(g) = \frac{d}{dt}|_{t=0}\ \psi(g + tf)$$

$$N_t\psi(g) = \frac{d}{ds}|_{s=0}\ \psi(e^{s\chi_{[0,t]}}g)$$

we write N(s,t) for $N_t - N_s$.

The W(f) are unitary operators on $\mathcal{H}$ satisfying the CCR

$$W(f)W(g) = e^{-\frac{\|f\|^2}{2} - <f,g>}\psi(f+h)$$

- $f_{t]} = \chi_{[0,t]}f$; $f_{[t} = \chi_{[t,\infty)}f$
- H_o a complex Hilbert space , called the initial space.
- $\mathcal{H} = H_o \otimes \Gamma(L^2(\mathbf{R}_+)$
- $\mathcal{H}_{t]} = H_o \otimes \Gamma(L^2([0,t]) \otimes \Phi_{[t}$
- $\mathcal{B} = \mathcal{B}(\mathcal{H}) = \mathcal{B}\left(H_o \otimes \Gamma(L^2(\mathbf{R}_+)\right)$
- $\mathcal{B}_{t]} = \mathcal{B}(H_o \otimes \Gamma(L^2([0,t])) \otimes 1_{t]}$
- $\mathcal{B}_{[t} = \mathcal{B}(1_{H_o} \otimes 1_{t]} \otimes \Gamma(L^2([t,\infty))$
- θ_t the shift on $L^2(\mathbf{R}_+)$.
- $\sigma_t = \Gamma(\theta_t)$ the shift on $\Gamma(L^2(\mathbf{R}_+)$
- $u_t^o = \imath_o \otimes \sigma_t$ the free time shift on $\mathcal{B}$. where $\imath_o$ is the identity map on $\mathcal{B}(H_o)$.

The objects described above provide a simple and , for a certain class of models , canonical example of a quantum Markov process (in fact also of a quantum independent increment process in the sense of [2]) and the Feynman-Kac formula allows to perturb such structures by means of unitary cocycles (for the free time shift) giving rise to new quantum processes [1]. In particular, the generator L of the quantum Markovian semigroup canonically associated to the new process is the sum of the generator L_o of the semigroup associated to the original process and of an additional perturbative piece, denoted L_I. The problem with the above class of quantum Markov processes is that the free time shift u_t^o acts trivially on the initial algebra and therefore the corresponding semigroup is zero so that, as remarked in [1], in this case one is in fact dealing with FK perturbations of the identity semigroup. As a consequence of this one looses one of the most attractive analytical advantages of the classical FK formula, namely the possibility of dealing with perturbations L_I so singular that the operator $L_o + L_I$ is not well defined (a typical example is the possibility of giving a meaning, via the FK formula, to the formal generator $-\triangle + V$ where $\triangle$ is the Laplacian on $\mathbf{R}^n$ and V is a highly singular potential). By analogy with the classical case we would like to have a time shift v_t^o wich shifts also the initial random variables (observables) and not only the increments. Moreover, in order to be able to apply the quantum FK perturbation technique, the free time shift should be such that the structure of the associated unitary cocycles should be determined quite explicitly, which is rarely the case if this time shift is itself a Feynman-Kac perturbation of u_t^o. This problem was posed by A. Meyer during the Obervolfach meeting and in the following I want to outline a possible general scheme for a solution and illustrate it with an example.

(2.) A POSSIBLE SCHEME FOR A SOLUTION

Let us look for a time shift v_t^o of the form

$$v_t^o = j_t \otimes \sigma_t : \mathcal{B} = \mathcal{B}(\mathcal{H}) = \mathcal{B}\left(H_o \otimes \Gamma(L^2(\mathbf{R}_+))\right) \cong \mathcal{B}(H_o) \otimes \mathcal{B}\left(\Gamma(L^2(\mathbf{R}_+))\right) \longrightarrow \mathcal{B} \tag{2.1}$$

where

$$j_t : \mathcal{B}(H_o) \cong \mathcal{B}(H_o) \otimes 1 \longrightarrow \mathcal{B}_{t]} \tag{2.2}$$

is a $*$-homomorphism. For all $a_o \in \mathcal{B}(H_o)$, $b \in \mathcal{B}(\Gamma(L^2(\mathbf{R}_+))$ one has

$$v_s^o v_t^o(a_o \otimes b) = j_s \otimes \sigma_s(j_t(a_o) \otimes \sigma_t(b)) = (j_s \otimes \sigma_s)(j_t(a_o)) \cdot (j_s \otimes \sigma_s)(1_o \otimes \sigma_t(b)) \tag{2.3}$$

$$= (j_s \otimes \sigma_s)(j_t(a_o)) \cdot (1_{t]} \otimes \sigma_{s+t}(b))$$

and since we want v_t^o to be a 1-parameter semi-group of $*$-endomorphisms, it follows that, for any $a_o \in \mathcal{B}(H_o)$, $b \in \mathcal{B}(\Gamma(L^2(\mathbf{R}_+))$ the right hand side of (2.3) must be equal to

$$v_{s+t}^o(a_o \otimes b) = j_{s+t}(a_o) \otimes \sigma_{s+t}(b) \tag{2.4}$$

Thus v_s^o will be a 1-parameter semi-group if and only if

$$(j_s \otimes \sigma_s)(j_t(a_o)) = j_{s+t}(a_o) \qquad \forall a_o \in \mathcal{B}(H_o) \tag{2.5}$$

Here we give an example of a j_t satisfying condition (2.5) above. First shall we give the expression of j_t in unbounded form and then shall write the corresponding bounded form.

In the notations of Section (1) choose $H_o = L^2(\mathbf{R})$ with a_o, a_o^+ denoting the usual annihilation and creation operators. Define

$$j_t(a_o^\varepsilon) = a_o^\varepsilon + X_{[0,t]} \qquad ; \qquad a_o^\varepsilon = a_o \text{ or } a_o^+ \tag{2.6}$$

with $(s,t) \mapsto X_{[0,t]}$ a σ-homogeneous normal additive process, i.e.

$$X_{[0,t]} = X_{[0,t]}^+ \hat{\in} 1_{H_o} \otimes \mathcal{B}\left(\Gamma(L^2([0,t]))\right) \qquad ; \qquad [X_{[0,t]}, X_{[0,t]}^+] = 0 \tag{2.7}$$

$$X_{[r,s]} + X_{[s,t]} = X_{[r,t]} \qquad ; \qquad r < s < t \tag{2.8}$$

$$\sigma_r(X_{[s,t]}) = X_{[s+r,t+r]} \tag{2.9}$$

$$[X_{[r,s]}, X_{[u,t]}] = 0 \qquad if\ (u,t)\cap(r,s) = \emptyset \tag{2.10}$$

where [. , .] denotes , as usual , the commutator . In bounded form and under the additional assumption that $X_{[0,t]}$ is self-adjoint, j_t can be defined on the Weyl operators on H_o by :

$$j_t(W_o(z)) = j_t(\exp i(za_o^+ + z^+a_o)) = \exp i(zj_t(a_o^+) + z^+j_t(a_o)) = \tag{2.11}$$

$$= e^{i(za_o^+ + z^+a_o) + (2Rez)X_{[0,t]}} = W_o(z)e^{i(2Rez)X_{[0,t]}}$$

An example of $X_{[0,t]}$ sastisfying the required conditions is the momentum operator $P(\chi_{[0,t]})$. Another example is $X_{[0,t]} = W_t - W_o$, which gives the usual free shift in Wiener space (not only in the increment space cf. Meyer ' s contribution to these proceedings) . Other examples could be obtained using the position or number processes or mixtures of them i.e., Weyl shifts of the form :

$$j_t(W_o(z)) = W_o(z)W(z\chi_{[0,t]}; e^{itz\chi_{[0,t]}}) \tag{1.12}$$

(cf. the remark at the end of this note).

(3.) THE SEMI-GROUP ASSOCIATED TO THE CHOICE $X_{[0,t]} = P(\chi_{[0,t]})$

The semi-group P_o^t , associated to the "free" evolution v_t^o is

$$P_o^t = E_{o]}(v_t^o(a_o)) \qquad ; \quad a_o \in \mathcal{B}(H_o) \tag{3.1}$$

$$E_{o]} = i_o \otimes < \Phi, (\ .\ )\Phi > : \ \mathcal{B}(H_o)\otimes\mathcal{B} \longrightarrow \mathcal{B}(H_o) \cong \mathcal{B}(H_o)\otimes 1 \tag{3.2}$$

In our case, choosing $b = W_o(z)$ $(z \in \mathbf{C})$ and $X_{[0,t]} = P(\chi_{[0,t]})$, one finds

$$P_o^t(W_o(z)) = E_{o]}(v_t^o(W_o(z))) = E_{o]}\left(W_o(z)e^{i(2Rez)P(\chi_{[0,t]})}\right) = W_o(z)e^{-2(Rez)^2t} \tag{3.3}$$

Hence the Weyl operators are in the domain of the generator of P_o^t and one has:

$$P_o^t = \exp tL \tag{3.4}$$

with

$$L(W_o(z)) = -2(Rez)^2W_o(z) \qquad ; \qquad z \in \mathbf{C}) \tag{3.5}$$

The explicit form of the generator can be obtained with the following semiheuristic, considerations:

$$W_o(z) = \exp i(za_o^+ + z^+a_o) \tag{3.6}$$

therefore

$$\frac{\partial}{\partial a_o^+}W_o(z) = (iz)W_o(z) \qquad ; \qquad \frac{\partial}{\partial a_o}W_o(z) = (iz^+)W_o(z)$$

hence

$$\left(\frac{\partial}{\partial a_o^+} + \frac{\partial}{\partial a_o}\right)W_o(z) = i(2Rez)W_o(z)$$

and therefore

$$\left[\frac{1}{\sqrt{2}}\left(\frac{\partial}{\partial a_o^+} + \frac{\partial}{\partial a_o}\right)\right]^2 - r \tag{3.7}$$

Now, from

$$[a_o\ ,\ a_o^+] = 1 \tag{3.8}$$

we deduce

$$[a_o\,,\,.\,] = \frac{\partial}{\partial a_o^+} \qquad ; \qquad [a_o^+\,,\,.\,] = -\frac{\partial}{\partial a_o} \tag{3.9}$$

In conclusion

$$L = \left[\frac{1}{\sqrt{2}}\left(\frac{\partial}{\partial a_o^+} + \frac{\partial}{\partial a_o}\right)\right]^2 = \left[\frac{1}{\sqrt{2}}\left([a_o\,,\,.\,] + [a_o^+\,,\,.\,]\right)\right]^2 = \frac{1}{2}[a_o - a_o^+\,,\,.\,]^2 \tag{3.10}$$

and since $\frac{1}{\sqrt{2}}[a_o - a_o^+\,,\,.\,]$

$$L = -[p\,,\,.\,]^2 = -[p\,,\,[p\,,\,.\,]\,] \tag{3.11}$$

So the free semigroup is a quasifree semigroup of the type considered by Lindblad in [4]. Notice moreover that, if f is a smooth function and M_f is multiplication by f in $L^2(\mathbf{R})$, then with the identification $p = \frac{1}{i}\frac{\partial}{\partial x}$ one has

$$[p, M_f] = \frac{1}{i} M_{(\frac{\partial}{\partial x} f)} \tag{3.12}$$

hence

$$-[p\,,\,[p\,,\,M_f\,]\,] = M_{(\frac{\partial^2}{\partial x^2} f)} \tag{3.13}$$

which gives the right answer when we restrict our attention to the classical Wiener process.

(4.) v_t^o- Markovian cocycles : an example

Consider the stochastic differential equation (SDE)

$$dU = \Big((L_o + X_{[0.t]})dA^+ - (L_o^+ + X_{[0.t]})dA + Zdt\Big)U \tag{4.1}$$

the unitarity condition (using the Fock Ito table for dA , dA^+) is

$$Z = iH - \frac{1}{2} \mid L_o^+ + X_{[0.t]} \mid^2 \tag{4.2}$$

with H self-adjoint. By shifting the equation (4.1) , with H = 0 , with the free shift v_s^o, we obtain the equation for $v_s^o(U_t)$ namely

$$dv_s^o(U_t) = \Big((L_o + X_{[0.t+s]})dA_s^+(t) - (L_o^+ + X_{[0.t+s]})dA_s(t) - \frac{1}{2} \mid L_o^+ + X_{[0.t+s]} \mid^2 dt\Big)v_s^o(U_t) \tag{4.3}$$

where we have used the notation

$$dA_s(t) = A(s + t + dt) - A(s + t) \tag{4.4}$$

which means that, by definition

$$\int_0^T Y_{t+s} dA_s(t) := \int_s^{s+T} Y_t dA(t) \tag{4.5}$$

Now, written in integral form, the equations (4.1) and (4.3) look respectively like

$$U_t = 1 + \int_0^t \Big((L_o + X_{[0.r]})dA^+(r) - (L_o^+ + X_{[0.r]})dA(r) - \frac{1}{2} \mid L_o^+ + X_{[0.r]} \mid^2 dr\Big)U_r \tag{4.6}$$

$$v_s^o(U_t) = 1 + \int_0^t \Big((L_o + X_{[0.s+r]})dA^+(s+r) - (L_o^+ + X_{[0.s+r]})dA(s+r) - \frac{1}{2} \mid L_o^+ + X_{[0.s+r]} \mid^2 dr\Big)v_s^o(U_r) \tag{4.7}$$

So that $v_s^o(U_t) \cdot U_s$ and U_{s+t} satisfy the same SDE (in the t-variable) with the same initial condition i.e. U_s at t = 0. Therefore , if the $X_{[0,t]}$-process is regular enough to assure the existence and uniqueness of the solutions of the above SDE , it will follow that

$$v_s^o(U_t) \cdot U_s = U_{s+t} \tag{4.8}$$

which means that U_t is a v_t^o - Markovian cocycle . The formal unitarity of follows from (4.2) and in many interesting cases the unitarity can be effectively proved . Having the unitary cocycle, we can apply the FK perturbation scheme to the free semigroup associated to v_t^o . Denoting $\mathcal{L}_o$ the generator of this semigroup, a simple calculation shows that the formal generator of the perturbed semigroup

$$P^t(b_o) := E_{o]}\left(U_t^* \cdot v_t^o(b_o) \cdot U_t\right) \tag{4.9}$$

will be

$$\mathcal{L}_o + L_o^+ b + bL_o + L_o^+ bL_o \tag{4.10}$$

If the operators $\mathcal{L}_o$, L_o are unbounded, the expression (4.10) will not be in general a well defined operator However, for $X_{[0,t]}$ as in Section (2) , the operators $L_o + X_{[0,t]}$; $L_o^+ + X_{[0,t]}$ are always well defined and therefore equation (1) makes sense and in some cases the conditions for the existence of a solution of this equation are much weaker than those which allow to realize $\mathcal{L}_o + L_o^+ b + bL_o + L_o^+ bL_o$ as a well defined operator.

Applying the considerations above to the additive functional $X_{[0,t]} = P(\chi_{[0,t]})$, for which the regularity problems mentioned above can be solved with standard techniques, one can produce singular perturbations of the noncommutative Laplacian in full analogy with the classical case.

The physical meaning of the operator $X_{[0,t]}$ in (4.1) can be understood in terms of Barchielli' s analysis [3] : $X_{[0,t]}$ is the input field (the number process in Barchielli 's paper) which interacts with an apparatus described by the operators L_o, Z in (4.1) . The free evolution of the system is given by the time shift v_t^o and equation (4.1) describes the interaction cocycle giving the evolution of the observables of the coupled system (input field + apparatus) according to the scheme proposed in [1] . The advantage of the present approach with respect to [3] is that, due to the v_t^o-cocycle property of the solution of (4.1), the interacting evolution $x \mapsto v_t^o(U_t^+ x U_t)$ will now be a 1-parameter automorphism group, in agreement with the basic principles of quantum physics.

BIBLIOGRAPHY

1) ACCARDI L. Quantum stochastic processes in : Statistical Physics and Dynamical Systems (Rigorous results) J.Fritz, A.Jaffe,D.Szasz (eds.) Birkhauser 1985

2) ACCARDI L., SCHURMANN M., von WALDENFELS W.: Quantum independent increment processes on super algebras. Heidelberg preprint 1987

3) BARCHIELLI A. Input and output channels in quantum systems and quantum stochastic differential equations. These Proceedings.

4) LINDBLAD G. Brownian motion of a quantum harmonic oscillator. Reports on Math. Phys. 10, (1976), 393-406.

4) MEYER P.A. A note on shifts and cocycles. These Proceedings.

STOCHASTIC INTEGRATION

Luigi Accardi Dipartimento di Matematica Universita' di Roma II

Franco Fagnola Dipartimento di Matematica Universita'di Trento

(0.) Introduction.

The programme of developing a "representation free" stochastic calculus was outlined in [1] . In the present note, which is part of a joint work in preparation with J.Quagebeur, we concentrate on the first step of this programme: the definition of stochastic integral. We have chosen to take as our starting point the axiomatic definition of semimartingale adopted by Letta [4] and based on Dellacherie's characterization of the classical semimartingales (cf. [5] and Definition (3.1) below) which has the advantage of looking almost the same in the classical and in quantum case and of not depending on the detailed structure of the Hilbert spaces on which the operators act or of the algebras to which these operators are affiliated. It turns out that in the quantum , as in the classical, case a semimartingale is the most general object for which a stochastic integral with meaningful properties can be defined. However, being at the moment very far from having anything like a quantum version of Dellacherie's theorem, the best we can hope for is to find some sufficient condition, for an additive process to be a semimartingale, which is at the same time easily applicable and sufficiently general to cover all the known cases (and at least some new ones). In Theorem (3.3) such a sufficient condition is proposed; in Section (4.) we show the connection between our notion of stochastic integrals and the Hilbert space valued stochastic integrals in the sense of Metivier and Pellaumail [6] ; in Section (5.) we show that the basic integrators appearing in the varoius quasi-free representation of the CCR or CAR over $L^2(\mathbf{R}_+)$ are semi-martingales in the sense of Definition (3.1); finally, in Section (6.) , we prove an existence and uniqueness theorem for stochastic differential equations.
As mentioned in Remark (3.), after Definition (3.1), the sufficient condition introduced in definition (3.3), is not yet general enough to include all the examples and applications we have in mind. However in the same Remark(3.) we show how this condition has to be modified in order to achieve this goal. It can be proved , but it is not done in this note, that, even with this modification the conclusions of both Theorem (3.3) and Theorem (6.2) continue to hold . In view of this we feel that the present approach is adequate for the development of a representation free notion of stochastic integral.

(1.) NOTATIONS

Let $\mathcal{H}$ be a complex separable Hilbert space. We write $\mathcal{B}(\mathcal{H})$ to denote the vector space of all bounded operators on $\mathcal{H}$. Let $\mathcal{A}$ be a Von Neumann sub algebra of $\mathcal{B}(\mathcal{H})$ and let $(\mathcal{A}_{t]})_{t\geq 0}$ be an increasing family of Von Neumann subalgebras of $\mathcal{A}$. We write $\mathcal{A}'$ and $\mathcal{A}'_{t]}$ to denote the commutant of $\mathcal{A}$ and $\mathcal{A}_{t]}$ in $\mathcal{B}(\mathcal{H})$. Let $\mathcal{E}$ be a subset of $\mathcal{H}$ such that the set $\mathcal{A}' \cdot \mathcal{E}$, that we shall denote by $\mathcal{D}$ is dense in $\mathcal{H}$. $\mathcal{L}(\mathcal{D};\mathcal{H})$ will denote the set of pairs (F, F^+) of linear operators on $\mathcal{H}$ with domain containing $\mathcal{D}$ such that

$$< \eta, F\xi >=< F^+\eta, \xi >$$

for all elements η , $\xi \in \mathcal{D}$. The pair (F, F^+) will be denoted $F^\cdot$ or, if no confusion can arise, simply F. One easily verifies that $\mathcal{L}(\mathcal{D};\mathcal{H})$ is a vector space. We shall consider two topologies on $\mathcal{L}(\mathcal{D};\mathcal{H})$: the strong-* topology on $\mathcal{D}$, defined by the semi-norms

$$—A \longrightarrow || A\xi || + || A^+\xi || \qquad , \qquad \xi \in \mathcal{D}$$

and the weak topology on $\mathcal{D}$ defined by the semi-norms

$$A \longrightarrow |<\xi, A\xi>| \quad , \quad \xi \in \mathcal{D}$$

If X is a linear operator on $\mathcal{H}$ we write D(X) to indicate the domain of X. For all $t \geq 0$ we say that an element A of $\mathcal{L}(\mathcal{D};\mathcal{H})$ is affiliated with $\mathcal{A}_{t]}$ and write $A \widehat{\in} \mathcal{A}$ if $AA' \supseteq A'A$ for all element A' of $\mathcal{A}'_{t]}$. A **stochastic process** in $\mathcal{H}$ is a family $(F_t)_{t\geq 0}$ of elements of $\mathcal{L}(\mathcal{D};\mathcal{H})$. Two stochastic processes are said to be equivalent if they coincide on $\mathcal{A}' \cdot \mathcal{E}$ The process (F_t) is **strongly-*** (resp. **weakly**) **measurable** if, for all elements $\xi \in \mathcal{D}$ the maps $t \mapsto \| F_t\xi \|$, $\| F_t^+\xi \|$(resp. $t \mapsto <\xi, F_t\xi>$) are measurable with respect to Lebesgue measure . The stochastic process (F_t) is called **adapted** (to the filtration $(\mathcal{A}_{t]})$ if, for all $t \geq 0$, the operators F_t and F_t^+ are affiliated with $\mathcal{A}$. A process is called an elementary predictable process if it can be written in the form

$$\sum_{k=1}^{n} \chi_{(t_k, t_{k+1}]} \otimes F_{t_k}$$

where $0 \leq t_0 < t_1 < ... < t_n < \infty$ and F_{t_k} is affiliated with $\mathcal{A}_{t_k]}$ (for all k). If moreover F_{t_k} is an element of $\mathcal{A}_{t_k]}$ then we say that (F_t) is a bounded elementary predictable process.

(2). SIMPLE STOCHASTIC INTEGRALS

DEFINITION (2.1) An **additive process** on $\mathcal{H}$ is a family $(X(s,t))$ $(0 \leq s < t)$ of elements of $\mathcal{L}(\mathcal{D};\mathcal{H})$ such that: (i) for all s,t with $s < t$, $X(s,t)$ is affiliated with $\mathcal{A}_{t]}$. (ii) for all r,s,t with $r < s < t$ we have

$$X(r,t) = X(r,s) + X(s,t)$$

on D. For all additive processes X we shall denote by S(X) the set of all adapted processes (F_t) which can be written in the form

$$F_t = \sum_{k=1}^{n} \chi_{(t_k, t_{k+1}]}(t) F_{t_k} \tag{2.2}$$

where:

$$0 \leq t_0 < t_1 < ..... < t_n < \infty \tag{2.3}$$

$$F_{t_k}(\mathcal{D}) \subseteq \bigcap_{t_k \leq r < s} D(X(r,s)) \tag{2.4}$$

$$\bigcup_{t_k \leq r < s} X(r,s)^+ D(X(r,s)^+) \subseteq D(F_{t_k}^+) \tag{2.5}$$

for all integers k with $0 \leq k \leq n$. Given an element F of S(X) one can define the **left stochastic integral**

$$\int dX_s F_s = \sum_{k=1}^{n} X(t_k, t_{k+1}) F_{t_k} \tag{2.6}$$

and the right stochastic integral

$$\int F_s^+ dX_s^+ = \sum_{k=1}^{n} F_{t_k}^+ X(t_k, t_{k+1})^+ \tag{2.7}$$

PROPOSITION (2.2) In the above notations
(i) The left and right stochastic integrals are independent of the representation of F in the form (2.2)
(ii) The pair $(\int dX_s F_s, \int F_s^+ dX_s^+)$ is an element of $\mathcal{L}(\mathcal{D};\mathcal{H})$
(iii) S(X) is a vector space and for all elements F , G of S(X) we have

$$\int dX_s(F_s + G_s) = \int dX_s F_s + \int dX_s G_s \tag{2.8}$$

$$\int (F_s^+ + G_s^+) dX_s^+ = \int F_s^+ dX_s^+ + \int G_s^+ dX_s^+ \tag{2.9}$$

(iv) For all elements F of S(X), a' of $\mathcal{A}'$ and ξ of $\mathcal{E}$ we have

$$\int dX_s F_s a'\xi = \overline{a^+} \int dX_s F_s \xi \quad ; \quad \int F_s^+ dX_s^+ a'\xi = a' \int F_s^+ dX_s^+ \xi \tag{2.10}$$

PROOF. (i) Let F be an element of S(X) and let

$$\sum_{k=1}^{n} \chi_{(t_k, t_{k+1}]}(t) F_{t_k} \quad ; \quad \sum_{h=1}^{m} \chi_{(s_h, s_{h+1}]}(t) F_{s_h}$$

be two representations of F in the form (2.2). We have then

$$\sum_{k=1}^{n} X(t_k, t_{k+1}) F_{t_k} = \sum_{k=1}^{n} \sum_{h=1}^{m} X(t_k \vee s_h, t_{k+1} \wedge s_{h+1}) F_{t_k} = \sum_{h=1}^{m} \sum_{k=1}^{n} X(t_k \vee s_h, t_{k+1} \wedge s_{h+1}) F_{s_h} =$$

$$= \sum_{h=1}^{m} X(s_h, s_{h+1}) F_{s_h}$$

Similar equalities hold for the simple right stochastic integral. (ii) Let F be an element of S(X) which can be written in the form (2.2); we have then (for all $\xi, \eta \in \mathcal{D}$)

$$< \eta, \sum_{h=1}^{m} X(s_h, s_{h+1}) F_{s_h} \xi > = < \sum_{h=1}^{m} F_{s_h}^+ X^+(s_h, s_{h+1}) \eta, \xi >$$

(iii) and (iv) are obvious.

NOTATION The pair

$$\left(\int dX_r F_r \chi_{(s,t]}(r), \int F_r^+ \chi_{(s,t]}(r) dX_r^+\right)$$

will be also denoted

$$\left(\int_s^t dX_r F_r, \int_s^t F_r^+ dX_r^+\right)$$

or simply, when no confusion can arise,

$$\int_s^t dX_r F_r$$

(3.) STOCHASTIC INTEGRALS

We shall denote by $\mathcal{P}_s(X)$ (resp. $\mathcal{P}_w(X)$) the vector space of all processes F such that there exists a sequence $F^{(n)}$ in S(X) with the following properties:

(i) for all t, the operator $F_t^{(n)}$ converges *-strongly (resp. weakly) on $\mathcal{D}$ to F .
(ii) for all $t \geq 0$, $\xi \in \mathcal{E}$ and all integers n one has

$$\sup_{s \leq t} \| F_s^{(n)} \xi \| \leq c_{t,\xi} \qquad \left(resp. \ \sup_{s \leq t} \| F_s^{(n)} \xi \| \leq c_{t,\xi} \right)$$

where $c_{t,\xi}$ is a constant. We say that a sequence $(F^{(n)})$ $(n \geq 1)$ of elements of $\mathcal{P}_s(X)$ (resp. $\mathcal{P}_w(X)$) converges in $\mathcal{P}_s(X)$ (resp. $\mathcal{P}_w(X)$) to F if the two conditions above are fulfilled.

DEFINITION (3.1) An additive process X is a **strong-*** (resp. **weak**) **semimartingale** if for all sequences $F^{(n)}$ in S(X) converging to zero in $\mathcal{P}_s(X)$ (resp. $\mathcal{P}_w(X)$) and for all $t \geq 0$ the simple stochastic integrals

$$\int dX_r F_r^{(n)}, \int F_r^{(n)+} dX_r^+$$

converge to zero strongly (resp. weakly) in $\mathcal{L}(\mathcal{D};\mathcal{H})$.

REMARK(1.) Identifying, as usual, a scalar (real or complex valued) classical process with the associated multiplication operators on the L^2-space of the process , the content of Dellacherie' s theorem mentioned in the introduction is that a scalar process X is a semimartingale in the sense of Definition (3.1) if and only if it admits a decomposition of the form

$$X = M + A$$

where M is a local martingale and A is the difference of two increasing processes. In fact, in Dellacherie' s formulation of condition (i) above, convergence in probability is substituted for *-strong convergence but, due to condition (ii) and to the fact that for classical processes the *-strong convergence reduces to L^2-convergence, in that case the two conditions are equivalent since a norm bounded sequence in L^2 which converges to zero in probability converges to zero also in L^2.

REMARK(2.) The following example shows that condition (ii) in the definition of convergence in $\mathcal{P}_s(X)$ (resp. $\mathcal{P}_w(X)$) is necessary to have a good notion of stochastic integral. Let

$$X(s,t) = (t-s)\cdot 1$$

and

$$F^{(n)} = 1 \cdot n \cdot \chi_{(0,1/n]}$$

Then, for any element ξ of $\mathcal{E}$ we have for all $t > 0$

$$\| F_t^{(n)}\xi \|^2 = n^2 \cdot \chi_{(0,1/n]}(t) \cdot \| \xi \|^2 \longrightarrow 0$$

but for all n

$$\| \int_0^1 dX_t F_t^{(n)} \xi \|^2 = \| \xi \|^2$$

REMARK(3.) If we want a larger class of semimartingales we must require that the continuity property expressed in Definition (3.1) hold for a stronger topologyes on a smaller space of integrands . In particular the topologies defined by the seminorms

$$A \mapsto \| CA\xi \| + \| A^+C^+\xi \| \qquad ; \qquad A \mapsto < \xi, A^+CA\xi >$$

where C is a process arise naturally in several applications. For example, if A is an element of $\mathcal{L}(\mathcal{D};\mathcal{H})$ not bounded and affiliated with $\mathcal{A}_{0]}$, then

$$X(s,t) = (t-s)A$$

is not, in general, an semimartingale in the sense of Definition (3.1) because it may not be true that if $F_t^{(n)} \cdot \xi \longrightarrow 0$ then $AF_t^{(n)} \cdot \xi \longrightarrow 0$ However in the present paper we shall only consider the case C = 1 (cf. the remark at the end of the introduction).

We can now define the strong-* integral with respect to a semimartingale X . Let F be an element of $\mathcal{P}_s(X)$ and $(F^{(n)}\cdot)$ be a sequence of elements of S(X) converging to F in $\mathcal{P}_s(X)$. For all elements ξ of $\mathcal{E}$ (hence also for all $\eta \in \mathcal{D}$ and all $t \geq 0$, the sequences $(F^{(n)}\cdot\xi)$ are Cauchy in $\mathcal{H}$. Moreover in view of the property of the semimartingale X the limits are independent of the particular sequence. One can therefore define

$$\int_0^t dX_s F_s \xi = \lim_{n\to\infty} \int_0^t dX_s F_s^{(n)} \xi$$

$$\int_0^t F_s^+ dX_s^+ \xi = \lim_{n\to\infty} \int_0^t F_s^{(n)+} dX_s^+ \xi$$

Similarly one defines the weak sthocastic integral. The following elementary properties of the stochastic integral are easily checked :

PROPOSITION (3.2) Let X be a weak semimartingale , then :

(i) For any element F of $\mathcal{P}_w(X)$ and for all $t \geq 0$, the pair $(\int_0^t dX_s F_s\ ,\ \int_0^t F_s^+ dX_s^+)$ is an element of $\mathcal{L}(\mathcal{D};\mathcal{H})$.

(ii) $\mathcal{P}_w(X)$ is a vector space and for all elements F , G of $\mathcal{P}_w(X)$ the relations (2.8) and (2.9) hold for all $t \geq 0$.

(iii) For all elements F of $\mathcal{P}_w(X)$, a' of A' and ξ of $\mathcal{E}$ and for all $t \geq 0$ the relations (2.10) hold.

Moreover the same statements hold for $\mathcal{P}_s(X)$ when X is a strong semimartingale .

PROOF. (i) Let $F^{(n)\cdot}$ a sequence of elements of S(X) converging to $F^{\cdot}$ in $\mathcal{P}_s(X)$ (resp. $\mathcal{P}_s(X)$). Then for all elements $\xi, \eta \in \mathcal{D}$ using (2.8) (ii) we have

$$< \int_0^t F_s^+ dX_s^+ \eta, \xi >= \lim_{n\to\infty} < \int_0^t F_s^{(n)+} dX_s^+ \eta, \xi >= \lim_{n\to\infty} < \eta, \int_0^t dX_s F_s^{(n)} \xi >=< \eta, \int_0^t dX_s F_s \xi >$$

The other statements can be proved in a similar way.

DEFINITION (3.3) An additive process X is called a regular semimartingale for the set $\mathcal{E}$. if it satisfies the following condition : for all elements $\xi \in \mathcal{E}$ there exist two positive funtions $g_\xi^{\cdot} \in L^1_{loc}(\mathbf{R}_+)$ such that, for all elements F of S(X) and all $t \geq 0$ we have:

$$\| \int_0^t dX_s F_s \xi \|^2 \leq c_{t.\xi} \cdot \int_0^t \| F_s \xi \|^2 g_\xi(s) ds \tag{3.2a}$$

$$\| \int_0^t F_s^+ dX_s^+ \xi \|^2 \leq c_{t.\xi} \cdot \int_0^t \| F_s^+ \xi \|^2 g_\xi^+(s) ds \tag{3.2b}$$

where $c_{t.\xi}$ is a positive constant.

THEOREM (3.4) Any regular semi-martingale is a strong-* semimartingale .

PROOF. Let $(F^{(n)\cdot})_{n\geq 1}$ be a sequence in S(X) converging to zero in $\mathcal{P}_s(X)$ then, for all $t \geq 0$ we have

$$\lim_{n\to\infty} \| F_s^{(n)\cdot} \xi \| = 0$$

$$\sup_{s\leq t} \| F_s \xi \| \leq c'_{t.\xi}$$

where $c'_{t.\xi}$ is a positive constant. But the conditions (3.5) combined with Lebesgue's theorem implies then that

$$\lim_{n\to\infty} \int_0^t dX_s F_s^{(n)} \xi = \lim_{n\to\infty} \int_0^t F_s^{(n)+} dX_s^+ \xi = 0$$

If X is a regular semimartingale then ,as shown by the following theorem , we can extend the stochastic integral (with respect to X) to a class of processes larger than $\mathcal{P}_s(X)$.

THEOREM (3.5) Let X be an additive process satisfying condition (3.4) and let $(F_t)_{t\geq 0}$ be a measurable adapted process such that, for all $\xi \in \mathcal{E}$ and all $t \geq 0$

$$\int_0^t \left(\| F_s\xi \|^2 g_\xi(s) + \| F_s^+ \xi \|^2 g_\xi^+(s) \right) ds < \infty \tag{3.3}$$

$$F_t(\mathcal{D}) \subseteq \bigcap_{s>r\geq t} D(X(r,s)) \tag{3.4}$$

$$\bigcup_{s>r\geq t} [X^+(r,s)\xi] \subseteq D(F_t^+) \tag{3.5}$$

then we can define the stochastic integral of F with respect to X . Moreover, for all element $\xi \in \mathcal{E}$ the inequalities (3.5) hold.

PROOF. Suppose first that for all element $\xi \in \mathcal{E}$ the functions $s \mapsto F_s\xi$ are continuous and consider then the sequence of elements of S(X)

$$F_t^{(n)\cdot} = \sum_k \chi_{[\frac{k}{n}, \frac{k+1}{n}]}(t) F_{\frac{k}{n}}$$

Then we can show that $F^{(n)\cdot}$ converges to $F^\cdot$ in $P_s(X)$. In fact for all $\xi \in \mathcal{E}$, all $t \geq 0$ and all $\epsilon > 0$ there exists a $\delta > 0$ such that, if

$$| r - s | < \delta \quad ; \quad 0 \leq r, s \leq t$$

then

$$\| F_r\xi - F_s\xi \| < \epsilon$$

Thus for all n such that $1/n < \delta$ we have

$$\| F_s^{(n)\cdot}\xi - F_s\xi \| < \epsilon \quad ; \quad \sup_{s\leq t} \| F_r^{(n)\cdot}\xi \| \leq \sup_{s\leq t} \| F_s\xi \| = c'_{t,\xi}$$

Now let $(F_t)_{t\geq 0}$ be a measurable adapted process satisfying conditions (3.3), (3.4), (3.5) and let $(\phi_n)_{n\geq 1}$ be the sequence of positive measurable functions

$$\phi_n(t) = n\chi_{[0,\frac{1}{n}]}(t)$$

Let us consider the processes

$$F_s^{(n)\cdot}\xi = \int_0^s \phi_n(u) F_{s-u}^\cdot du$$

which is strongly continuous on $\mathcal{E}$ and adapted. Then, for all $\xi \in \mathcal{E}$,

$$\| F_s^{(n)\cdot}\xi - F_s\xi \|^2 = \| \int_0^t \phi_n(u)[F_{s-u}\xi - F_s\xi] du \|^2$$

and therefore

$$\int_0^t \| F_s^{(n)}\xi - F_s\xi \|^2 g_\xi(s) ds \leq \int_0^t g_\xi(s) ds \int_0^s \phi_n(s-u) \| F_u\xi - F_s\xi \|^2 du$$

$$= \int_0^t g_\xi(s) ds \int_0^s \phi_n(u) \| F_{s-u}\xi - F_s\xi \|^2 du$$

$$= \int_0^t \phi_n(u) \int_u^t \| F_{s-u}\xi - F_s\xi \|^2 g_\xi(s)ds$$

$$= n \int_0^{\frac{1}{n}} du \int_u^t \| F_{s-u}\xi - F_s\xi \|^2 g_\xi(s)ds$$

$$\leq n \int_0^{\frac{1}{n}} du \int_0^t \| F_{s-u}\xi - F_s\xi \|^2 g_\xi(s)ds \longrightarrow 0$$

as $n \to \infty$. Similarly

$$\int_0^t \| F_s^{(n)+}\xi - F_s^+\xi \|^2 g_\xi^+(s)ds \leq n \int_0^{\frac{1}{n}} du \int_0^t \| F_{s-u}^+\xi - F_s^+\xi \|^2 g_\xi^+(s)ds$$

And so

$$\lim_{n\to\infty} \int_0^t \| F_s^{(n)}\xi - F_s\xi \|^2 g_\xi^+(s)ds = 0 \tag{3.6}$$

Therefore the sequences

$$\int_0^t dX_s F_s^{(n)}\xi \quad , \quad \int_0^t F_s^{(n)+} dX_s^+\xi$$

are Cauchy in $\mathcal{H}$. Moreover these limits are the same for any sequence satisfying (3.6) so we can define

$$\int_0^t dX_s F_s\xi = \lim_{n\to\infty} \int_0^t dX_s F_s^{(n)}\xi$$

$$\int_0^t F_s^+ dX_s^+\xi = \lim_{n\to\infty} \int_0^t F_s^{(n)+} dX_s^+\xi$$

And we have moreover

$$\| \left(\int_0^t dX_s F_s \xi \right) \|^2 \leq c_{t,\xi} \int_0^t \| F_s\xi \|^2 g_\xi(s)ds$$

(4.) CLASSICAL STOCHASTIC INTEGRALS

In this section we shall show that, under quite general conditions, the definition of quantum stochastic integral of the preceeding section includes the classical one . We shall deal with the following objects:

- H - separable Hilbert space
- $(\Omega, \mathcal{F}, P)$ a probability space
- $(\mathcal{F}_t)$ $(t \geq 0)$ a filtration
- (x_t) $(t \geq 0)$ a locally integrable H-valued semimartingale

We will suppose, for simplicity, that x has a decomposition of the form

$$x_t = m_t + \int_0^t b(s)ds \tag{4.1}$$

where: - $b : \mathbf{R}_+ \times \Omega \to H$ is adapted and , for all $t \geq 0$:

$$\int_0^t E(\| b(s) \|^2 ds < \infty$$

- m is a locally square integrable martingale with quadratic variation $< m >$ of the form

$$< m >_t = \int_0^t c(s)ds$$

with $c : \mathbf{R}_+ \times \Omega \to H$ adapted and such that , for all $t \geq 0$:

$$\int_0^t E(\| c(s) \| ds < \infty$$

One easily sees (cf. Metivier-Pellaumail [6]) that for all bounded adapted processes G with values in $\mathcal{B}(\mathcal{H})$ such that, for all $t \geq 0$ one has

$$\int_0^t E\Big(\| G_s \|^2_{\mathcal{B}(H)} \big(| c(s) |_H + | b(s) |_H\big)\Big) ds < \infty \tag{4.2}$$

we can define the classical stochastic integral

$$\int_0^t G_s dx_s$$

Now let us consider

$$\mathcal{H} = L^2(\Omega, \mathcal{F}, P) \otimes H$$

$$\mathcal{A} = L^\infty(\Omega, \mathcal{F}, P) \otimes \mathcal{B}(H) \subseteq \mathcal{B}(\mathcal{H})$$

$$\mathcal{A}_{t]} = L^\infty(\Omega, \mathcal{F}_{t]}, P) \otimes \mathcal{B}(H) \subseteq \mathcal{B}(\mathcal{H})$$

(the algebra $L^\infty(\Omega, \mathcal{F}, P)$ acts on $L^2(\Omega, \mathcal{F}, P)$ by multiplication). Fix $h \in H$. For all $t \geq 0$ the vector $\xi = 1 \otimes h$ is cyclic for $\mathcal{A}_{t]}$ in $L^2(\Omega, \mathcal{F}, P) \otimes H$. Then , by the BT theorem [7] , for each t there exists an operator X_t affiliated with $\mathcal{A}_{t]}$ such that

$$x_t = X_t(1 \otimes h) \tag{4.3}$$

Notice that any classical $\mathcal{B}(H)$-valued adapted process F satisfying (4.2) can be identified with a quantum process in $\mathcal{H}$ (by identifying it with $1 \otimes F$) . Therefore if there exists a dense subset $\mathcal{D}$ of $\mathcal{H}$ with $\xi \in \mathcal{D}$ such that, for all $t \geq 0$, the stochastic integral

$$\int_0^t F_s dX_s \quad ; \quad \int_0^t dX_s^+ F_s^+ \tag{4.4}$$

have a meaning as operators in $\mathcal{L}(\mathcal{D}; \mathcal{H})$, then the identity (4.3) can be interpreted as stating that the operator $\int_0^t F_s dX_s$ applied to the vector ξ produces the vector $\int_0^t F_s dx_s$. In fact , if F is an element of S(X) , then the stochastic integrals (4.4) are well defined as operators and the identity

$$\int_0^t F_s dX_s = \int_0^t F_s dx_s$$

takes place. Moreover, if F is any bounded adapted process satisfying (4.2) and $(F_t^{(n)})$ is a sequence of elements of S(X) converging to F in $P_s(X)$ then, from the equality

$$\| F_s^{(n)} \cdot \xi \|^2_{\mathcal{H}} = E\Big(\| F_s^{(n)} \cdot \xi\Big) \|^2_H\Big) \tag{4.5}$$

it follows that

$$\int_0^t F_s dx_s = \lim_{n \to \infty} \int_0^t F_s^{(n)} dx_s = \lim_{n \to \infty} \int_0^t F_s^{(n)} dX_s \xi = \int_0^t F_s dX_s \xi \tag{4.6}$$

where the limits are meant strongly in $L^2(\Omega, \mathcal{F}_t, P) \otimes H$.

(5.) SOME QUANTUM STOCHASTIC INTEGRALS

In this section we shall see that the notion of stochastic integral introduced above , when particularized to suitable representations of the CCR or CAR, reduces to known examples of quantum stochastic integrals .

(5.1) BOSON FOCK STOCHASTIC INTEGRALS OVER $L^2(\mathbf{R}_+$

Denote :
- $\mathcal{H} = \Gamma(L^2(\mathbf{R}_+)$ the Fock space over the one-particle space $L^2(\mathbf{R}_+)$
- $\mathcal{E} = \{\psi(f) : f \in L^2(\mathbf{R}_+)\}$ the set of exponential vectors in $\mathcal{H}$.
- $\Phi = \psi(0)$ the vacuum state in $\mathcal{H}$

$$\mathcal{H}_{t]} = \Gamma(\chi_{[0,t]})\mathcal{H} = \Gamma(L^2([0,t]) \otimes \Phi_{[t}$$

where $t \geq 0$ and $\Gamma(\chi_{[0,t]})$ is the orthogonal projector defined by

$$\Gamma(\chi_{[0,t]})\psi(f) = \psi(\chi_{[0,t]}f)$$

- $\mathcal{A}_{t]} = \mathcal{B}(\mathcal{H}_{t]})$
- W(f) ($f \in L^2(\mathbf{R}_+)$ the Weyl operator characterized by the property

$$W(f)\psi(g) = e^{-\frac{\|f\|^2}{2} - <f,g>}\psi(f+g)$$

- A , A^+ , N the annihilation, creation and number fields defined, on $\mathcal{E}$ by the relations :

$$A(f)\psi(g) = < f, g > \psi(g)$$

$$A^+(f)\psi(g) = \frac{d}{dt}|_{t=0}\ \psi(g + tf)$$

$$N_t\psi(g) = \frac{d}{ds}|_{s=0}\ \psi(e^{s\chi_{[0,t]}}g)$$

we write N(s,t) for $N_t - N_s$.

The W(f) are unitary operators on $\mathcal{H}$ satisfying the CCR

$$W(f)W(g) = e^{-Im<f,g>}W(f+h) \tag{5.2}$$

With the notation

$$f_{t]} = \chi_{[0,t]}f \qquad ; \qquad f_{[t} = \chi_{[t,\infty]}f$$

one has that $W(f_{t]})$ is an element of $\mathcal{A}_{t]}$ and $W(f_{[t})$ an element of $\mathcal{A}_{[t}$. Moreover for any $\mathcal{A}_{t]}$-adapted process F_t and for any $f \in L^2_{loc}(\mathbf{R}_+)$ the following identities hold on $\mathcal{E}$:

$$[F_t, W(f_{[t})] = 0 \tag{5.3}$$

$$[F_t, A^{\cdot}(\chi_{[t,u]}f)] = 0 \tag{5.4}$$

$$[F_t, N(t,u)] = 0 \tag{5.5}$$

for all $u > t$.

<u>*PROPOSITION*</u> For all $f \in L^2_{loc}(\mathbf{R}_+)$, the additive processes $N(s,t)$ and $A^{\cdot}(\chi_{[t,u]}f)$ are regular semimartingales for the set $\mathcal{E}$.

<u>*PROOF*</u> We shall use the following identities (whose validity on the domain $\mathcal{E}$ is easily verified)

$$W(g)^* A(\chi_{[s,t]}f)W(g) = A(\chi_{[s,t]}f) + \int_s^t \bar{f}(r)g(r)dr \tag{5.6a}$$

$$W(g)^* A^+(\chi_{(s,t]} f) W(g) = A^+(\chi_{(s,t]} f) + \int_s^t f(r)\bar{g}(r)dr \tag{5.6b}$$

$$W(g)^* N((s,t]) W(g) = N((s,t]) + A^+(\chi_{(s,t]} g) + A(\chi_{(s,t]} g) + \int_s^t |g(r)|^2 dr \tag{5.6c}$$

If $\psi(g)$ is an exponential vector and F is an element of S(A(f)) written in the form

$$F_s = \sum_{k=1}^n F_{t_k} \chi_{(t_k,t_{k+1}]}(s)$$

where $0 \le t_1 < t_2 < \ldots < t_{k+1} = t$ is a partition of [0,t], then

$$A(\chi_{(s,t]} f) F_s \psi(g) = e^{\frac{1}{2}\|g\|^2} A(\chi_{(s,t]} f) F_s W(g)\Phi = e^{\frac{1}{2}\|g\|^2} A(\chi_{(s,t]} f) W(g_{[s}) F_s W(g_{s]})\Phi$$

and applying (5.5) this becomes equal to

$$e^{\frac{1}{2}\|g\|^2} W(g_{[s}) A(\chi_{(s,t]} f) F_s W(g_{s]})\Phi + e^{\frac{1}{2}\|g\|^2} \int_s^t \bar{f}(r) g(r) dr \; W(g_{[s}) F_s W(g_{s]})\Phi = \left(\int_s^t \bar{f}(r) g(r) dr\right) \cdot F_s \psi(g)$$

From this we deduce that

$$\|\sum_{k=1}^n A(\chi_{(t_k,t_{k+1}]} f) F_{t_k} \psi(g)\|^2 = \|\int_s^t \bar{f}(r) g(r) F_r \psi(g) dr\|^2 \le \tag{5.7}$$

$$\le \left(\int_s^t |g(r)|^2 dr\right) \cdot \int_s^t \|F_s \psi(g)\|^2 |f(s)|^2 ds$$

Moreover

$$\|\sum_{k=1}^n F_{t_k}^+ A^+(\chi_{(t_k,t_{k+1}]} f)\psi(g)\|^2 = \|\sum_{k=1}^n A^+(\chi_{(t_k,t_{k+1}]} f) F_{t_k}^+ \psi(g)\|^2 =$$

$$= \sum_{h,k=1}^n < F_{t_k}^+ \psi(g), A(\chi_{(t_k,t_{k+1}]} f) A^+(\chi_{(t_h,t_{h+1}]} f) F_{t_h}^+ \psi(g) >$$

and using the CCR the above expression becomes

$$\|\sum_{k=1}^n A(\chi_{(t_k,t_{k+1}]} f) F_{t_k}^+ \psi(g)\|^2 + \sum_{k=1}^n \int_{t_k}^{t_{k+1}} \|F_s^+ \psi(g)\|^2 |f(s)|^2 ds$$

hence , using (5.7)

$$\|\sum_{k=1}^n F_{t_k}^+ A^+(\chi_{(t_k,t_{k+1}]} f)\psi(g)\|^2 \le \left(1 + \int_s^t |g(r)|^2 dr\right) \cdot \int_s^t \|F_r^+ \psi(g)\|^2 |f(r)|^2 dr \tag{5.8}$$

This proves that that the additive process $A^{\cdot}(\chi_{(t_k,t_{k+1}]} f)$ $0 \le s < t$ is a regular semimartingale for the set $\mathcal{E}$.

To show that also the additive process N(s,t) $(0 \le s < t)$ is a regular semimartingale (for the set $\mathcal{E}$) first compute

$$N(s,t) F_s \psi(g) = e^{\frac{1}{2}\|g\|^2} N(s,t) W(g_{[s}) F_s W(g_{s]})\Phi$$

Applying twice (5.6c) this becomes equal to

$$e^{\frac{1}{2}\|g\|^2} W(g_{[s}) N(s,t) F_s W(g_{s]})\Phi + W(g_{[s}) A^+(\chi_{(s,t]} g) F_s W(g_{s]})\Phi +$$

$$+ W(g_{[s}) A(\chi_{(s,t]} g) F_s W(g_{s]})\Phi + \int_s^t F_r |g(r)|^2 \psi(g) dr$$

The first and the third term vanish because of (5.4) ,(5.5) and applying (5.6b) to the second term we obtain

$$N(s,t) F_s \psi(g) = A^+(\chi_{(s,t]} g) F_s \psi(g)$$

from which, using again (5.8) , we have finally

$$\|\sum_{k=1}^n N((t_k, t_{k+1}]) F_{t_k} \psi(g)\|^2 \le \left(1 + \int_s^t |g(r)|^2 dr\right) \cdot \int_s^t \|F_r \psi(g)\|^2 |g(r)|^2 dr$$

(5.2) BOSON UNIVERSALLY INVARIANT STOCHASTIC INTEGRALS OVER $L^2(\mathbf{R}_+)$

We keep the notations of (5.8) with the following modifications : all the objects relative to the Fock space $\Gamma(L^2(\mathbf{R}_+)$ will be denoted with a subscript " o " and all the objects relative to the conjugate Fock space $\bar{\Gamma}(L^2(\mathbf{R}_+)$ with a superscript " - " i.e., for example, $\psi_o(g)$, $W_o(g)$ $(g \in L^2(\mathbf{R}_+))$ denote respectively the exponential vectors and Weyl operators introduced in (5.1) ; $\bar{\psi}_o(f)$ denotes the exponential vector $\psi_o(f)$ considered as an element of $\bar{\Gamma}(L^2(\mathbf{R}_+)$ and, if T is a linear operator on $\Gamma(L^2(\mathbf{R}_+))$, we write $\bar{T}$ to indicate the natural action of the linear operator T on $\bar{\Gamma}(L^2(\mathbf{R}_+))$. With these notations we write :

$$\mathcal{H} = \Gamma(L^2(\mathbf{R}_+)) \otimes \bar{\Gamma}(L^2(\mathbf{R}_+))$$

$$\Phi = \Phi_o \otimes \bar{\Phi}_o$$

and for a given positive number $\sigma^2 > 1$

$$\lambda^2 = \frac{1}{2}(\sigma^2 + 1) \qquad ; \qquad \mu^2 = \frac{1}{2}(\sigma^2 - 1)$$

The W^*- algebra generated by the operators of the form

$$W(f) = W_o(\lambda\chi_{[s,t]}f) \otimes \bar{W}_o(-\mu\chi_{[s,t]}f) \tag{5.9a}$$

$$A(f) = \lambda A_o(f) \otimes 1 + \mu 1 \otimes \bar{A}_o^+(f) \tag{5.9b}$$

$$A^+(f) = \lambda A_o^+(f) \otimes 1 + \mu 1 \otimes \bar{A}_o(f) \tag{5.9c}$$

It is well known that :

- Φ is cyclic and separating for A.
- $\mathcal{A}'$ is the W^*-algebra generated by the operators (5.9)
- for all t, $\mathcal{A}_{t]}$ is the W^*-algebra generated by the operators

$$W'(f_{t]}) \otimes A$$

(with $A \in \mathcal{B}(\Gamma(L^2((t,\infty)) \otimes \bar{\Gamma}(L^2((t,\infty)))$

PROPOSITION The additive process $A(s,t) = A(\chi_{[s,t]})$ is a regular semimartingale for the set $\mathcal{E} = \Phi$ and, in particular, a Φ-regular martingale.

PROOF Let F be an element of S(A) written in the form

$$F_s = \sum_{k=1}^{n} F_{t_k}\chi_{[t_k,t_{k+1}]}(s)$$

where $0 \le t_o < t_1 < ... < t_n = t$ is a partition of [0,t] , then we can compute

$$\| \sum_{k=1}^{n} F_{t_k} A(t_k, t_{k+1})\Phi \|^2 = \| \sum_{k=1}^{n} F_{t_k}\Phi_{t_k} \otimes A(t_k,t_{k+1})\Phi_{[t_k,t_{k+1}]} \otimes \Phi_{[t_{k+1}} \|^2 =$$

$$= \| \sum_{k=1}^{n} F_{t_k}\Phi_{t_k} \otimes \left(\mu\Phi_{o,[t_k,t_{k+1}]} \otimes \bar{A}_o^+(t_k,t_{k+1})\Phi_{[t_k,t_{k+1}]}\right) \otimes \Phi_{[t_{k+1}} \|^2 =$$

$$= \int_0^t \| F_s\Phi \|^2 \mu ds == \| \sum_{k=1}^{n} F_{t_k}\Phi_{t_k} \otimes \left(\mu \otimes \chi_{[t_k,t_{k+1}]}\right)\Phi_{[t_k,t_{k+1}]} \otimes \Phi_{[t_{k+1}} \|^2 =$$

$$\int_0^t \| F_s\Phi \|^2 \mu ds = \sum_{k=1}^{n} \| F_{t_k}\Phi_{t_k} \|^2 \mu^2(t_{k+1} - t_k) = \int_0^t \| F_s\Phi_s \|^2 \mu^2 ds$$

A similar computation shows that

$$\| \sum_{k=1}^{n} F_{t_k}^+ A^+(t_k,t_{k+1})\Phi \|^2 = \int_0^t \| F_s\Phi \|^2 \lambda^2 ds$$

which proves that A(s,t) is a regular semimartingale for the set consisting of the single vector Φ.

(5.3) CLIFFORD STOCHASTIC INTEGRALS OVER $L^2(\mathbf{R}_+)$

Denote
- $\mathcal{H}$ the antisymmetric Fock space $\Lambda(L^2(\mathbf{R}_+))$
- Φ the vacuum state in $\mathcal{H}$
- A , A^+ annihilation and creation fields

$$\psi_t = \psi(u\chi_{(0,t]}) = A^+(u\chi_{(0,t]}) + A(u\chi_{(0,t]}) \quad ; \quad u \in L^2(\mathbf{R}_+)$$

- $\mathcal{A}$ the W^*-algebra generated by the ψ_t
- $\mathcal{A}_{t]}$ the W^*-algebra generated by the ψ_s with $s \leq t$

Then Φ is cyclic separating for $\mathcal{A}$ and one easily verifies ([2] Th. (3.5) (c)) the following

PROPOSITION For any simple ψ_t-adapted process F we have:

$$\| \int_0^t F_s d\psi_s \|^2 = \int_0^t \| F_s \Phi \|^2 | u(s) |^2 ds$$

in particular ψ_t is a regular semimartingale for the set consisting of the single vector Φ.

(6.) STOCHASTIC DIFFERENTIAL EQUATIONS

We shall work in the simplest situation.

DEFINITION ((6.1) We say that a process Y (resp. Y^+) satisfies the left (resp. right) stochastic differential equation

$$dY_t = dX_t Y_t \quad ; \quad Y_o \hat{\in} \mathcal{A}_{o]} \tag{6.1}$$

respectively

$$dY_t^+ = Y_t^+ dX_t^+ \quad ; \quad Y_o \hat{\in} \mathcal{A}_{o]} \tag{6.2}$$

if, for all $t \geq 0$ one has

$$Y_t = Y_o + \int_0^t dX_s Y_s \quad ; \quad \text{resp.} \quad Y_t^+ = Y_o^+ + \int_0^t Y_s^+ dX_s^+ \tag{6.3}$$

THEOREM ((6.2) Let X be a regular semimartingale. The left and right SDE (6.3) have a unique solution $Y^\cdot$ in $\mathcal{L}(\mathcal{D}, \mathcal{H})$ which satisfy the estimate

$$\| (Y_t^\cdot - Y_o^\cdot)\xi \|^2 \leq \| Y_o^\cdot \xi \|^2 \cdot exp\left(c_{t,\xi} \int_0^t g_\xi(s) ds \right) \tag{6.4}$$

where $c_{t,\xi}$ and $g_\xi(s)$ are determined by X through (3.2a) and (3.2b) . Moreover Y and Y^+ are adjoint of each other on $\mathcal{D}$

PROOF. Let us define the sequence $(Z_t^{(n)\cdot})$ in $\mathcal{L}(\mathcal{D}, \mathcal{H})$ by

$$Z_t^{(o)\cdot} = Y_o^\cdot \quad ;$$

$$Z_t^{(n+1)} = \int_0^t dX_s Z_s^{(n)}$$

$$Z_t^{(n+1)+} = \int_0^t Z_s^{(n)+} dX_s^+$$

Then for any element ξ of $\mathcal{E}$ and any integer n we have

$$\| Z_t^{(n)}\xi \|^2 \le \frac{c_{t,\xi}^n}{n!} \| Y_0\xi \|^2 \cdot \left(\int_0^t g_\xi(s)ds \right)^n \tag{6.5a}$$

$$\| Z_t^{(n)+}\xi \|^2 \le \frac{c_{t,\xi}^n}{n!} \| Y_0^+\xi \|^2 \cdot \left(\int_0^t g_\xi^+(s)ds \right)^n \tag{6.5b}$$

Let us verify, for example, (6.5). This inequality is trivially satisfied when n = 1. If it is satisfied for $k \le n$ then

$$\| Z_t^{(n+1)}\xi \|^2 = \| \int_0^t dX_s Z_s^{(n)}\xi \|^2 \le c_{t,\xi} \cdot \left(\int_0^t \| Z_s^{(n)}\xi \|^2 g_\xi(s)ds \right)$$

$$\le \| Y_0\xi \|^2 c_{t,\xi} \frac{c_{t,\xi}^n}{n!} \int_0^t g_\xi(s) \Big(\int_0^s g_\xi(r)dr \Big)^n ds$$

$$= \| Y_0\xi \|^2 \frac{c_{t,\xi}^{n+1}}{(n+1)!} \Big(\int_0^t g_\xi(t)ds \Big)^{n+1} ds$$

From (6.5) and (6.5) it follows that, for all $t \ge 0$ the operators

$$Y_t = \sum_{n=1}^{\infty} Z_t^{(n)} \quad ; \quad Y_t^+ = \sum_{n=1}^{\infty} Z_t^{(n)+}$$

are well defined in the sense that the series converges uniformly on bounded intervals in the strong topology on $\mathcal{E}$. Moreover, for all $\xi \in \mathcal{E}$,

$$\int_0^t \| Y_s\xi \|^2 g_\xi(s)ds \le \| Y_0\xi \|^2 \cdot exp\left(c_{t,\xi} \int_0^t g_\xi(s)ds \right) \tag{6.6}$$

and so the stochastic integral of Y with respect to X is well defined. We now show that (Y_t) is a solution of equation (6.2). Fix t and $\xi \in \mathcal{E}$, then for any integer N we have

$$\sum_{n=1}^{N+1} Z_t^{(n)} = \int_0^t dX_s (\sum_{n=0}^{N} Z_s^{(n)}) \tag{6.7}$$

$$\sum_{n=1}^{N+1} Z_t^{(n)+} = \int_0^t (\sum_{n=0}^{N} Z_s^{(n)}) dX_s^+$$

Moreover, for all N,

$$\lim_{N\to\infty} \| \int_0^t dX_s Y_s \xi - \int_0^t dX_s \sum_{n=1}^{N} Z_s^{(n)}\xi \|^2 \le \lim_{N\to\infty} c_{t,\xi} \cdot \int_0^t g_\xi(s) \| Y_s\xi - \sum_{n=1}^{N} Z_s^{(n)}\xi \|^2 ds = 0$$

and

$$\lim_{N\to\infty} \| \int_0^t Y_s^+ dX_s^+ \xi - \int_0^t \sum_{n=1}^{N} Z_s^{(n)+} dX_s^+ \xi \|^2 \le \lim_{N\to\infty} c_{t,\xi} \cdot \int_0^t g_\xi^+(s) \| Y_s^+\xi - \sum_{n=1}^{N} Z_s^{(n)+}\xi \|^2 ds = 0$$

Therefore the limits, as $N \to \infty$ of both sides of the equations (6.7) exist , are equal to

$$Y_t\xi - Y_0\xi = \int_0^t dX_s Y_s \xi \quad ; \quad Y_t^+\xi - Y_0^+\xi = \int_0^t Y_s^+ dX_s^+ \xi$$

and clearly satisfy the estimate (6.6). Moreover by the same iteration argument as above, it is easy to verify that every solution of the equations (6.3) satisfies the estimate (6.4) . From this uniqueness immediately follows.

BIBLIOGRAPHY

[1] Accardi L. Quantum stochastic calculus . Proccedings IV Vilnius Conference on Probability and mathematical Statistics VNU Science Press 1987
[2] Barnett C., Streater R., Wilde I.F. The Ito-Clifford integral J.Funct. Anal.48(1982)172-212
[3] Hudson R.L., Parthasarathy K.R. Quantum Ito' s formula and stochastic evolutions. Comm. Math. Phys.93(1984)301-323
[4] Letta G. Martingales et integration stochastique Quaderni Scuola Normale Superiore 1984
[5] Metivier M., J. Pellaumail Stochastic Integration Academic Press 1980
[6] Meyer P.A. Caracterisation des semimartingales d' apres Dellacherie Sem. Prob. XIII, Springer LNM 721 (1979) 620-623
[7] Sakai S. C^*-Algebras and W^*-Algebras . Springer 1970

QUANTUM STOCHASTIC PARALLEL TRANSPORT ON NON-COMMUTATIVE VECTOR BUNDLES

David Applebaum
Department of Mathematics
University of Nottingham
University Park
Nottingham NG7 2RD
England

Work carried out while the author was supported by SERC Grant No GR/D51292.

ABSTRACT

A concept of quantum stochastic parallel transport is formulated in a finitely generated projective module over a "smooth" subalgebra of a C*-algebra where the noise arises from a family of semi-martingales in Fock space. The main examples studied are Heisenberg modules over the non-commutative torus where we find instances in which quantum parallel transport exists, but there is no diffusion on the underlying algebra of which it is the horizontal lift.

INTRODUCTION

The non-commutative differential geometry of A Connes ([5] to [9]) provides us with an apparatus to quantise the notion of "smoothness". In this framework an appropriate norm dense *-subalgebra of a C*-algebra plays the role of a smooth manifold and finitely generated projective modules over this algebra are "non-commutative vector bundles". Connes has generalised many familiar geometrical notions to this setting, such as connection and Chern character.

In a similar fashion, the subject of quantum probability (see, for example, the volumes containing [1] and [18]) aims to develop ideas of classical probability theory within a non-commutative framework. In particular, the quantum stochastic calculus of Hudson and Parthasarathy (see e.g. [18] and [19]) extends the classical theory initiated by Itô. As in Connes' geometry, their theory is C*-algebraic in character, with the probabilistic input arising from the properties of quasi-free states on the (Weyl) C*-algebra of the canonical commutation relations ([1], [2]).

Our aim in this paper is to combine these two theories within the context of a "quantum stochastic differential geometry". One should bear in mind, of course, that the interaction of the two commutative "classical" theories is a well developed subject [13] which has found applications to e.g. the heat equation on manifolds [20], path integral solutions of the Schrödinger equation [25] and index theory [4].

The first two sections of this paper are designed to be a gentle initiation into those aspects of Connes' geometry which we will be using in the sequel. In particular, care has been taken to motivate all the definitions through their classical analogues.

The subsequent sections are devoted to a discussion of the main topic of this paper, namely the concept of quantum stochastic parallel transport. We propose that this be a unitary operator valued process acting on the tensor product of an "initial" Hilbert space, which contains the geometric information, with an appropriate Fock space to carry "quantum noise". This process should, moreover, satisfy a certain quantum stochastic differential equation whose coefficients are determined by a suitable connection.

We investigate two examples of this concept, these being the classical case and the case of Heisenberg modules over the non-commutative two-torus algebra (see e.g. [7] and [9]). In the latter case, we discover a feature which is peculiar to the non-commutative framework, i.e. solutions of the parallel transport equation which are not the horizontal lifts of a diffusion process on the underlying smooth algebra.

We will employ the following notation, $(X, X^\dagger)$ denote mutually adjoint, densely defined operators on some specified domain. $X^\#$ should be read "X and $X^\dagger$". If V_1 and V_2 are inner product spaces, $V_1 \underline{\otimes} V_2$ denotes their algebraic tensor product.

We use the Einstein summation convention throughout with respect to upper and lower repeated indices so that e.g.

$$a^i b_i = a^1 b_1 + a^2 b_2 + \ldots + a^n b_n .$$

It is a great pleasure to thank Robin Hudson for a number of valuable discussions and Mauro Spera, Mark Evans and John Klauder for several stimulating comments.

1. PROBABILITY AND GEOMETRY IN OPERATOR ALGEBRAS

The following discussion of quantum probability will be very familiar to most readers. Let $(\Omega, F, \mathbf{P})$ be a probability space, then $L^\infty(\Omega, F, \mathbf{P})$ is a commutative von Neumann algebra and expectations of random variables $f \in L^\infty(\Omega, F, \mathbf{P})$ are determined by the state $\mathbf{E}$ where

$$\mathbf{E}(f) = \int_\Omega f \, d\mathbf{P}.$$

The philosophy of quantum probability is to replace the pair $(L^\infty(\Omega, F, \mathbf{P}), \mathbf{E})$ by (N, ω) where N is an arbitrary von Neumann algebra and ω is a state on N. We then investigate the extent to which the wealth of mathematical structure in $(L^\infty(\Omega, F, \mathbf{P}), \mathbf{E})$ which we call "probability theory" can be understood as special cases of more general results in (N, ω).

In a number of recent papers ([5] to [9]), A Connes has initiated a similar programme to the above for differential geometry. The remainder of this section will be concerned with describing his concept of "non-commutative manifold".

Let V be a real, compact, finite dimensional C^{∞}-manifold. The space $C^{\infty}(V)$ of smooth (complex valued) functions on V is a norm-dense *-subalgebra of the commutative C*-algebra $C(V)$ of continuous functions on V. Connes defines a *"non-commutative manifold"* to be a "smooth" norm dense *-algebra A^{∞} of a C*-algebra A. A precise definition of "smooth" is given in [6]; in the present paper we will only be concerned with a specific class of examples, which we will now describe.

Let G be a connected Lie group and H a closed subgroup so that G acts by left translation on the homogeneous manifold $V = G/H$. We obtain a representation α of G in $Aut(C(V))$ by the prescription

$$\alpha_g(f)(x) = f(g^{-1}x) \tag{1.1}$$

where $f \in C(V)$, $g \in G$, $x \in V$. Thus $(C(V), G, \alpha)$ is a C*-dynamical system and we may characterise $C^{\infty}(V)$ by

$$C^{\infty}(V) = \{f \in C(V),\ g \mapsto \alpha_g(f) \text{ is } C^{\infty}\}. \tag{1.2}$$

So given a C*-dynamical system (A, G, α) where G is a connected Lie group and A contains an identity I, a *"non-commutative homogeneous manifold"* is the norm dense *-subalgebra ([5], [7]) A^{∞} of A given by

$$A^{\infty} = \{X \in A,\ g \mapsto \alpha_g(X) \text{ is } C^{\infty}\}. \tag{1.3}$$

Example 1.1 *The non-commutative 2-torus* ([22], [7]).

Let T denote the torus $\{z \in \mathbb{C},\ |z| = 1\}$ and for fixed $\theta \in \mathbb{R}$, let A_θ be the transformation group C*-algebra $C(T) \times_\beta \mathbb{Z}$ [12] where the action β of $\mathbb{Z}$ on T is by rotation through integral multiples of $2\pi\theta$. Alternatively, A_θ may be characterised [22] as the unique C*-algebra generated by any pair of unitary operators $\{U_i,\ i = 1, 2\}$ satisfying

$$U_2 U_1 = e^{2\pi i\theta} U_1 U_2. \tag{1.4}$$

We obtain a C*-dynamical system (A_θ, T^2, α) where the action of T^2 on A_θ is given by

$$\alpha(z)U_j = z_j U_j \qquad (j = 1,2) \tag{1.5}$$

for $z = (z_1, z_2) \in T^2$.

It can be shown [7] that the algebra A_θ^∞, defined as in (1.3), is precisely those elements $X \in A_\theta$ which can be written in the form

$$X = \sum_{n,m\in\mathbb{Z}} s(m,n)U_1^m U_2^n \tag{1.6}$$

where s is a rapidly decreasing function on $\mathbb{Z}^2$, i.e. $\lim_{n,m\to\infty} |(m^p + n^q)s(m,n)| = 0$ for all $p,q \geqslant 0$.

2. GEOMETRY OF PROJECTIVE MODULES

Let E be a complex smooth vector bundle whose base is a real compact smooth manifold V and whose fibres are of fixed (finite) dimension. It was shown by Swan in [24] that the space of smooth sections of E, which we denote by $\Gamma(E)$, is a finitely generated projective $C^\infty(V)$-module, where the operations are given by

$$(r+s)(x) = r(x) + s(x)$$

$$(fs)(x) = f(x)s(x) \tag{2.1}$$

$r, s \in \Gamma(E)$, $f \in C^\infty(V)$, $x \in V$. Conversely, given any finitely generated projective $C^\infty(V)$-module M, there exists such a vector bundle E with base V such that $M = \Gamma(E)$.

Now let A^∞ be a smooth norm dense *-subalgebra of a C*-algebra A. Connes [5] has defined a "*non-commutative vector bundle*" to be a finitely generated, projective A^∞-module.

We now describe those geometric properties of such modules, as developed in [9], which we will require in later sections. Let (A, G, α) be a C*-dynamical system so that A^∞ is given by (1.3). By differentiating in (1.3) we obtain an action of the Lie algebra $\mathbf{L}$ of G, by derivations on A^∞. We denote this action by δ, so that for each $X \in \mathbf{L}$ and $a, b \in A^\infty$,

$$\delta_X(ab) = \delta_X(a)b + a\delta_X(b). \tag{2.1}$$

Now let $\mathcal{Z}$ be a finitely generated, projective (right) A^∞-module. We will make the following assumptions.

(*a*) A^∞ is equipped with a faithful δ-invariant trace τ, so that

$$\tau(\delta_X(a)) = 0 \text{ for all } X \in \mathbf{L} \text{ and } a \in A^\infty.$$

(*b*) $\mathbb{Z}$ is hermitian in that there exists an A^∞-valued inner product $\langle\ ,\ \rangle_A$ on $\mathbb{Z}$.

From (*a*) and (*b*) we see that a $\mathbb{C}$-valued inner product on $\mathbb{Z}$ is given by $\tau(\langle\ ,\ \rangle_A)$. We denote by h_0 the corresponding Hilbert space completion of $\mathbb{Z}$.

A *connection* on $\mathbb{Z}$ is a linear map $\nabla: \mathbb{Z} \longrightarrow \mathbb{Z} \otimes \mathbf{L}^*$ which satisfies the Leibnitz identity

$$\nabla_X(\xi a) = \nabla_X(\xi)a + \xi\delta_X(a) \tag{2.2}$$

for all $\xi \in \mathbb{Z}$, $a \in A^\infty$ and $X \in \mathbf{L}$.

∇ is said to be *compatible* with $\langle\ ,\ \rangle_A$ if

$$\langle\nabla_X\xi, \mu\rangle_A + \langle\xi, \nabla_X\mu\rangle_A = \delta_X(\langle\xi, \mu\rangle_A) \tag{2.3}$$

for all $\xi, \mu \in \mathbb{Z}$ and $X \in \mathbf{L}$.

We denote by $CC(\mathbb{Z})$ the space of compatible connections on $\mathbb{Z}$. From (*a*), (*b*) and (2.3) we see that for each $\nabla \in CC(\mathbb{Z})$ and $X \in \mathbf{L}$, ∇_X is a skew-hermitian operator on h_0 with invariant domain $\mathbb{Z}$.

The *curvature* R of a connection ∇ on $\mathbb{Z}$ is a 2-form on $\mathbf{L}$ taking values in $B(h_0)$ [9] given by the formula

$$R(X, Y) = [\nabla_X, \nabla_Y] - \nabla_{[X, Y]} \tag{2.4}$$

for $X, Y \in \mathbf{L}$.

Example 2.1 *Classical Vector Bundles*

Let E be a complex smooth vector bundle whose base is the compact homogeneous manifold $V = G/H$. In this case τ is integration on V and (*a*) is the requirement that the volume form be left G-invariant. General conditions under which this is satisfied can be found in [14]. If V is itself a Lie group then (*a*) is always satisfied, the required volume element being given by the wedge product of the Maurer-Cartan forms on V [11].

Since E has compact base, we know there exists [10] a Hermitian structure on E, i.e. a C^∞ assignment of inner products in the fibres of E. Assumption (*b*) is then satisfied as follows; for each $r, s \in \Gamma(E)$, we obtain $\langle r, s\rangle \in C^\infty(V)$ by

$$\langle r, s\rangle(x) = \langle r(x), s(x)\rangle_{E_x}$$

where $x \in V$ and $\langle\ ,\ \rangle_{E_x}$ denotes the inner product in the fibre E_x.

(2.2) to (2.4) are the usual definitions of connection and curvature in this framework ([10]).

Example 2.2 *Heisenberg Modules Over the Non-commutative 2-Torus* ([7], [9], [23])

Let A_θ^∞ be the non-commutative 2-torus algebra defined in example 1.1 and let $\mathcal{E}$ be the space $S(\mathbb{R}, K)$ of Schwartz functions taking values in a finite dimensional complex Hilbert space K, so $f \in S(\mathbb{R}, K)$ provided $f \in C^\infty(\mathbb{R}, K)$ and

$$\lim_{x \to \infty} \left\| x^m \frac{d^n f}{dx^n}(x) \right\|_K = 0$$

for all $m, n > 0$ where $\|\ \|_K$ denotes the norm in K.

Now choose $a, b \in \mathbb{Z}$ with $b > 0$, such that a and b are relatively prime. Let W_i $(i = 1, 2)$ be unitary operators on K satisfying

$$W_i^b = I, \qquad i = 1, 2,$$

$$W_2 W_1 = \exp\left(-\frac{2\pi a i}{b}\right) W_1 W_2. \tag{2.5}$$

For $i = 1, 2$ we define the operators V_i on $S(\mathbb{R})$ by

$$(V_1 \xi)(s) = \xi(s - \lambda)$$

$$(V_2 \xi)(s) = e^{2\pi i s} \xi(s) \tag{2.6}$$

for $\xi \in S(\mathbb{R})$, $s \in \mathbb{R}$ where $\lambda = \frac{a}{b} - \theta$.

It is not difficult to verify that $\mathcal{E}$ becomes a right A_θ^∞-module where the action of A_θ^∞, as operators on the right, is given by

$$U_i = V_i \otimes W_i. \tag{2.7}$$

Furthermore [23], $\mathcal{E}$ is finitely generated and projective.

For $X \in A_\theta^\infty$ of the form (1.6), the faithful trace τ on A_θ^∞ is given by

$$\tau(X) = s(0, 0). \tag{2.8}$$

The Lie algebra of T^2 is $\mathbb{R}^2$ and a basis for the representation of $\mathbb{R}^2$ in $\mathrm{Der}(A_\theta^\infty)$ is given by $\{\delta_k, k = 1,2\}$ where

$$\delta_k(U_j) = 2\pi i U_j \qquad (j = k),$$

$$\delta_k(U_j) = 0 \qquad (j \neq k). \tag{2.9}$$

Using (2.1), (2.9) and (2.8) in (1.6), it is easily verified that assumption (*a*) is satisfied.

A hermitian structure on $\mathcal{Z}$ is given by

$$\langle \xi, \eta \rangle_A = \sum_{m,n \in \mathbb{Z}} \langle \xi, \eta \rangle(m,n) U_1^m U_2^n$$

where for $\xi, \eta \in \mathcal{Z}$,

$$\langle \xi, \eta \rangle(m,n) = \int_{-\infty}^{\infty} \langle W_2^n W_1^m \xi(s - m\varepsilon), \eta(s) \rangle_K e^{-2\pi i n s}\, ds. \tag{2.10}$$

Applying (2.8) in (2.10), we see that $h_0 = L^2(\mathbb{R}, K)$.

We postpone a discussion of connections in $\mathcal{Z}$ to §4.

3. QUANTUM STOCHASTIC PARALLEL TRANSPORT IN PROJECTIVE MODULES

Let H denote the symmetric Fock space over $L^2(\mathbb{R}^+, \mathbb{C}^d)$ where $d \geqslant 1$ and let D be the dense subspace of H comprising finite, linear combinations of exponential vectors $\{\psi(g), g \in L^2(\mathbb{R}^+, \mathbb{C}^d)\}$. Annihilation and creation operators $\{a^\#(f), f \in L^2(\mathbb{R}^+, \mathbb{C}^d)\}$ are densely defined and mutually adjoint on D and yield a representation of the canonical commutation relations

$$[a(f), a(g)] = 0, \qquad [a(f), a^\dagger(g)] = \langle f, g \rangle I \tag{3.1}$$

for all $f, g \in L^2(\mathbb{R}^+, \mathbb{C}^d)$.

For $\alpha \in L^\infty(\mathbb{R}^+, M_d(\mathbb{C}))$ its differential second quantisation $d\Gamma(\alpha)$ is densely defined on E as the infinitesimal generator of the one parameter group which maps each $\psi(f)$ to $\psi(e^{it\alpha} f)$.

Let $\{e^j, 1 \leqslant j \leqslant d\}$ be the natural basis in $\mathbb{C}^d$, so each $e^j = (0,\ldots,0,\overset{(j)}{1},0,\ldots,0)$ and let $\{E_j^i, 1 \leqslant i,j \leqslant d\}$ be that basis of $M_d(\mathbb{C})$ wherein each E_j^i is the matrix whose only non-zero entry is the (i,j)th, this being 1.

Identifying $L^2(\mathbb{R}^+, \mathbb{C}^d)$ with $L^2(\mathbb{R}^+) \otimes \mathbb{C}^d$ and $L^\infty(\mathbb{R}^+, M_d(\mathbb{C}))$ with $L^\infty(\mathbb{R}^+) \otimes M_d(\mathbb{C})$, the three fundamental families of processes in H are those of *annihilation*, *creation* and *conservation*, defined, respectively, by

$$A^j(t) = a(\chi_{[0,t)} \otimes e^j)$$

$$A^{\dagger j}(t) = a^\dagger(\chi_{[0,t)} \otimes e^j)$$

$$\Lambda^i_j(t) = d\Gamma(\chi_{[0,t)} \otimes E^i_j) \tag{3.2}$$

for $1 \leqslant i, j \leqslant d$, $t \geqslant 0$.

We refer the reader to the basic paper [19] and the article by M Evans and R L Hudson in this volume for the details of how a stochastic calculus in H may be constructed using the processes (3.2) as integrators.

Let $h = h_0 \otimes H$ where h_0 is the completion of a finitely generated, projective A^∞-module $\mathbb{Z}$, as described in §2. We will be concerned with semi-martingales $M = (M(t), t \geqslant 0)$ in h of the form

$$M(t) = M(0) + \int_0^t (F_j \, dA^{\dagger j} + G^j_k \, d\Lambda^k_j + H_j \, dA^j + L \, dt) \tag{3.3}$$

where each of the coefficient processes are adapted and square integrable. We further require that each of the operators $F_j(t)$, $G^j_k(t)$, $H_j(t)$ and $L(t)$ has domain $\mathbb{Z} \underline{\otimes} D$ and range $\mathbb{Z} \otimes H$.

If M_i ($i = 1, 2$) are semi-martingales, their product is again a semi-martingale whose coefficients are determined by the quantum Ito formula [19]

$$d(M_1 M_2) = dM_1 . M_2 + M_1 . dM_2 + dM_1 . dM_2 \tag{3.4}$$

where $dM_1 . dM_2$ is evaluated by bilinear extension of the following table.

	$d\Lambda^p_q$	$dA^{\dagger r}$	dA_r	dt
$d\Lambda^i_j$	$\delta^p_j \Lambda^i_q$	$\delta^r_j dA^{\dagger i}$	0	0
dA_k	$\delta^p_k dA_q$	$\delta^r_k dt$	0	0
$dA^{\dagger k}$	0	0	0	0
dt	0	0	0	0

Now let ∇ be a connection on $\mathbb{Z}$ and $\{X_1, \ldots, X_n\}$ be a fixed orthonormal basis for the Lie algebra $\mathbb{L}$. A *quantum stochastic parallel transport* in h is a unitary operator valued process $U = (U(t), t \geqslant 0)$ satisfying the stochastic differential equation

$$dU = U(\nabla_{X_j} dM^j + \tfrac{1}{2}(\nabla_{X_j} dM^j)^2)$$

$$U(0) = I \tag{3.5}$$

where $(M_1, \ldots, M_n)$ are semi-martingales of the form (3.3).

The existence of a solution to (3.8) is guaranteed by the theory of [19]. However, as was shown therein, the requirement of unitarity imposes stringent restrictions upon the choice of coefficients in (3.3).

The *deterministic* form of (3.5) is obtained when all pure noise terms vanish and is hence of the form

$$dU = U\nabla_{X_j} L^j \, dt \tag{3.6}$$

and the unitarity requirement is clearly satisfied whenever, $(1 \leqslant j \leqslant n)$

$$\nabla_{X_j} L^j - L^{j*}\nabla_{X_j} = 0. \tag{3.7}$$

For example, if each $L^j = c^j I$ where c^j is an integrable real valued function on $\mathbb{R}^+$, we may write (3.6) as

$$dU = U\nabla_{c(t)} \, dt \tag{3.8}$$

where, for $t \in \mathbb{R}^+$, $c(t) = c^j(t)X_j$ describes the velocity along the curve $\exp \int_0^t c(s)\, ds$ in G. The existence of solutions to (3.8) may be established by considering the convergence of its Dyson series.

We now consider an example of (3.6) where the noise terms are present.

Example 3.1 *Classical Stochastic Parallel Transport*

Let ∇ be a connection on the complex vector bundle E of example 2.1. Let each semi-martingale M_p $(1 \leqslant p \leqslant n)$ in (3.8) be of the form (3.3) with each $G^j_k = 0$ and each $F_j = H_j$, thus for $1 \leqslant p \leqslant n$, $t > 0$

$$M_p(t) = M_p(0) + \int_0^t (H_j dQ^j + L \, dt) \tag{3.9}$$

where $Q^j = A^j + A^{\dagger j}$ $(1 \leqslant j \leqslant d)$.

Let $C = \{\omega\colon \mathbb{R}^+ \longrightarrow \mathbb{R}^d;\ \omega$ is continuous with $\omega(0) = 0\}$ and let P denote Wiener measure on C. We identify H with the space $L^2(C,P)$ by means of the duality transformation which associates exponential vectors in H to exponential martingales [19]. Under this correspondence, the processes $\{Q^j, 1 \leqslant j \leqslant d\}$ are identified with the components of d-dimensional Brownian motion on C.

(3.5) is now the Ito form of the usual stochastic parallel transport equation in E [21]. Because the Ito integral does not transform nicely under local changes of co-ordinates, (3.5) is usually written in Stratonovitch form

$$dU = U\nabla_{X_j} \circ dM^j,$$

$$U(0) = I. \tag{3.10}$$

If a unitary solution to (3.10) exists, it generates horizontal lifts of a diffusion process $\eta = (\eta(t), t \geqslant 0)$ on V, which satisfies the stochastic differential equation

$$d\eta(t) = X_j(\eta(t)) \circ dM^j$$

$$\eta(0) = x \tag{3.11}$$

where $x \in V$. The lifting of η is a process $\pi_x = (\pi_x(t), t \geqslant 0)$ where each $\pi_x(t)$ is (almost surely) an isomorphism between the fibre at $\eta(t)$ and the fibre at x, given by

$$\pi_x(t)(Y(\eta(t))) = (U(t)Y)(x) \tag{3.12}$$

for each $Y \in \Gamma(E)$. (Note that (3.11) holds almost surely [21].)

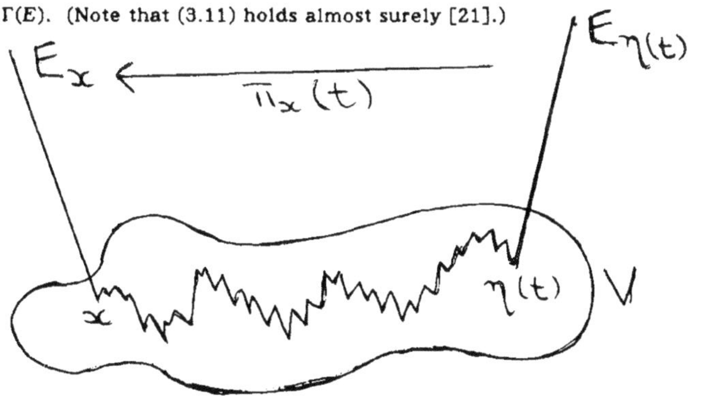

Remark: The above discussion of classical stochastic parallel transport is critically dependent upon the Hermitian structure on E, indeed the isomorphisms $\pi_x(t)$ will preserve the inner products on the individual fibres. A detailed comparison of this approach with those given in [3] and [13] is clearly required.

For the remainder of this article, we will concentrate on exploring a non-commutative example.

4. QUANTUM STOCHASTIC PARALLEL TRANSPORT IN HEISENBERG MODULES

Let $\mathbb{Z} = S(\mathbb{R}, K)$ be a Heisenberg module over A_θ^∞ as constructed in example 2.2. Connes and Rieffel in [9] have defined a Yang-Mills action functional on the space $CC(\mathbb{Z})$. We will not explore the details of this notion here (the interested reader should consult [9]) but note that this functional is minimised by the connection with components ∇_i $(i = 1, 2)$, corresponding to the derivations δ_i $(i = 1, 2)$, given by

$$(\nabla_1\xi)(s) = 2\pi i\left(\frac{s}{\lambda}\right)\xi(s), \qquad (\nabla_2\xi)(s) = \frac{d\xi}{ds}(s). \tag{4.1}$$

The curvature of ∇ is constant and is determined, from (2.4), by

$$[\nabla_1, \nabla_2] = -\frac{2\pi i}{\lambda}I. \tag{4.2}$$

If we choose a, b and θ so that $\frac{2\pi}{\lambda} = \hbar^{-1}$ and write $q = -i\hbar\nabla_1$, $p = -i\hbar\nabla_2$, then q and p are the familiar quantum mechanical position and momentum operators on $\mathbb{Z}$ and (4.2) is the Heisenberg commutation relation.

We will study an example of quantum stochastic parallel transport in $\mathbb{Z}$ in which ∇ is given by (4.1), $d = 1$ and the M_i's $(i = 1, 2)$ are linear combinations of annihilation and creation processes with time independent coefficients, i.e.

$$M_i = \alpha_i A + \beta_i A^\dagger \tag{4.3}$$

where $\alpha_i, \beta_i \in \mathbb{C}$ $(i = 1, 2)$. (3.5) now becomes

$$\begin{aligned} dU = U[&(\alpha_1\nabla_1 + \alpha_2\nabla_2)dA + (\beta_1\nabla_1 + \beta_2\nabla_2)dA^\dagger \\ &+ \tfrac{1}{2}(\alpha_1\beta_1\nabla_1^2 + \alpha_1\beta_2\nabla_1\nabla_2 + \alpha_2\beta_1\nabla_2\nabla_1 + \alpha_2\beta_2\nabla_2^2)\, dt]. \end{aligned} \tag{4.4}$$

A necessary condition [19] for (4.4) to yield a unitary solution is that (4.4) can be written in the form

$$dU = U[L^\dagger dA - L dA^\dagger + (iH - \tfrac{1}{2}L^\dagger L)\, dt] \tag{4.5}$$

where L and H are operators on h_0 with common invariant domain $\mathbb{Z}$ such that H is self-adjoint.

The only way to recover (4.5) from (4.4) is to choose $\beta_i = \overline{\alpha_i}$ $(i = 1, 2)$ in (4.3), whence $H = 0$ and $L = \alpha_1\nabla_1 + \alpha_2\nabla_2$.

We may now proceed along the same lines as the analysis of [18], i.e. we introduce annihilation and creation operators on h_0,

$$a = -\frac{i\hbar}{\sqrt{2}}(\nabla_1 + i\nabla_2), \qquad a^\dagger = -\frac{i\hbar}{\sqrt{2}}(\nabla_1 - i\nabla_2)$$

which satisfy the commutation relation

$$[a, a^\dagger] = \hbar I \tag{4.6}$$

so that $L = \hbar^{-1}(\varepsilon a + \eta a^\dagger)$ where

$$\varepsilon = \frac{1}{\sqrt{2}}(\alpha_2 + i\alpha_1) \quad \text{and} \quad \eta = \frac{1}{\sqrt{2}}(\alpha_2 - i\alpha_1).$$

Now, by means of gauge transformations and linear canonical transformations of the pair $(a, a^\dagger)$ and gauge transformations of $(A, A^\dagger)$ we may transform the pair of coefficients (ε, η) to either of the three forms (ρ, ρ), $(\rho, 0)$ and $(0, \rho)$ where ρ is a non-negative real number [18].

We hence obtain three canonical forms of (4.5)

$$dU = U(\tilde{\rho}\nabla_1 dP + \tfrac{1}{2}\tilde{\rho}^2\nabla_1^2\, dt) \tag{4.7}$$

$$dU = U(\tilde{\rho}a^\dagger dA - \tilde{\rho}a dA^\dagger - \tfrac{1}{2}\tilde{\rho}^2 a^\dagger a\, dt) \tag{4.8}$$

$$dU = U(\tilde{\rho}a dA - \tilde{\rho}a^\dagger dA^\dagger - \tfrac{1}{2}\tilde{\rho}^2 a a^\dagger\, dt) \tag{4.9}$$

where $\tilde{\rho} = \rho\hbar^{-1}$.

In (4.7) we may realise the process $P = -i(A - A^\dagger)$ as a classical Brownian motion on Wiener space, and write the solution explicitly as

$$U(t) = e^{\tilde{\rho}\nabla_1 P(t)} \tag{4.10}$$

for $t \geqslant 0$.

(4.8) and (4.9) are the quantum Ornstein-Uhlenbeck and anti-Ornstein-Uhlenbeck processes, respectively [18]. Unitarity of the solution of the former was rigorously established in [17].

We may now apply the results of [18] to obtain dilations of semigroups associated with the processes generated by U in each of the cases (4.7) to (4.9), e.g. in (4.8) we take vacuum expectations to obtain, for $t \geqslant 0$,

$$\langle\psi(0), U(t)\psi(0)\rangle = e^{tS} \tag{4.11}$$

where $S = \tfrac{1}{4}\rho^2(\nabla_1^2 + \nabla_2^2 + \hbar^{-1}I)$.

5. THE CONSTRUCTION OF HORIZONTAL LIFTS

A natural geometrical question which arises from the analysis of §4 is whether our parallel transport operators induce horizontal lifts of a diffusion process in the algebra A_θ^∞. We begin by returning to the classical case as discussed at the end of §3. Here U gives rise to a horizontal lift of η through (3.12). (Pictorially, the lift is a curve in the total space of E traced out by the process π_x.) The following result gives us a hint as to how to reformulate this notion in our non-commutative framework.

Proposition 1. U gives rise to a horizontal lift of η in the bundle E if and only if

$$U(t)(fY)(x) = f(\eta(t))U(t)(Y)(x) \tag{5.1}$$

holds almost surely, for $t \geqslant 0$, $f \in C^\infty(V)$, $Y \in \Gamma(E)$, $x \in V$.

Proof. By the module structure of $\Gamma(E)$, (3.12) holds if and only if

$$\pi_x(t)(fY)(\eta(t)) = U(t)(fY)(x) \quad a.s.$$

$$\Longleftrightarrow f(\eta(t))\pi_x(t)(Y(\eta(t)) = U(t)(fY)(x) \ a.s.$$

$$\Longleftrightarrow f(\eta(t))(U(t)Y)(x) = U(t)(fY)(x) \ a.s.$$

as required. □

To make our non-commutative generalisation, we need the concept of a *quantum diffusion process* on a smooth algebra A^∞. This has been defined by R L Hudson (see [15], [16] and his article with M Evans in this volume) to be a family of injective *-homomorphisms $(j_t,\ t \geqslant 0)$ of A^∞ into $B(h)$ such that for all $X_0 \in A^\infty$, the process $X = (X(t),\ t \geqslant 0)$, where each $X(t) = j_t(X_0)$, satisfies a stochastic differential equation of the form

$$dX = \alpha_j(X)dA^{\dagger j} + \lambda_j^i(X)d\Lambda_i^j + \alpha_j^\dagger(X)dA^j + \tau(X)dt \tag{5.2}$$

where each of α_j, λ_j^i, $\alpha_j^\dagger$ and τ are mappings from A^∞ into itself with $a_j^\dagger(X_0) = (\alpha_j(X_0^*))^*$ for all $X_0 \in A^\infty$.

Generalising (5.1), we say that a quantum stochastic parallel transport U induces *horizontal lifts* if there exists a quantum diffusion process $(j_t,\ t \geqslant 0)$ on a^∞ such that

$$U(t)(\xi a) \otimes \psi(f) = (U(t)\xi \otimes \psi(f))j_t(a) \tag{5.3}$$

for all $t \geqslant 0$, $\xi \in \mathbb{Z}$, $a \in A^\infty$, $f \in L^2(\mathbb{R}^+, \mathbb{C}^d)$. [Note that on the right hand side of (5.3), we are regarding h as a right $B(h)$-module.] If it exists, we will call the process $(j_t, t \geqslant 0)$ arising in (5.3), the *underlying diffusion*. Given a quantum stochastic parallel transport process U, it follows from the results of [18] and [19] that the process $AdU = (Ad\,U(t),\, t \geqslant 0)$ where

$$(Ad\,U(t))X = U(t)XU(t)^{\dagger} \quad (t \geqslant 0,\ X \in B(h_0))$$

is a diffusion process on the "flat" algebra $B(h_0)$.

Proposition 2. A quantum stochastic parallel transport process U induces horizontal lifts if and only if $Ad\,U$ is a diffusion process on A^{∞}, in which case $Ad\,U$ is the unique underlying diffusion.

Proof. For the duration of this proof, we will regard $\mathbb{Z}$ as a left A^{∞}-module so that we may have the convenience of writing all our operators on the left. (5.3) now takes the form

$$U(t)a\xi \otimes \psi(f) = j_t(a)U(t)\xi \otimes \psi(f).$$

Now, since $\mathbb{Z}$ is dense in h_0 and exponential vectors are total in Fock space, we have, by linearity and boundedness

$$U(t)a = j_t(a)U(t)$$

and the result follows. □

Corollary 3. U induces horizontal lifts if and only if the coefficients in the stochastic differential of $Ad\,U$ leave A^{∞} globally invariant.

Proof. This is immediate from Proposition 2 and the definition of quantum diffusion process. □

Corollary 3 yields the means for testing whether horizontal lifts exist. As we will see in the examples below, this is by no means always the case.

We now specialise to the case of Heisenberg modules over the torus algebra A_θ^{∞}. Recall that each of our three canonical forms of parallel transport are of the form (4.5). Hence ([18], [19]) we obtain a quantum diffusion process, as described above, on $B(h_0)$ by

$$j_t(X_0) = U(t)X_0U(t)^{\dagger} \tag{5.4}$$

with

$$dX = [L^{\dagger}, X]dA - [L, X]\,dA^{\dagger} + N(X)\,dt \tag{5.5}$$

where N is the Lindblad generator

$$N(X) = -\tfrac{1}{2}(L^{\dagger}LX - 2L^{\dagger}XL + XL^{\dagger}L). \tag{5.6}$$

Now let $X_0 \in A_\theta^\infty$ in (5.4). Clearly $\{j_t,\, t \geqslant 0\}$ is a family of *-homomorphisms of A_θ^∞ into $B(h)$. If this is to be a diffusion process, the restriction of (5.5) to A_θ^∞ must yield an expression of the form

$$dX = \alpha^\dagger(X)\,dA + \alpha(X)\,dA^\dagger + \tau(X)\,dt \tag{5.7}$$

where ([15], [16]), α and τ are constrained to satisfy the cohomological relation

$$\tau(ab) - a\tau(b) - \tau(a)b = \alpha^\dagger(a)\alpha(b) \tag{5.8}$$

for $a, b \in A_\theta^\infty$.

We examine each of the three cases (4.7) to (4.9) in turn.

Case I - equation (4.7)

In this case, for $X_0 \in B(h_0)$, (5.5) becomes

$$dX = \tilde{\rho}[\nabla_1, X]\,dP + \tfrac{1}{2}\tilde{\rho}^2[\nabla_1, [\nabla_1, X]]\,dt. \tag{5.9}$$

If we restrict $X_0 \in A_\theta^\infty$, and apply (2.2), we obtain

$$dX = \tilde{\rho}\delta_1(X)\,dP + \tfrac{1}{2}\tilde{\rho}^2\delta_1^2(X)\,dt \tag{5.10}$$

and (5.8) is satisfied with $\alpha = \alpha^\dagger = \tilde{\rho}\delta_1$ and $\tau = \frac{1}{2}\tilde{\rho}^2\delta_1^2$ (c.f. [16]). Hence (4.10) induces horizontal lifts of the process given by (5.10).

Remark: Note that the parallel transport equation

$$dU = U(\tilde{\rho}\nabla_2\,dQ + \tfrac{1}{2}\tilde{\rho}^2\nabla_2^2\,dt) \tag{5.11}$$

is reducible to the canonical form (4.7) via the gauge transformations $a \longrightarrow ia$, $A \longrightarrow -iA$.

Furthermore, (5.11) yields a horizontal lift of

$$dX = \tilde{\rho}\delta_2(X)\,dQ + \tfrac{1}{2}\tilde{\rho}^2\delta_2^2(X)\,dt \tag{5.12}$$

so there is no asymmetry in our treatment of δ_1 and δ_2.

Case II - equation 4.7

$L = \frac{i\rho}{\sqrt{2}}(\nabla_1 - i\nabla_2)$, thus by (2.2) we find $\alpha = \frac{i\rho}{\sqrt{2}}(\delta_1 - i\delta_2)$. However, by the analysis of [16], we know that for this α we cannot find a τ such that (5.8) is satisfied. Hence, in this case, we have an example of a quantum parallel transport process which cannot be the horizontal lift of a diffusion process on the underlying smooth algebra.

Indeed, by (5.5) we see that if such a τ were to exist, it would be the restriction of N to A_θ^∞. However, a direct computation yields

$$N(a)\psi = -i(\nabla_1\psi)\delta_2(a) + i(\nabla_2\psi)\delta_1(a) + \psi(\delta_1^2(a) + \delta_2^2(a)) \tag{5.13}$$

for $a \in A_\theta^\infty$, $\psi \in \mathcal{Z}$. Clearly, A_θ^∞ is not preserved by the action of N.

Case III - equation 4.8

$L = \frac{i\rho}{\sqrt{2}}(\nabla_1 + i\nabla_2)$, $\alpha = \frac{i\rho}{\sqrt{2}}(\delta_1 + i\delta_2)$and by a similar argument to that of case II, we find that there is no diffusion on A_θ^∞ corresponding to this parallel transport.

REFERENCES

1. D Applebaum, *Quasi-free Stochastic Evolutions* in Quantum Probability and Applications II, ed L Accardi and W von Waldenfels (Springer LNM 1136), 46-57 (1985).

2. C Barnett, R F Streater, I F Wilde, *Quasi-free Quantum Stochastic Integrals for the CAR and CCR*, J Funct Anal **52**, 19-47 (1983).

3. J M Bismut, *Mécanique Aléatoire* (Springer LNM Vol 866) (1981).

4. J M Bismut, *The Atiyah-Singer Index Theorem for Families of Dirac Operators: Two Heat Equation Proofs*, Invent Math **83**, 91-151 (1986).

5. A Connes, *C*-algèbres et Géométrie Différentielle*, C R Acad Sc Paris t **290**, Série A, 599-604 (1980).

6. A Connes, *An Analogue of the Thom isomorphism for Crossed Products of a C*-algebra by an Action of* ℝ, Adv Math **39**, 31-55 (1981).

7. A Connes, *A Survey of Foliations and Operator Algebras*, Proc Symp Pure Math **38** (Amer Math Soc Providence), 521-628 (1982).

8. A Connes, *Non-commutative Differential Geometry*, (*Parts I and II*), IHES Publ Math **62**, 41-144 (1985).

9. A Connes, M Rieffel, *Yang Mills for Non-commutative Two-Tori* (preprint).

10. S S Chern, *Complex Manifolds Without Potential Theory*, Van Nostrand (1967).

11. C Chevalley, *Theory of Lie Groups Volume I*, Princeton University Press (1946).

12. E G Effros, F Hahn, *Locally Compact Transformation Groups and C*-algebras*, Mem Amer Math Soc (1967).

13. K D Elworthy, *Stochastic Differential Equations on Manifolds*, Cambridge University Press (1982).

14. S Helgason, *Differential Geometry and Symmetric Spaces*, Academic Press (1962).

15. R L Hudson, *Algebraic Theory of Quantum Diffusions* (preprint).

16. R L Hudson, *Quantum Diffusions and Cohomology of Algebras*, to appear in Proceedings of First World Congress of the Bernouilli Society.

17. R L Hudson, P D F Ion, K R Parthasarathy, *Time-orthogonal Unitary Dilations and Non-commutative Feynman-Kac Formulae*, Commun Math Phys **83**, 261-80 (1982).

18. R L Hudson, K R Parthasarathy, *Construction of Quantum Diffusions* in Quantum Probability and Applications, ed L Accardi, A Frigerio, V Gorini (Springer LNM 1055), 173-99 (1984).

19. R L Hudson, K R Parthasarathy, *Quantum Ito's Formula and Stochastic Evolution*, Commun Math Phys **93**, 301-23 (1984).

20. N Ikeda, S Watanabe, *Stochastic Differential Equations and Diffusion Processes*, North Holland/Kodansha (1981).

21. H Kunita, *Some Extensions of Ito's Formula*, in Seminaire de Probabilités XV, ed J Azéma, M Yor (Springer LNM 850), 118-42 (1980).

22. M A Rieffel, *C*-algebras Associated with Irrational Rotations*, Pacific J Math, **93**, 415-29 (1981).

23. M A Rieffel, *The Cancellation Theorem for Projective Modules Over Irrational Rotation C*-algebras*, Proc London Math Soc, **47**, 285-302 (1983).

24. R G Swan, *Vector Bundles and Projective Modules*, Trans Amer Math Soc, **105**, 264-77 (1962).

25. C de Witt-Morette, K D Elworthy, B L Nelson, G S Sammelson, *A Stochastic Scheme for Constructing Solutions of the Schrödinger Equation*, Ann Inst H Poincaré, Vol XXXII, 327-41 (1980).

INPUT AND OUTPUT CHANNELS IN QUANTUM SYSTEMS AND QUANTUM STOCHASTIC DIFFERENTIAL EQUATIONS

Alberto Barchielli
Dipartimento di Fisica, Università di Milano
Istituto Nazionale di Fisica Nucleare, Sezione di Milano
Via Celoria 16, 20133 Milano, Italy

1. Noise in quantum open systems and quantum stochastic calculus.

The aim of this paper is to show how quantum stochastic differential equations (QSDE's) [1-3] can be used for treating input and output channels in quantum systems [4,5]. A typical example could be an atom driven by a laser (the input channel) and emitting fluorescence light (the output channel) [6]. However, quantum stochastic calculus (QSC) was developed as a mathematical theory of quantum noise and to apply it to the description of quantum channels needs a change of point of view with respect to the original motivations. In order to stress these differences, first the well known application of QSDE's to the treatment of noise in quantum open systems is shortly reviewed (Sec.1); only afterwards the problem of how quantum channels can be described by using QSC is considered (Sec.2). Finally (Sec.3), a concrete physical example taken from single ion spectroscopy is given. The emphasis will be on the physical concepts, not on the mathematical formalism. Also when very general mathematical results are available, only the simplest cases are presented.

Let Γ be the symmetric Fock space over $L^2(\mathbb{R})$ and $\psi(f)$, $f \in L^2(\mathbb{R})$, be the exponential vectors in Fock space,

$$\psi(f) = \left(1, f, \ldots, (n!)^{-1/2} f\otimes\ldots\otimes f, \ldots\right). \tag{1.1}$$

$\{\psi(f),\ f \in L^2(\mathbb{R})\}$ is a total family in Γ, $\psi(0)$ is the Fock vacuum and we have

$$\langle\psi(g), \psi(f)\rangle = \exp\langle g,f\rangle\ , \qquad f,g \in L^2(\mathbb{R}). \tag{1.2}$$

Denote by $A_t^\dagger$, A_t, Λ_t the creation, annihilation and gauge (or number) processes, defined by

$$A_t^\dagger\psi(f) = \frac{d}{d\varepsilon}\, \psi(f+\varepsilon\chi_{[0,t]})\Big|_{\varepsilon=0}, \qquad A_t\psi(f) = \int_0^t ds\ f(s)\ \psi(f), \tag{1.3a}$$

$$\Lambda_t\psi(f) = \frac{d}{d\varepsilon}\, \psi(e^{\varepsilon\chi_{[0,t]}} f)\Big|_{\varepsilon=0}, \tag{1.3b}$$

where $f \in L^2(\mathbb{R})$ and $\chi_{[0,t]}$ is the indicator function of the interval $[0,t]$. By using the heuristic notation usually adopted in theoretical physics, we can say that we have a Bose field $a(t)$, $a^\dagger(t)$ (in Fock representation), satisfying CCR's

$$[a(t),a^{\dagger}(s)] = \delta(t-s), \qquad [a(t),a(s)] = 0, \qquad [a^{\dagger}(t),a^{\dagger}(s)] = 0. \tag{1.4}$$

Then we can write

$$A_t = \int_0^t ds\ a(s) \equiv a(\chi_{[0,t]}), \qquad A_t^{\dagger} = \int_0^t ds\ a^{\dagger}(s) \equiv a^{\dagger}(\chi_{[0,t]}), \tag{1.5a}$$

$$\Lambda_t = \int_0^t ds\ a^{\dagger}(s)\ a(s). \tag{1.5b}$$

A QSC of Itô type, based on the integrators $dA_t^{\dagger}$, dA_t, $d\Lambda_t$ (and dt), has been developed by Hudson and Parthasarathy (see, for instance, Ref.[2]). This calculus obeys very simple formal rules, which can be summarized in the fact that the fundamental integrators "point into the future" and commute with adapted processes (Ref.[2], Def.3.1 and Theor.4.5) and in the fact that these integrators satisfy simple multiplication rules (Ref.[2], eq.(7.1)).

The Hudson and Parthasarathy's theory allows, in particular, to give meaning to "stochastic Schrödinger equations", in which $dA_t^{\dagger}$, dA_t play the role of a noncommutative analogue of white noise. Let h_0 (initial space) be an Hilbert space (representing the Hilbert space of some system S) and consider the following QSDE for operators in $h_0 \otimes \Gamma$

$$dU_t = [-R^{\dagger}dA_t + RdA_t^{\dagger} - (\tfrac{1}{2} R^{\dagger}R + iH)dt]U_t, \qquad U_0 = \mathbb{1}, \tag{1.6}$$

where $R,H \in B(h_0)$ (bounded operators on h_0) and $H=H^{\dagger}$. Here and in the following we identify R with $R \otimes \mathbb{1}$, A_t with $\mathbb{1} \otimes A_t$ and so on. The solution U_t of this equation exists and is unique; $\{U_t,\ t \geq 0\}$ is an adapted process and, for any $t \geq 0$, U_t is a unitary operator on $h_0 \otimes \Gamma$ (Ref.[2], Sec.7).

Let now θ_t be the shift on Γ defined by

$$\theta_t \psi(f) = \psi(f_t), \qquad f_t(s) = f(t+s). \tag{1.7}$$

By identifying θ_t with $\mathbb{1} \otimes \theta_t$ and setting

$$\bar{U}_t := \theta_t U_t, \quad t \geq 0, \qquad \bar{U}_t := \bar{U}_{|t|}^{\dagger}, \quad t < 0, \tag{1.8}$$

then $\{\bar{U}_t,\ t \in \mathbb{R}\}$ is a one-parameter group of unitary operators on $h_0 \otimes \Gamma$ [7].

By these results, U_t can be interpreted as the evolution operator for the system S interacting with the quantum noise described by A_t, $A_t^{\dagger}$, in the interaction picture with respect to the free dynamics θ_t of the noise. The operator H in eq.(1.6) represents the "free" Hamiltonian for system S and the expression $i(RdA_t^{\dagger} - R^{\dagger}dA_t)$ gives the interaction between system S and noise. The term $-\frac{1}{2} R^{\dagger}Rdt$ is a correction term due to the fact that we are using a stochastic calculus of Itô type. Formally, it is possible to introduce also a QSC of Stratonovich type [4]; in this case no correction term would appear in the equation for the evolution operator.

Equation (1.6) is a stochastic equation generalizing Schrödinger equation. Also stochastic equations of Heisenberg type for systems operators can be introduced [1].

For any $X \in B(h_0)$ define

$$X_t := U_t^\dagger X U_t \tag{1.9}$$

and, similarly,

$$H_t := U_t^\dagger H U_t, \qquad R_t := U_t^\dagger R U_t; \tag{1.10}$$

then X_t satisfies the following QSDE

$$dX_t = i[H_t,X_t]dt + [R_t^\dagger,X_t]dA_t + [X_t,R_t]dA_t^\dagger + \frac{1}{2}\left([R_t^\dagger,X_t]R_t + R_t^\dagger[X_t,R_t]\right)dt. \tag{1.11}$$

This equation can be easily derived from eq.(1.6) for U_t and its adjoint equation for $U_t^\dagger$, by using the formal rules of QSC mentioned before.

Consider now the vacuum conditional expectation E_0: $B(h_0 \otimes \Gamma) \to B(h_0)$, defined by

$$\langle u,E_0(J)v\rangle = \langle u\otimes\psi(0),Jv\otimes\psi(0)\rangle, \quad u,v \in h_0, \quad J \in B(h_0\otimes\Gamma). \tag{1.12}$$

Then we have the following result (Ref.[2], Theor.8.1): the family $\{T_t, t \geq 0\}$, where

$$T_t X := E_0(U_t^\dagger X U_t), \qquad X \in B(h_0), \tag{1.13}$$

is a uniformly continuous one-parameter semigroup of completely positive maps on $B(h_0)$, whose infinitesimal generator L' is given by

$$L'X = i[H,X] - \frac{1}{2}(R^\dagger RX + XR^\dagger R) + R^\dagger XR, \qquad X \in B(h_0). \tag{1.14}$$

One can say that U_t gives a unitary dilation of the dynamical semigroup T_t on $B(h_0)$. By using many independent noises A_t^j, $A_t^{j\dagger}$, dilations of more general semigroups can be obtained [3].

In a language more usual in the physical literature, the result above can be restated in the following way. Let ρ be the initial state for system S (ρ is a normalized, positive, trace-class operator on h_0) and let the initial state for the noise be the Fock vacuum. The expression

$$\rho(t) := \mathrm{Tr}_\Gamma\{U_t(\rho\otimes|\psi(0)\rangle\langle\psi(0)|)U_t^\dagger\} \tag{1.15}$$

(where Tr_Γ is the partial trace over Fock space) represents the reduced dynamics of system S. Then $\rho(t)$ satisfies the <u>quantum master equation</u>

$$\frac{d}{dt}\rho(t) = L\rho(t), \tag{1.16a}$$

$$L\rho = i[H,\rho] + \frac{1}{2}\left([R,\rho R^\dagger] + [R\rho,R^\dagger]\right); \tag{1.16b}$$

L is a map on the trace class on h_0 defined by eq.(1.16b) and it is usually called the Liouville operator; the map L', given by eq.(1.14), is its dual in $B(h_0)$. Formally, eqs.(1.16) can be obtained by taking the quantum expectation of eq.(1.11) with respect to the state $\rho\otimes|\psi(0)\rangle\langle\psi(0)|$. Another important result is the so called quantum regression theorem, which says that certain time-ordered multi-time correlation functions for system operators can be reexpressed by means of the reduced dynamics exp(Lt) (Ref.[4], Sec.IV.D, and Ref.[7]).

In a pictorial language, what we have done in this section is the following. We have an open system S interacting with some reservoir. For describing this situation we make certain approximations. First, the physical reservoir is replaced by a Bose field (or, in general, by many Bose fields). Then, the interaction between system S and fields is chosen so singular that, when the degrees of freedom of the fields are eliminated, the reduced dynamics of system S obeys exactly to a quantum master equation; moreover, the quantum regression theorem holds, not as an approximation, but as an exact result. The approximations usually needed for eliminating the "memory" terms from the equation for the reduced dynamics are made directly on the dynamics of the total system (system S + reservoir). Note that in this situation we are not interested on what happens to the reservoir, but only on system S. If we want to get information on S, we must perform some measurement directly on it.

2. Input and output channels.

Consider now a different situation, described by the following scheme: an input channel, a system S (possibly interacting with some reservoir), an output channel, a detector. No direct observation is made on S, but the detector performs some measurement on the output channel and from this measurement we extract information on S. As before, we approximate the reservoir by some Bose fields; moreover, we replace also the physical channels by other Bose fields of the type described in the previous section. This is a strong approximation; we expect it to be a good approximation as far as "memory" contributions are not relevant. For simplicity, we continue to use a single field; therefore, there is no other external noise than that one due to the channel itself.

We call A_t, $A_t^\dagger$ the "input field" (it represents the field before the interaction with system S); this field is regarded now as a certain approximation of a physical field (typically the electromagnetic field). Some intuitive justification of this approximation is given in Ref.[4]. In this order of ideas, we can take as initial state for the field any state in Fock space. To change this initial state means that we are changing the preparation of the input channel. For instance, by taking as initial state for the field a (normalized) exponential vector, we can describe a coherent input signal (e.g., a laser field). We shall make use of this possibility only in the next section.

Consider now the output channel: we can describe it by using the fields after the interaction with system S. Therefore, we introduce the "output fields" (the fields in the Heisenberg picture) and the "output gauge process" by

$$A_t^{out} := U_t^\dagger A_t U_t, \qquad A_t^{out\dagger} := U_t^\dagger A_t^\dagger U_t, \qquad \Lambda_t^{out} := U_t^\dagger \Lambda_t U_t. \tag{2.1}$$

These output fields enjoy some nice properties. First, by using the localization properties of the input fields and of U_t, it is possible to show [5] that

$$A_t^{out} = U_T^\dagger A_t U_T, \qquad \forall T \geq t, \tag{2.2}$$

and similar equations for $A_t^{out\dagger}$ and Λ_t^{out}. In particular, by taking T sufficiently large, we see that the output fields are obtained from the input fields by a single unitary transformation and all the commutation rules are preserved. Secondly, the output fields satisfy the simple QSDE's [5]

$$dA_t^{out} = dA_t + R_t dt, \qquad dA_t^{out\dagger} = dA_t^\dagger + R_t^\dagger dt, \tag{2.3}$$

$$d\Lambda_t^{out} = d\Lambda_t + R_t^\dagger dA_t + R_t dA_t^\dagger + R_t^\dagger R_t dt, \tag{2.4}$$

where R_t is defined in eq.(1.10). Equations (2.3) and (2.4) are simple consequences of eqs.(2.1) and (1.6) and of the formal rules of QSC. From these equations one explicitly sees that the output fields contain information on system S: R is the system operator appearing in the interaction term in eq.(1.6).

Now we have to discuss the action of the detector. There are many possibilities. The detector can measure the "amplitude" of the output field, or can count the number of quanta in the output, or can make some more complicate measurement [5]. Here, we consider only counting processes: the detector counts the quanta in the output. The number of quanta up to time t in the input is described by the number operator Λ_t and the number of quanta in the output is given by Λ_t^{out}. The operators Λ_t^{out}, $t \geq 0$, are a family of commuting selfadjoint operators (from the definition (1.3b) and eqs.(2.1), (2.2) we have $[\Lambda_t,\Lambda_s] = [\Lambda_t^{out},\Lambda_s^{out}] = 0$) and, therefore, it exists a projection-valued measure describing the joint measurement of all these observables. Because infinitely many selfadjoint operators are involved, it is easier to work with the Fourier transform of this projection-valued measure.

For any $\phi \in L^\infty_{loc}(\mathbb{R})$ (locally bounded, measurable, real functions on $\mathbb{R}$) consider the unitary operator on Fock space

$$V_t[\phi] := \exp\Big\{ i\int_0^t \phi(s)\, d\Lambda_s \Big\}. \tag{2.5}$$

One can give a rigorous meaning to this expression, for instance by defining the action of $V_t[\phi]$ on the exponential vectors and by extending it by linearity to the whole space Γ. Precisely, we have

$$V_t[\phi]\psi(g) = \psi(U_t^\phi g), \qquad (U_t^\phi g)(s) := \exp\,[i\phi(s)\chi_{[0,t]}(s)]\, g(s). \tag{2.6}$$

By considering the Weyl operators (Ref.[2], eq.(2.1)) $W(f,U)$, $f \in L^2(\mathbb{R})$, U unitary operator on $L^2(\mathbb{R})$, we have

$$V_t[\phi] = W(0, U_t^\phi). \tag{2.7}$$

The operators $V_t[\phi]$ satisfy the QSDE (Ref.[8], eq.(4.11), and Ref.[2], eq.(6.1))

$$dV_t[\phi] = \{[e^{i\phi(t)}-1]d\Lambda_t\}V_t[\phi], \qquad V_0[\phi] = 1 . \tag{2.8}$$

In order to understand the meaning of $V_t[\phi]$, choose p functions $g_j \in L^\infty_{loc}(\mathbb{R})$ and set

$$V_t(\vec{\eta}|\{g_j\}) := V_t[\sum_{j=1}^{p} \eta_j g_j], \qquad \vec{\eta} \in \mathbb{R}^p. \tag{2.9}$$

Equation (2.9) defines a strongly continuous unitary representation of $\mathbb{R}^p$ in Γ. This proposition follows from the properties of Weyl operators (Ref.[2], Sec.2). Therefore, $V_t(\cdot|\{g_j\})$ is the p-dimensional Fourier transform of a projection-valued measure on $\mathbb{R}^p$ (the joint projection-valued measure of the commuting operators $\int_0^t g_j(s)d\Lambda_s$). In particular, by choosing in eq.(2.9)

$$g_j(s) = \chi_{[t_{j-1},t_j]}(s), \qquad 0\equiv t_0<t_1<\ldots<t_p\equiv t, \tag{2.10}$$

we get the Fourier transform of the projection-valued measure associated with the operators

$$\Lambda_{t_j} - \Lambda_{t_{j-1}} \equiv \int_{t_{j-1}}^{t_j} d\Lambda_t. \tag{2.11}$$

All these operators have a pure discrete spectrum (0,1,2,...) and, therefore, the projection-valued measure we have obtained is supported by the integers, as it must be for counting processes. If we denote by $P_n(t_{j-1},t_j)$ the eigenprojection of the operator (2.11) associated with the eigenvalue n, then the expectation value of

$$P_{n_1}(t_0,t_1)P_{n_2}(t_1,t_2)\ldots P_{n_p}(t_{p-1},t_p)$$

on some state in Fock space is interpreted as the probability (in this state) of n_1 counts in the interval $(t_0,t_1]$, n_2 counts in $(t_1,t_2]$, and so on. All the projection-valued measures one gets from eq.(2.9) are consistent and can be obtained from a unique projection-valued measure on a suitable measurable space. This can be seen by applying Minlos theorem to $\Phi_u[\phi] = \langle u|V_t[\phi]u\rangle$, where u is a normalized vector in Γ and ϕ a C^∞-function with compact support in (0,t). By the properties of $V_t[\phi]$, $\Phi_u[\phi]$ is the characteristic functional of a generalized stochastic process in the sense of Gel'fand [9] for any u; then, it can be shown that there exists a projection-valued measure on a suitable σ-algebra, whose restriction to cylinder sets is the projection-valued measure associated with $V_t(\vec{\eta}|\{g_j\})$ (see Ref.[10], Sec.2, for similar results, although in a different context).

Up to now we have not considered the interaction between system S and fields. If we want to describe the counts of the quanta in the output field, we have only to replace Λ_t by Λ_t^{out} and this can be done by defining (cf. eq.(2.2))

$$V_t^{out}[\phi] := U_t^\dagger V_t[\phi] U_t. \tag{2.12}$$

Then, by construction, the quantity

$$\Phi_t[\phi] = \mathrm{Tr}_{h_0 \otimes \Gamma}\{V_t^{out}[\phi](\rho \otimes |\psi(0)\rangle\langle\psi(0)|)\} \tag{2.13}$$

is the characteristic functional of a counting process in $(0,t)$ for any state ρ for system S.

In order to obtain explicit expressions for this counting process it is useful to eliminate the degrees of freedom of the fields by a procedure similar to that one used for obtaining the master equation (1.16). First, define an operator $G_t[\phi]$ on the trace class on h_0 by

$$\mathrm{Tr}_{h_0}\{X\, G_t[\phi]\rho\} = \mathrm{Tr}_{h_0 \otimes \Gamma}\{U_t^\dagger(X \otimes V_t[\phi])U_t(\rho \otimes |\psi(0)\rangle\langle\psi(0)|)\}, \qquad \forall\, X \in B(h_0), \tag{2.14}$$

so that the characteristic functional (2.13) can be written as

$$\Phi_t[\phi] = \mathrm{Tr}_{h_0}\{G_t[\phi]\rho\}. \tag{2.15}$$

Then, by exploiting eqs.(1.6), (2.8) and (2.14) and by using the rules of QSC, one gets

$$\frac{d}{dt} G_t[\phi] = \{L + K(\phi(t))\}\, G_t[\phi], \qquad G_0[\phi] = \mathbb{1}, \tag{2.16}$$

where L is given by eq.(1.16b) and $K(\phi)$ by

$$K(\phi(t))\rho = [e^{i\phi(t)} - 1]\, R\rho R^\dagger. \tag{2.17}$$

It can be shown [8] that $G_t[\phi]$ is the Fourier transform of an <u>operation-valued measure</u> (or <u>instrument</u>) [11] describing a certain continuous (in time) observation on system S [12]. By starting from more general operators $V_t[\phi]$ and more general evolution operators U_t, a very general class of continuous measurements can be obtained, by a procedure similar to that one sketched in this paper [5,8,10,13].

In our case the characteristic functional (2.15) can be rewritten in such a way that the structure of a counting process is apparent. By setting

$$\bar{L}\rho := -i[H,\rho] - \frac{1}{2}(R^\dagger R\rho + \rho R^\dagger R), \qquad J\rho := R\rho R^\dagger, \tag{2.18}$$

one has

$$L + K(\phi(t)) = \bar{L} + e^{i\phi(t)} J. \tag{2.19}$$

Now, the solution of eq.(2.16) can be written as the Dyson series

$$G_t[\phi] = e^{\bar{L}t} + \sum_{m=1}^{\infty} \int_0^t dt_m \int_0^{t_m} dt_{m-1} \cdots \int_0^{t_2} dt_1\, \exp[i \sum_{r=1}^{m} \phi(t_r)] \times$$

$$\times\ e^{\bar{L}(t-t_m)}\, J\, e^{\bar{L}(t_m-t_{m-1})}\, J \ldots e^{\bar{L}(t_2-t_1)}\, J\, e^{\bar{L}t_1} . \tag{2.20}$$

By inserting eq.(2.20) into eq. (2.15), we get

$$\Phi_t[\phi] = P_t(0|\rho) + \sum_{m=1}^{\infty} \int_0^t dt_m \int_0^{t_m} dt_{m-1} \ldots \int_0^{t_2} dt_1 \exp\Big[i \sum_{r=1}^{m} \phi(t_r)\Big] \times$$

$$\times\ p_t(t_m, t_{m-1}, \ldots, t_1|\rho), \tag{2.21}$$

where

$$P_t(0|\rho) = \mathrm{Tr}_{h_0}\{e^{\bar{L}t}\rho\}, \tag{2.22}$$

$$p_t(t_m, t_{m-1}, \ldots, t_1|\rho) = \mathrm{Tr}_{h_0}\{e^{\bar{L}(t-t_m)}\, J\, e^{\bar{L}(t_m-t_{m-1})}\, J \ldots e^{\bar{L}(t_2-t_1)}\, J\, e^{\bar{L}t_1}\rho\}. \tag{2.23}$$

The structure (2.21) is typical for the characteristic functional of a regular point process (roughly speaking, a point process for which the probability density of two or more counts at the same time vanishes). The quantity (2.22) can be interpreted as the probability of no counts in the interval $(0,t]$ and the quantity (2.23) as the probability density of a count at time t_1, a count at time t_2, ..., a count at time t_m and no other count in the interval $(0,t]$. The quantities (2.23) are called elementary probability densities or exclusive probability densities [14,15].

The example of counting process we have constructed (eqs.(2.16)-(2.23)) is a quantum stochastic process in the sense of Davies [8,11,14-16]. An essential ingredient in the approach of Davies are the quantities $N_t(m)$, $m=0,1,\ldots$, defined by [14]

$$N_t(0) := \exp(\bar{L}t), \tag{2.24}$$

$$N_t(m) := \int_0^t dt_m \int_0^{t_m} dt_{m-1} \ldots \int_0^{t_2} dt_1\, e^{\bar{L}(t-t_m)}\, J \ldots e^{\bar{L}(t_2-t_1)}\, J\, e^{\bar{L}t_1}, \quad m \geq 1. \tag{2.25}$$

Note that we have

$$\sum_{m=0}^{\infty} N_t(m) = \exp(Lt), \tag{2.26}$$

where $\exp(Lt)$ is the quantum dynamical semigroup with generator (1.16b). The operators $N_t(m)$ give the probabilities of counts. Precisely, the probability of m counts in the time interval $(0,t]$ is given by

$$P\{m,(0,t]|\rho\} = \mathrm{Tr}_{h_0}\{N_t(m)\rho\}, \tag{2.27}$$

the probability of m counts in $(0,t]$ and n counts in $(t,t+s]$ by

$$P\{m,(0,t];n,(t,t+s]|\rho\} = \mathrm{Tr}_{h_0}\{N_s(n)N_t(m)\rho\}, \tag{2.28}$$

and so on. The relationships between these probabilities and the quantities (2.22) and (2.23) are obvious.

Finally, note that, by using the projections $P_n(s,t)$ introduced above and the vacuum conditional expectation E_0 {eq.(1.12)}, we have

$$\mathrm{Tr}_{h_0}\{X\, N_t(m)\rho\} = \mathrm{Tr}_{h_0}\{E_0[U_t^{\dagger}(X\otimes P_m(0,t))U_t]\rho\}, \tag{2.29}$$

for any $X\in B(h_0)$ and any trace-class operator ρ. This equation summarizes the essential steps of our construction: i) we measure something on the system living in Fock space (in eq.(2.29) this measurement is represented by the projection-valued measure $P_m(0,t)$); ii) the measured system and the system of interest (system S) interact (U_t); iii) by eliminating the degrees of freedom of the system living in Fock space (E_0), an indirect measurement (represented by $N_t(m)$) on S is obtained.

3. The electron shelving effect.

In this section we want to apply the formalism of the previous sections to a concrete system. Our main purpose is not to obtain essentially new physical results, but simply to show how QSC can help in modelling physical systems.

Consider a three level atom. Following Ref. [17], call the states $|g\rangle$ (ground), $|B\rangle$ (blue), and $|R\rangle$ (red) and the corresponding transitions the blue transition ($B\to g$) and the red one ($R\to g$). Assume the blue transition be very strong and the red one very weak. When both transitions are driven by two suitably tuned lasers, we expect the atom emits blue fluorescence light. But sometimes, when the atom absorbs a red photon, the electron goes into the red state, which has a long lifetime (a fraction of a second), and the fluorescence light stops up to when the red state decays. So, we expect to observe bright and dark periods, randomly distributed. In a pictorial language, we say that during a dark period the electron is "shelved" in the red state. Indeed, experimentalists succeeded in observing this effect [18]. Let us stress that for observing dark and bright periods one has to make the experiment on a single ion, because this effect is completely masked when many emitters are involved. A good theoretical explanation of the electron shelving effect was given in Ref.[17].

For the dynamics of the system described above we can use an evolution equation of the type of eq.(1.6), but now we need two fields (one for red and one for blue photons). We take

$$dU_t = \Big\{\sum_{j=1}^{2}\big(-R_j^{\dagger}dA_t^j + R_j dA_t^{j\dagger} - \tfrac{1}{2}R_j^{\dagger}R_j dt\big) - iHdt\Big\}\,U_t, \qquad U_0 = \mathbb{1}\,, \tag{3.1}$$

where

$$R_j = \sqrt{\Gamma_j}\;|0\rangle\langle j|, \qquad H = \sum_{j=1}^{2}\omega_j\,|j\rangle\langle j|. \tag{3.2}$$

Here we have made the identification $|0\rangle\equiv|g\rangle$ (ground state), $|1\rangle\equiv|B\rangle$ (blue state), $|2\rangle\equiv|R\rangle$ (red state). The parameters Γ_j ($\Gamma_j>0$) are called transition rates and $\omega_j \equiv$

$\equiv(E_j-E_0)/\hbar$ are the transition frequencies. The first term in eq.(3.1) describes the absorption of a photon of type j and the transition of the electron from the ground state to the excited state $|j\rangle$ (j=1,2), the second term describes emission of a photon and decay of the electron into the ground state, the third term is the Itô correction and the fourth one is the free Hamiltonian of the three level system containing the energy differences among the levels. The evolution equation (3.1) can be justified in the so called rotating wave approximation [4].

The Hilbert space for the fields is now the symmetric Fock space over $\mathbb{C}^2 \otimes L^2(\mathbb{R})$ and the exponential vectors depend on two functions. In order to describe the two lasers, we take as initial state for the fields a coherent vector

$$\Psi := \frac{\psi(f_1,f_2)}{\|\psi(f_1,f_2)\|} \equiv \exp\{-\tfrac{1}{2}\left(\|f_1\|^2+\|f_2\|^2\right)\}\ \psi(f_1)\otimes\psi(f_2), \tag{3.3}$$

where $\psi(f_j)$ is given by eq.(1.1) and

$$f_j(t) = i\lambda_j \exp(-i\bar{\omega}_j t)\ \chi_{[t_i,t_f]}(t). \tag{3.4}$$

The parameters $\bar{\omega}_j$ are the laser frequencies and $[t_i,t_f]$ is the time interval in which the lasers are on; after all, we shall take $t_i \to -\infty$ and $t_f \to +\infty$. Moreover, we set

$$\Delta_j := \bar{\omega}_j - \omega_j, \qquad \Omega_j := 2|\lambda_j|\sqrt{\Gamma_j}; \tag{3.5}$$

the quantities Δ_j are called detuning parameters and the quantities Ω_j are called Rabi frequencies.

Now we want to describe the counting of the fluorescence photons emitted by the atom, not of all the photons in the fields. For this purpose, we change eq.(2.5) by taking

$$V_t[\phi] := \exp\{i\int_0^t \phi(s) \sum_{j=1}^{2} d\Pi_s^j\}, \tag{3.6}$$

$$d\Pi_t^j = d\Lambda_t^j - f_j(t)dA_t^{j\dagger} - f_j^*(t)dA_t^j + |f_j(t)|^2 dt. \tag{3.7}$$

This choice is due to the fact that the number operators Λ_t^j give the number of photons in the two fields, but we have to subtract from them the contribution of the laser fields. Indeed, the operators Π_t^j, in the heuristic notation of eqs.(1.4)-(1.5b), can be written as

$$\Pi_t^j = \int_0^t ds\ (a_j^\dagger(s) - f_j^*(s))(a_j(s) - f_j(s)), \tag{3.8}$$

from which the subtraction is apparent. The operators Π_t^j are simple examples of quantum Poisson processes and can be obtained from the operators Λ_t^j by a unitary transformation (see, for instance, the contribution by Frigerio in these proceedings). The sum in eq.(3.6) is due to the fact that we are counting both types of photons without distinguishing between them.

As in Sec.2, we introduce the operators $V_t^{out}[\phi]$ by eq.(2.12) and the characteristic functional by eq.(2.13). However, in this last equation, we replace the Fock vacuum $\psi(0)$ by the coherent vector (3.3). An operator $G_t[\phi]$ could be defined as in eq.(2.14); however, it is more convenient to eliminate the explicit time dependence coming in from the initial state Ψ for the fields (eqs.(3.3) and (3.4)). Therefore, we define $G_t[\phi]$ by

$$\mathrm{Tr}_{h_0}\{X\, G_t[\phi]\rho\} = \mathrm{Tr}_{h_0\otimes\Gamma}\{U_t^{\dagger}(X(t)\otimes V_t[\phi])U_t(\rho\otimes|\Psi\rangle\langle\Psi|)\}, \tag{3.9}$$

where

$$X(t) = e^{-iH_0 t}\, X\, e^{iH_0 t}, \qquad H_0 = \sum_{j=1}^{2} \bar{\omega}_j |j\rangle\langle j| . \tag{3.10}$$

In spite of this change in the definition of $G_t[\phi]$, the characteristic functional is always given by eq.(2.15).

By the same computations leading to eqs.(1.16) or (2.16) and (2.17), we obtain (cf. eqs.(2.16)-(2.19))

$$\frac{d}{dt} G_t[\phi] = \{\bar{L} + e^{i\phi(t)}J\}\, G_t[\phi], \qquad G_0[\phi] = \mathbb{1}, \tag{3.11}$$

$$\bar{L}\rho = -i\bar{H}\rho + i\rho\bar{H}^{\dagger}, \tag{3.12}$$

$$J\rho = \sum_{j=1}^{2} R_j\rho R_j^{\dagger} = \sum_{j=1}^{2} \Gamma_j |0\rangle\langle j|\rho|j\rangle\langle 0|, \tag{3.13}$$

$$\bar{H} = H - H_0 - \frac{i}{2}\sum_{j=1}^{2} R_j^{\dagger}R_j = \sum_{j=1}^{2}\left[\frac{1}{2}\Omega_j(|0\rangle\langle j| + |j\rangle\langle 0|) - (\Delta_j + \frac{i}{2}\Gamma_j)|j\rangle\langle j|\right]. \tag{3.14}$$

In these equations we have made the replacement $\exp(i \arg \lambda_j)|j\rangle \to |j\rangle$. Apart from the new expressions of the operators $\bar{L}$ and J, we are exactly in the situation discussed in the previous section, so that the probability of no counts in the interval $(0,t]$ is given by eq.(2.22) and the exclusive probabilities densities by eq.(2.23).

The quantity of main interest in the case of the shelving effect is the probability of no emission in the interval $(0,t]$, given the emission of a photon at time zero. It is given by

$$P_0(t) := \lim_{\tau\to 0+} \frac{\int_0^{\tau} dt'\, p_t(t'|\rho)}{\int_0^{\tau} dt'\, p_\tau(t'|\rho)} = \lim_{\tau\to 0+} \mathrm{Tr}_{h_0}\{N_{t-\tau}(0)N_\tau(1)\rho\}/\mathrm{Tr}_{h_0}\{N_\tau(1)\rho\} =$$

$$= \mathrm{Tr}_{h_0}\{e^{\bar{L}t} J\rho\}/\mathrm{Tr}_{h_0}\{J\rho\} = \mathrm{Tr}_{h_0}\{e^{\bar{L}t}(|0\rangle\langle 0|)\} = P_t(0||0\rangle\langle 0|). \tag{3.15}$$

By setting

$$c_j(t) := \langle j|e^{-i\bar{H}t}|0\rangle, \qquad j = 0,1,2, \tag{3.16}$$

we can write

$$P_0(t) = \sum_{j=0}^{2} |c_j(t)|^2. \tag{3.17}$$

By taking into account eq.(3.14), we get for the quantities (3.16) the equations

$$\begin{aligned}
\dot{c}_0(t) &= -\frac{i}{2}\Omega_1 c_1(t) - \frac{i}{2}\Omega_2 c_2(t)\\
\dot{c}_1(t) &= -\frac{i}{2}\Omega_1 c_0(t) + \left(i\Delta_1 - \frac{1}{2}\Gamma_1\right)c_1(t)\\
\dot{c}_2(t) &= -\frac{i}{2}\Omega_2 c_0(t) + \left(i\Delta_2 - \frac{1}{2}\Gamma_2\right)c_2(t)
\end{aligned} \tag{3.18}$$

with the initial condition $c_j(0) = \delta_{j0}$. These equations can be solved by the Laplace transform technique [6]. For $\Omega_j \neq 0$ and $\Gamma_j \neq 0$, the real parts of the three roots of the characteristic equation of system (3.18) are negative, so that eq.(3.17) implies

$$\lim_{t\to+\infty} P_0(t) = 0. \tag{3.19}$$

Moreover, by taking the time derivative of the expression (3.15), we get

$$\frac{d}{dt}P_0(t) = -\sum_{j=1}^{2} \Gamma_j |c_j(t)|^2 \tag{3.20}$$

and, therefore, eqs.(3.19) and (3.20) give

$$P_0(t) = \int_t^{+\infty} ds \sum_{j=1}^{2} \Gamma_j |c_j(s)|^2. \tag{3.21}$$

Now, the idea of a weak and a strong transition can be realized by asking

$$\Gamma_1 \gg \Gamma_2, \quad \Gamma_1 \gg \Omega_2, \quad \Omega_1 \gg \Omega_2, \quad \Omega_1^2 \gg \Gamma_1\Gamma_2, \quad \Delta_1 = 0; \tag{3.22}$$

the last condition means simply that we are considering a perfectly tuned blue laser. Then, the three roots of the characteristic equation of system (3.18) are approximately given by

$$z_\pm \approx -\frac{\Gamma_1}{4} \pm \frac{1}{2}\sqrt{\frac{\Gamma_1^2}{4} - \Omega_1^2}, \tag{3.23}$$

$$z_0 = -\frac{\Gamma_2}{2} + i\Delta_2 - \zeta, \tag{3.24a}$$

$$\zeta \approx \frac{\Omega_2^2}{2} \cdot \frac{\Gamma_1\Omega_1^2 + 2i\Delta_2(\Omega_1^2 - 4\Delta_2^2 - \Gamma_1^2)}{(\Omega_1^2 - 4\Delta_2^2)^2 + 4\Delta_2^2\Gamma_1^2}; \tag{3.24b}$$

$z_\pm$ are real or complex numbers according to the sign of $\Gamma_1^2 - 4\Omega_1^2$. By eqs.(3.22) we have that

$$|\mathrm{Re}\, z_0| \ll |\mathrm{Re}\, z_\pm|. \tag{3.25}$$

Following Ref.[17], we collect together all the terms with short decaying time in $P_0(t)$ (see Ref.[6] for the computations)

$$P_0(t) = \int_t^{+\infty} dt' \,[F_{short}(t') + F_{long}(t')], \tag{3.26}$$

where, by taking into account eq.(3.22), we have approximately

$$F_{short}(t) \approx \frac{\Gamma_1\Omega_1^2}{|\Gamma_1^2-4\Omega_1^2|} \left|e^{z_+t} - e^{z_-t}\right|^2, \tag{3.27}$$

$$F_{long}(t) = \Pi \,|2\mathrm{Re}(z_0)|\, \exp[2\mathrm{Re}(z_0)t], \tag{3.28a}$$

$$\Pi \approx \frac{\Omega_2^2}{(\Omega_1^2-4\Delta_2^2)^2+4\Delta_2^2\Gamma_1^2} \cdot \frac{1}{|2\mathrm{Re}(z_0)|} \times$$

$$\times \left\{\Gamma_1\Omega_1^2\Omega_2^2 \frac{\Gamma_1^2\Omega_1^4+4\Delta_2^2(\Omega_1^2-4\Delta_2^2-\Gamma_1^2)^2}{[(\Omega_1^2-4\Delta_2^2)^2+4\Delta_2^2\Gamma_1^2]^2} + \Gamma_2(\Gamma_1^2+4\Delta_2^2)\right\}. \tag{3.28b}$$

By taking t=0 in eq.(3.15), we have

$$P_0(0) = 1 \tag{3.29}$$

and, therefore, eqs.(3.26) and (3.28a) imply

$$\int_0^{+\infty} dt\, F_{long}(t) = \Pi, \qquad \int_0^{+\infty} dt\, F_{short}(t) = 1 - \Pi. \tag{3.30}$$

Moreover, eqs.(3.22) and (3.28b) give

$$\Pi \ll 1. \tag{3.31}$$

In Ref.[17] it is shown that the structure (3.26) for $P_0(t)$ implies the existence of bright and dark periods. The argument is as follows.

The quantity $1-P_0(t)$ is the probability of at least a count in $(0,t]$. Then the quantity

$$p(t) := \frac{d}{dt}[1 - P_0(t)] = F_{short}(t) + F_{long}(t) \tag{3.32}$$

is the probability density for the waiting time Δt between two counts. In passing, it is interesting to note that there is a link between $1-P_0(t)$ and the stop times in Fock space in the sense of Ref.[19]. Let $P_n^{\Pi}(t)$, $n=0,1,\ldots$, be the eigenprojections of the operator $\Pi_t^1+\Pi_t^2$. Then, the equations

$$P^{T_1}([0,t]) = \sum_{n=1}^{\infty} P_n^{\Pi}(t), \quad t < \infty, \qquad P^{T_1}(\{\infty\}) = |\Psi\rangle\langle\Psi|, \tag{3.33}$$

where Ψ is the state given by eq.(3.3), define a stop time in the sense of Ref.[19], Def.3.1 (cf. also Ref.[19], eqs.(3.3)). From eqs.(2.24)-(2.27), where now $\bar{L}$ and J are defined by eqs.(3.12) and (3.13), we have

$$1 - P_0(t) = \sum_{m=1}^{\infty} \mathrm{Tr}_{h_0}\{N_t(m)\rho_0\}, \qquad \rho_0 = |0\rangle\langle 0|. \tag{3.34}$$

Consider now eq.(2.28). In this section the initial state for the fields is Ψ and not the Fock vacuum and the operator $\Pi_t^1+\Pi_t^2$ is used instead of Λ_t. Therefore, for

$X \equiv \mathbb{1}$, eq.(2.29) must be replaced by

$$\mathrm{Tr}_{h_0}\{N_t(m)\rho\} = \mathrm{Tr}_{h_0\otimes\Gamma}\{U_t^{\dagger}(\mathbb{1}\otimes P_m^{\Pi}(t))U_t(\rho\otimes|\Psi\rangle\langle\Psi|)\}. \tag{3.35}$$

By eqs.(3.33)-(3.35) we have

$$1 - P_0(t) = \mathrm{Tr}_{h_0\otimes\Gamma}\{U_t^{\dagger}(\mathbb{1}\otimes P^{T_1}([0,t]))U_t(\rho_0\otimes|\Psi\rangle\langle\Psi|)\}. \tag{3.36}$$

This equation shows the connection between $1-P_0(t)$ and the stop time T_1 in Fock space.

We introduce now a time delay θ such as

$$|2\mathrm{Re}(z_-)|^{-1} << \theta << |2\mathrm{Re}(z_0)|^{-1} \tag{3.37}$$

and consider the time interval Δt between succesive counts as "short" if $\Delta t<\theta$ and as "long" if $\Delta t>\theta$. The probability of having a short waiting time between two counts is

$$P(\Delta t<\theta) = \int_0^{\theta} dt\ p(t) \simeq \int_0^{\theta} dt\ F_{short}(t) \simeq \int_0^{+\infty} dt\ F_{short}(t) = 1 - \Pi \tag{3.38}$$

and, similarly, the probability of a long waiting time is

$$P(\Delta t>\theta) \simeq \int_0^{+\infty} dt\ F_{long}(t) = \Pi. \tag{3.39}$$

Note that these probabilities are independent from θ as far as eq.(3.37) holds. We can say that the "short" waiting times are distributed with a probability density $(1-\Pi)^{-1}F_{short}(t)$ and the "long" ones with a probability density $\Pi^{-1}F_{long}(t)$. Therefore, the mean duration of short intervals is given by

$$T_{short} = \frac{1}{1-\Pi}\int_0^{+\infty} dt\ t\ F_{short}(t) \simeq \Gamma_1\Omega_1^{-2} + 2\ \Gamma_1^{-1}, \tag{3.40}$$

and the mean duration of the long intervals is

$$T_{long} = \frac{1}{\Pi}\int_0^{+\infty} dt\ t\ F_{long}(t) = |2\mathrm{Re}(z_0)|^{-1}. \tag{3.41}$$

When many photons are emitted, separated by short time intervals, the intensity of the signal registered by the detector is approximately constant, with small fluctuations due to shot noise, and we have a bright period. Instead, when two photons are emitted, separated by a long time interval, a dark period is registered. The mean duration τ_B of a bright period is given by $T_{short}\bar{N}$, where $\bar{N}$ is the average number of consecutive short intervals. The durations of the intervals between two emissions are uncorrelated variables, because after any emission the electron goes into the ground state (cf. the structure (3.13) of the operator J). Therefore, the probability of having N short intervals followed by a long one is $(1-\Pi)^N\Pi$. If we want at least a short interval in the sequence, we have to divide this probability by $(1-\Pi)$. Finally, we have

$$\bar{N} = \sum_{N=1}^{\infty} N(1 - \Pi)^{N-1}\Pi = 1/\Pi \tag{3.42}$$

and the mean duration τ_B of bright periods is

$$\tau_B = T_{short}/\Pi. \tag{3.43}$$

Analogously, the mean duration τ_D of dark periods is

$$\tau_D = T_{long}/(1-\Pi) \simeq T_{lcng}, \tag{3.44}$$

When eqs.(3.23), (3.24), (3.28b), (3.40)-(3.44) are particularized to the cases considered in Ref.[17] ($\Omega_1 << \Gamma_1$ or $\Omega_1 >> \Gamma_1$), exactly the same results are obtained. In Ref.[17] it is also shown that it is possible to get spectroscopic information on the weak transition by plotting the ratio τ_D/τ_B vs. the detuning Δ_2. Any other statistical property of the emitted light can be derived from eqs.(2.22), (2.23), (3.12)-(3.14).

References

1. R.L. Hudson and K.R. Parthasarathy, in "Quantum Probability and Applications to the Quantum Theory of Irreversible Processes", edited by L. Accardi, A. Frigerio, and V. Gorini, Lecture Notes in Mathematics, Vol. 1055 (Springer, Berlin, 1984), pp. 173-198.
2. R.L. Hudson and K.R. Parthasarathy, Commun. Math. Phys. 93, 301-323 (1984).
3. R.L. Hudson and K.R. Parthasarathy, Acta Appl. Math. 2, 353-378 (1984).
4. C.W. Gardiner and M.J. Collet, Phys. Rev. A 31, 3761-3774 (1985).
5. A. Barchielli, Phys. Rev. A 34, 1642-1649 (1986).
6. A. Barchielli, Quantum stochastic differential equations: An application to the electron shelving effect, preprint IFUM 324/FT, Milan Univ. (1986).
7. A. Frigerio, Publ. R.I.M.S. Kyoto Univ. 21, 657-675 (1985).
8. A. Barchielli and G. Lupieri, J. Math. Phys. 26, 2222-2230 (1985).
9. I.M. Gel'fand and N.Ya. Vilenkin, Generalized Functions, Vol.4, Applications of Harmonic Analysis (Academic, New York, 1964).
10. A. Barchielli, in "Stochastic Processes in Classical and Quantum Systems", edited by S. Albeverio, G. Casati, and D. Merlini, Lecture Notes in Physics, Vol.262 (Springer, Berlin, 1986), pp.14-23.
11. E.B. Davies, Quantum Theory of Open Systems (Academic, London, 1976).
12. A. Barchielli, L. Lanz, and G.M. Prosperi, Found. Phys. 13, 779-812 (1983).
13. K.R. Parthasarathy, Boll. U.M.I. 5-A, 391-397 (1986).
14. M.D. Srinivas and E.B. Davies, Opt. Acta 28, 981-996 (1981).
15. A.S. Holevo, Izv. Vuz. Mat. 26, 3-19 (1982) [Sov. Math. 26, 1-20 (1982)].
16. E.B. Davies, Commun. Math. Phys. 15, 277-304 (1969); 19, 83-105 (1970); 22, 51-70 (1971).
17. C. Cohen-Tannoudji and J. Dalibard, Europhys. Lett. 1, 441-448 (1986).
18. W. Nagourney, J. Sandberg, and H.G. Dehmelt, Phys. Rev. Lett. 56, 2797-2799 (1986); T. Sauter, W. Neuhauser, R. Blatt, and P.E. Toschek, Phys. Rev. Lett. 57, 1696-1698 (1986); J.C. Bergquist, R.G. Hulet, W.M. Itano, and D.J. Wineland, Phys. Rev. Lett. 57, 1699-1702 (1986).
19. K.R. Parthasarathy and K.B. Sinha, Stop times in Fock space stochastic calculus, preprint Indian Statistical Institute, New Delhi-110016 (1986).

Some noncommutative Radon-Nikodym Theorems for von Neumann algebras

by

Carlo Cecchini

Dipartimento di Matematica dell'Universita di Genova-
Via L. B. Alberti 4 - Genova - Italy

1. Introduction and notations. The aim of this paper is to give Radon-Nikodym type results for a von Neumann algebra M on which two normal faithful (n.f.) states w_1 and w_2 are defined. Our starting point will be the following theorem due to A. Connes [5].

1.1. Theorem. Let w_1, w_2 be n.f. weights on a von Neumann algebra M. Then the following two statements are equivalent:

a) $w_1 \leq \alpha w_2$ for some $\alpha > 0$

b) the mapping $t \to u_t = (Dw_1 : Dw_2)_t$ (the Connes cocycle for w_1 and w_2, see [6]) can be extended to a mapping which is bounded and continuous on the strip $\{z \in \mathbf{C} ; 0 \leq \operatorname{Im} z \leq \frac{1}{2}\}$ and analytic in its interior, and, for $a \in$ M, $w_1(a) = w_2(u_{\frac{i}{2}}\, a\, u^+_{\frac{i}{2}})$, with $u_{\frac{i}{2}}$ the value of this function for $z = \frac{i}{2}$.

First we shall consider the spaces $L(p, M, w_i)$ $(1 \leq p \leq +\infty , i = 1,2)$ defined and studied in [3], and construct explicitely a family $D^p(w_1,w_2)$ of positive linear isometries from $L(p, M, w_1)$ to $L(p, M, w_2)$ well behaved with respect to the products and dualities between those spaces established in [3]. In particular we shall show that $D^\infty(w_1,w_2)$ is equivalent to the mapping $a \to u_{\frac{i}{2}}\, a\, u^+_{\frac{i}{2}}$ if condition a) in theorem 1.1. holds.

In the last part of this paper a Radon-Nikodym theorem obtained in [4] for the w_i-conditional expectations from M to a von Neumann subalgebra M_0 (for definitions and properties see [1]) under a boundedness condition shall be extended to the general situation.

In the following M will be a von Neumann algebra with M' acting on a separable Hilbert space H. Let w(w') be a n.f. state on M(M'). We shall denote by H_w this Hilbert space of the standard (left) representation π_w of M with respect to w, with a cyclic and separating vector Ω such that

$$w(a) = < \Omega, \pi_w (a) \Omega > \qquad \text{for} \quad x \in M.$$

J_w will denote the isometrical involution on H_w, π'_w the right representation of M on H_w, Δ_w the modular operator and σ^t_w ($t \in \mathbf{R}$) the modular automorphism group on M associated to w.

Let:

$$D(H,w) = \{ \xi \in H : \| a \xi \|^2 \leq \alpha\, w(a^+a) \text{ for some } \alpha > 0 \text{ and all } a \in M \}.$$

and $R_w (\xi)\, a\, \Omega = a \xi$ for $\xi \in D(H,w)$. Then $|R_w(\xi)|^2 \in \pi'_w (M)$ and we set $\zeta_w(\xi) = \pi_w^{-1} (J_w |R_w (\xi)|^2 J_w)$. The linear span of the set $\{ \zeta_w (\xi) : \xi \in D(H,w) \}$ is weak operator dense in M.

We define, for all $\phi \in M_*$, $[q(\varphi,w)] (\xi) = \phi(\zeta_w(\xi))$ for $\xi \in D(H,w)$, and the mapping $q : \varphi \to q(\varphi,w)$ defines a linear bijection between M_* and $L(1;M,w) = \{ q(\varphi,w) : \phi \in M_* \}$. $L(1;M,w)$ is a Banach space with the norm $\|q(\varphi,w)\|_{L(1;M,w)} = \|\varphi\|_{M_*}$. We shall set, for $q(\varphi,w), \in L(1; M,w)$ $\int q(\varphi,w)dw = \varphi(I)$.

Also $L(\infty; M,w) = \{ q(a,w) : a \in M \}$, with $[q(a,w)] (\xi) = < \xi, a\xi >$ for $\xi \in D(H,w)$, is a Banach space with the norm $\| q(a,w) \|_{L(\infty;M,w)} = \|a\|_M$. So the mapping $a \to q(a,w)$ defines a linear isometry between M and $L(\infty;M,w)$. It turns out that $L(\infty;M,w)$ is a dense subset of $L(1;M,w)$ and that those spaces are a compatible couple in the sense of interpolation theory; we call $L(p;M,w)$ $(1 < p < \infty)$ the interpolation spaces for this couple. There is natural linear isometry between those spaces and the spaces $L^p(M,w')$ studied by Connes [7], Hilsum [8] and Terp [11], which has been established in [3] explicitely. In the following we shall denote by $q(T;p,w,w')$ the element of $L(p;M,w)$ corresponding through this mapping to $T \in L^p (M,w')$, and define $T(q;p,w,w')$ by setting $q(T(q;p,w,w') ; p,w,w') = q$ for all $q \in L(p;M,w)$.

We recall that for $T \in L^{p_1} (M,w')$ we have $T(q(T,p_1,w,w'), p_2,w,w') = d^{\frac{1}{2p_2} - \frac{1}{2p_1}}\, T d^{\frac{1}{2p_2} - \frac{1}{2p_1}}$ for $1 < p_2 < p_1 < +\infty$. It is also possible to define an $L(r;M,w)$ valued product $q_1(p_1;M,w)\, q_2(p_2;M,w)$ between two elements $q_1(p_1;M,w)$ of $L(p_1;M,w)$ and $q_2(p_2;M,w)$ of $L(p_2;M,w)$ for $p_1^{-1} + p_2^{-1} = r^{-1}$, $1 < p_1,p_2, r < +\infty$. In the particular case in which $r = 1$ the mapping

$$\left(q_1(p_1;M,w),\ q_2(p_2;\ M,w)\right) \rightarrow \int q_1(p_1;M,w)\ q_2(p_2;M,w)\ \ dw$$

implements the duality between $L(p_1;M,w)$ and $L(p_2;M,w)$.

We remark that for us in the following the product of unbounded linear operators will be the strong product (i.e. the closure of the composition of the two operators) whenever it exists; it will be the composition if not. Also by an abuse of notation, we will denote with the same symbol an unbounded operator and its closure, when it exists.

For a full treatment of the above matter and comprehensive references, see [2] and [3]; for the theory of spatial Radom-Nikodym derivatives and of $L^p(M,w')$ spaces we shall use the notations and the results of [7] and [8].

2. A Radon-Nikodym theorem for the spaces L(p;M,w). In the following $\frac{dw}{dw'}$ (or, shortly, d) will

denote the spatial Radon-Nikodym introduced by Connes in [7] or a n.f. state w on M with respect to a n.f. state w' on M'. When various states on M will be dealt with simultaneously, their index will be attached to all the related objects of the theory.

2.1. Lemma. Let $\xi \in D(H,w)$; then, for each $t \in \mathbf{R}$, $d^{it}\xi \in D(H,w)$ and $R_w(d^{it}\xi) = d^{it}\ R_w\ (\xi)\ \Delta_w^{it}$.

Proof. Let $a \in M$; then for $t \in \mathbf{R}$:

$$\| a\, d^{it}\xi \|^2 = \| d^{-it}\, a\, d^{it}\, \xi\|^2 = \| \sigma_w^{-t}(a)\, \xi \|^2 < \alpha\, w\left(\sigma_w^{-t}(a)^+\, \sigma_w^{-t}(a)\right) =$$

$$= \alpha\, w\left(\sigma_w^{-t}(a^+a)\right) = \alpha\, w(a^+a)$$

for some $\alpha > 0$; so $d^{it}\xi \in D(w,H)$. Moreover:

$$R_w\ (d^{it}\ \xi)\ \pi_w(a)\ \Omega = ad^{it}\ \xi = d^{it}\ \sigma_w^{-t}\ (a)\ \xi =$$

$d^{it}\ R_w(\xi)\ \pi_w\left(\sigma_w^{-t}(a)\right)\ \Omega = d^{it}\ R_w\ (\xi)\ \Delta_w^{it}\ \pi_w(a)\ \Omega$, which implies our statement.

2.2. Corollary. Let $x \in D(H,w)$. Then:

$$| R_w(d^{it}\ \xi)^+|^2 = \sigma_{w'}^t\ \left(|R_w\ (\xi)^+|^2\right)$$

$$\zeta_w(d^{it}\xi) = \sigma_w^t\left(\zeta_w(\xi)\right)$$

Proof. It follows from lemma 2.1 by straightforward computations, recalling that for each $\xi \in D(H,w)$, $|R_w(\xi)^+|^2 \in M'$, and the definition of $\zeta_w(\xi)$.

In the following we shall denote, as usual, by $(Dw_1 : Dw_2)_t$ (or, shortly, by $d_{1,2}(it)$) the Connes Radon-Nikodym cocycle (see [3]) in M with respect to the states w_1 and w_2. We shall also need the strip $S = \{ z \in \mathbb{C} = 0 \leq \operatorname{Re} z \leq 1\}$ and its interior $\mathcal{S}$.

2.3. Proposition. Let $z \in S$. Then, for $\xi \in D(H,w_2)$, $d_2^{-\frac{z}{2}}\xi$ is well defined and belongs to $\mathcal{D}(d_1^{\frac{z}{2}})$. The vector $d_1^{\frac{z}{2}}\, d_2^{-\frac{z}{2}}\xi$ does not depend on w' and the mapping $z \to d_1^{\frac{z}{2}}\, d_2^{-\frac{z}{2}}\xi$ is an extension on the mapping $it \to (Dw_1 : Dw_2)_t\,\xi$ which is bounded and continuous on S and analytic on $\mathcal{S}$ (in the weak topology).

Proof. As w is a state, $D(H,w_2) \subseteq \mathcal{D}(d_s^{-\frac{1}{2}}) \subseteq \mathcal{D}(d_2^{-\frac{z}{2}})$ for $z \in S$, by the definition of d_2 (see [7]); so $d_2^{-\frac{z}{2}}\xi$ is well defined.

Let now $\eta \in \mathcal{D}(d_1^{\frac{1}{2}})$. The mapping

$$\varphi : z \to \left< d_1^{\frac{\bar{z}}{2}}\eta,\ d_2^{-\frac{z}{2}}\xi \right> \text{ is well defined, bounded and continuous on S and analytic on } \mathcal{S}.$$

Moreover:

$$|\varphi(it)| = \left| \left< d_1^{-\frac{it}{2}}\eta,\ d_2^{-\frac{it}{2}}\xi \right> \right| \leq \|\eta\|\ \|\xi\|$$

and using lemma 2.1., corollary 2.2., the fact that $d_1^{\frac{1}{2}}\, d_2^{-\frac{1}{2}}\, D(H,w_2) = d_1^{\frac{1}{2}}\, D(H,w') = D(H,w_1)$ and that

$R_{w'}|(d^{-\frac{1}{2}}\xi)^+|^2 = \zeta_w(\xi)$ for $\xi \in D(M,w)$ (see [3]) we have

$$|\varphi(1+it)| = |<d_1^{\frac{1-it}{2}}\eta, d_2^{-\frac{1+it}{2}}\xi>| =$$

$$= |<d_1^{-\frac{it}{2}}\eta, d_1^{\frac{1}{2}}d_2^{-\frac{1+it}{2}}\xi>| < \|\eta\| \, \|d_1^{\frac{1}{2}}d_2^{-\frac{1+it}{2}}\xi\| =$$

$$= \|\eta\| \, w_1\left(|R_{w'}(d_2^{-\frac{1+it}{2}}\xi)^+|^2\right) = \|\eta\| \, w_1\left(\sigma_{w_2}^{\frac{1}{2}}(\zeta_{w_2}(\xi))\right) \leq \|\eta\| \, \|\zeta_{w_2}(\xi)\|.$$

The three lines principle yields, for $z \in S$:

$$|\varphi(z)| \leq \|\eta\| \, \|\xi\|^{1-\frac{\mathrm{Re}z}{2}} \|\zeta_{w_2}(\xi)\|^{\frac{\mathrm{Re}z}{2}} \; ; \; \text{so}$$

$d_2^{-\frac{z}{2}}\xi \in \mathcal{D}(d_1^{-\frac{z}{2}})$ for $z \in S$ and $\|d_1^{\frac{z}{2}}d_2^{-\frac{z}{2}}\xi\| < \|\xi\|^{1-\frac{\mathrm{Re}z}{2}} \|\zeta_{w_2}(\xi)\|^{\frac{\mathrm{Re}z}{2}}$.

For $\eta \in H$, choose a sequence η_n in $\mathcal{D}(d_1^{\frac{1}{2}})$ converging to η. The sequence

$$\varphi_n = z \to <\eta_n, d_1^{\frac{z}{2}}d_2^{-\frac{z}{2}}\xi>$$

converges uniformly on S to

$$\varphi = z \to <\eta_n, d_1^{\frac{z}{2}}d_2^{-\frac{z}{2}}\xi>;$$

so φ too is bounded and continuous on S and analytic on S.

If $z = it$, $t \in \mathbf{R}$, then

$$\varphi(z) = \varphi(it) = <\eta, d_1^{\frac{it}{2}}d_2^{-\frac{it}{2}}\xi> = <\eta, (Dw_1 : Dw_2)_t \, \xi>;$$

as the last member of our equality does not depend on w', for all $z \in S$, $d_1^{\frac{z}{2}}d_2^{-\frac{z}{2}}\xi$ does not depend on w' and our proof is complete. In the following we shall set $d_1^{\frac{z}{2}} d_2^{-\frac{z}{2}} = d_{1,2}(z)$ for $z \in S$. We have just proved that $\mathcal{D}(d_{1,2}(z)) \subseteq D(H,w_2)$ for $z \in S$.

2.4. Proposition. Let $z \in S$. Then $d_{1,2}(z)$ commutes with M', in the sense that, for $\xi \in D(H,w_2)$, $a' \in M'$, the expression $d_{1,2}(z)\, a' \xi$ makes sense and $d_{1,2}(z)a' \xi = a'd_{1,2}(z)\, \xi$.

Proof. In the above hypothesis $a' \xi \in D(H,w_2) \subset \mathcal{D}(d_{1,2}(z))$.
Let $\eta \in H$,

$$\varphi_1(z) = \langle \eta, a' d_{1,2}(z) \xi \rangle$$
$$\varphi_2(z) = \langle \eta, d_{1,2}(z) a' \xi \rangle$$

Then, for $t \in \mathbf{R}$

$$\varphi_1(it) = \langle \eta, a' (Dw_1 : Dw_2)_t \xi \rangle =$$
$$= \langle \eta, (Dw_1 : Dw_2)_t a' \xi \rangle = \varphi_2(it)$$

as $(Dw_1 : Dw_2)_t \in M$, and therefore by proposition 2.3 $\varphi_1(z) = \varphi_2(z)$ for all $z \in S$, which implies our statement.

2.5. Lemma. Let $q \in L(p;M,w_2)_+$ (i.e $q(\xi) \geq 0\ \forall\ \xi \in D(H,w_2)$) , $1 < p < \infty$. There are then a unique $D^p_{1,2}(q) \in L(p;m,w_1)_+$ and a unique positive (see [7]) form Q such that

a) $d_{1,2}(\frac{1}{p})\, D(H,w_2) \cup D(H,w_1)$ spans linearly $\mathcal{D}(Q)$

b) Q is lower semicontinuous in the norm $\xi \to \| d_1^{-\frac{1}{2p}} \xi \|$

c) $Q(\xi) = (D^p_{1,2}(q))(\xi)$ for $\xi \in D(H,w_1)$

$Q(d_{1,2}(\frac{1}{p})\, \xi) = q(\xi)$ for $\xi \in D(H,w_2)$

Let $q \in L(\infty; M,w_2)$. There are then a unique $D^{\infty}_{1,2}(q)$ in $L(\infty; M,w_1)_+$ and a unique continuous positive form Q on H such that $D^{\infty}_{1,2}(q)$ and q coincide with Q on $D(H,w_1)$ and $D(H,w_2)$ respectively.

<u>Proof.</u> Let $T \in L^p(M,w')$, $q = q(T;p,w_2,w') \in L(p;M,w_2)$, $p^{-1} + p'^{-1} = 1$. We have, as w' is a state:

$$D(H,w_1) \subseteq \mathcal{D}(d_1^{-\frac{1}{2p}}) \subseteq \mathcal{D}(d_1^{-\frac{1}{2p}})$$

moreover, if $\xi \in D(H,w_1)$, then $d_1^{-\frac{1}{2p}}\xi = d_1^{\frac{1}{2p}}(d_1^{\frac{1}{2}}\xi)$, and as $d_1^{-\frac{1}{2}}\xi \in D(H,w')$ (see [3]) and $T^{\frac{1}{2}}\, d_1^{\frac{1}{2p}} \in L^1(M,w')$, we get $D(h,w_1) \subseteq \mathcal{D}(T^{\frac{1}{2}}\, d_1^{-\frac{1}{2}})$ (cfr. [3]). Similarly, by the definition of $d_{1,2}(\frac{1}{p})$, $\mathcal{D}(d_1^{-\frac{1}{2p}}) \supset d_{1,2}(\frac{1}{p})\, D(H,w_2)$ and $d_1^{-\frac{1}{2p}}\, d_{1,2}(\frac{1}{p}) = d_2^{-\frac{1}{2p}}$. We can now prove as in the preceding case that $d_{1,2}(\frac{1}{p})\, D(H,w_2) \subseteq \mathcal{D}(T^{\frac{1}{2}}\, d_1^{\frac{1}{2p}})$. So $\mathcal{D}(T^{\frac{1}{2}}\, d_1^{\frac{1}{2p}})$ contains the linear span of $D(H,w_1) \cup d_{1,2}(\frac{1}{p})\, D(H,w_2)$. If we set

$$D^{p}_{1,2}(q) = q(T; p,w_1, w'),$$

$$Q(\xi) = \| T^{\frac{1}{2}}\, d_1^{-\frac{1}{2p}}\xi \|^2 \text{ for } \xi \text{ in the linear span of } D(H,w_1) \cup d_{1,2}\left(\frac{1}{p}\right) D(H,w_2),$$

conditions a) and b) in our statement are trivially satisfied. Moreover, for $\xi \in D(H,w_1)$:

$$\left[D^{p}_{1,2}(q)\right](\xi) = q(T;p,w_2,w')(\xi) =$$

$$= \| T^{\frac{1}{2}}\, d_1^{\frac{1}{2p'}}\, d_1^{-\frac{1}{2}}\xi \|^2 = \| T^{\frac{1}{2}}\, d_1^{-\frac{1}{2p}}\xi \| = Q(\xi);$$

for $\xi \in D(H,w_2)$:

$$q(\xi) = q(T,p,w_2,w')\,(\xi) = \| T^{\frac{1}{2}} d_2^{-\frac{1}{2p}} \xi \|^2 = \| T^{\frac{1}{2}} d_1^{-\frac{1}{2p}} d_{1,2}(\tfrac{1}{p})\, \xi \|^2 = Q(d_{1,2}(\tfrac{1}{p})\,\xi).$$

The unicity for $D^p_{1,2}(q)$ follows from the fact that an element of $L(p;M,w_1)$ is uniquely determined by its values on $D(H,w_1)$; so, by c), Q is also determined uniquely.

If $q \in L(\infty;M,w_2)$, then $q = q(a,w_1)$ for some $a \in M_+$. It is straightforward to check that $Q(\xi) = \| a^{\frac{1}{2}} \xi \|$ $(\xi \in H)$ and $D^\infty_{1,2}(q) = Q(a,w_1)$ satisfy the requirements in the second part of our statement.

2.6. Lemma. Let $q \in L(p;M,w)$, $1 \le p \le +\infty$. Then $D^p_{1,2}\, q$ does not depend on the particular w' used in its construction.

Proof. The statement is obvious for $p = +\infty$.
Let now $1 \le p < +\infty$. The proof of lemma 2.5 implies that

$D^p_{1,2}(q) = q\big(T(q;p,w_2\, w'),\, p,\, w_1\, w'\big)$; therefore $T\big(D^p_{1,2}(q),\, 1,\, w_1\, w'\big) =$

$$= d_1^{\frac{1}{2p'}}\, T(q;p,w_2\, w')\, d_1^{\frac{1}{2p}}.$$

If $q = q(a,w_2) \in L(\infty;M,w_2\, w')_+$ $(a \in M_+)$, this implies $T\big(D^p_{1,2}(q),\, 1, w_1, w'\big) =$

$= d_1^{\frac{1}{2p'}}\, d_2^{\frac{1}{2p}}\, a\, d_2^{\frac{1}{2p}}\, d_1^{\frac{1}{2p'}}$; so for $\xi \in D(H,w_2)$, we have $(D^p_{1,2}\, q)(\xi) = \| a^{\frac{1}{2}}\, d_2^{\frac{1}{2p}}\, d_1^{-\frac{1}{2p}}\, \xi \|^2$.

As in proposition 2.3 we have proved that the vector $d_2^{\frac{1}{2p}}\, d_1^{-\frac{1}{2p}}\, \xi$ does not depend on the particular w' used, we get our statement for $q \in L(\infty;M,w_2\, w')$ by linearity. Note now that

$\|D^{p}_{1,2}(q)\|_{L(p;\,M,w)} = \|T(q;p,w_2\, w')\|_{L^{p}(M,w')} = \|q\|_{L(p;M,w_2)}$; so our statement follows by the density of $L(\infty;\, M,w_2\, w')$ in $L(p;M,w_2\, w')$ (see [3]).

2.7. Theorem. The mapping $D^{p}_{1,2} : L(p;\, M,w_2)$ in $L(p;M,w_2) \to L(p;M,w_1)$ is a positive linear isometry. For $1 \le p_1, p_2, r \le \infty$, $p_1^{-1} + p_2^{-1} = r^{-1}$, $q_i \in L(p_i;\, M,w_2)$ $(i = 1,2)$:

$$[\, D^{p_1}_{1,2}(q_1) \quad (p_1, M,w_1)\,]\,[\, D^{p_2}_{1,2}(q_2) \quad (p_2,M,w_1)\,] = q_1(p_1,M,w_2)\, q_2(p_2,M,w_2).$$

if $q \in L(1,M,w_2)$, then

$$\int q\, dw_2 = \int D^{1}_{1,2}(q)\, dw_1$$

Proof. The linearity and positivity of $D^{p}_{1,2}$ are obvious; the isometricity has been shown in the proof of lemma 2.6. By [3] and lemma 2.5 and 2.6 we have, for all n.f. states w' on M' :

$$T\Big([D^{p_1}_{1,2}(q_1)\,]\,(p_1;M,w_1)\, [D^{p_2}_{1,2}(q_2)\,]\,(p_2;\, M,w_1)\, r,w_1,\, w'\Big) =$$

$$= T\left(D^{p_1}_{1,2}(q_1),\, p_1,w_1,w'\right) \quad T\left(D^{p_2}_{1,2}(q_2),\, p_2,w_1,w'\right) =$$

$$= T(q_1;\, p_1,w_2\, w')\, T(q_2;\, p_2,w_2\, w') =$$

$$= T(\, q_1(\, p_1;M,w_2)\;\; q_2(p_2;M,w_2)\;\; q_2(p_2;M,w_2),\, r,w_2,w'),$$

which implies by [3] our statement.

The last equality is clear for $q \le 0$, as in this case $\int q\, dw_2 = \|q\|_{L(1,M,w_2)} =$

$= \|D^{1}_{1,2}\, q\|_{L(1,M,w_1)} = \int D^{1}_{1,2}\, q\, dw_1$ and follows by linearity for the general $q \in L(1,M,w_1)$.

Note that the previous theorem implies that also the duality relations are preserved by the mappings $D^{p}_{1,2}$.

2.8. Remark. For $p = 1$ the mappings $D^p_{1,2}$ can be defined directly by setting for $q \in L(1,M,w_2)$, $\xi \in D(H,w_2)$:

$$D^1_{1,2}(q)\ \left(d_{1,2}(1)\,\xi\right) = q(\xi)\,.$$

2.9. Theorem. (chain rule) Let w_1, w_2, w_3 be n.f. states on M. Then, for $1 \leq p \leq \infty$,

$$D^p_{1,3} = D^p_{1,2}\ D^p_{2,3}\,.$$

Proof. We have already remarked that, for $1 < p \leq +\infty$, $q \in L(p;M,w_3) : D^p_{1,3}(q) =$

$= q\,(T\,(q;\ p, w_3\ w')\,,\ p,\ w_1,\ w')$.

On the other hand

$$D^p_{1,2}\ \left(D_{2,3}(q)\right) = D^p_{1,2}\left(\ q(T(q;p,w_3,w')\ ,\ p,w_2,w'\right) = q\left(T(q;\ p,w_3\ w'),\ p,w_3,w'\right).$$

2.10. Corollary. For $1 \leq p \leq +\infty$ $(D^p_{1,2})^{-1} = D^p_{2,1}$

2.11. Remark. With some more effort we could have proved the same result as above also by allowing the auxiliary w' to be a n.f. semifinite weight on M; also w_1 and w_2 could have been chosen to be n.f.s. wight on M with the following additional hypothesis (needed in the proof of lemma 2.3):

the mapping $t \to w_1\left(\sigma^t_{w_2}(a)\right)$ is bounded on **R** for all $a \in M$.

Let us now deal with two particular cases:

2.12. Theorem. $w_1 < \alpha\, w_2$ for some $\alpha > 0$ iff $d_{1,2}(1) \in M$. If so, the mapping $z \to d_{1,2}(z)$ from S to M is bounded and weakly continuous on S and weakly analytic on S.

Proof. Let $x \in D(H,w_2)$; then

$$\|d_{1,2}(1)\,\xi\|^2 = \|\,d_1^{\frac{1}{2}}\ d_2^{-\frac{1}{2}}\xi\,\|^2 = w_1\ \left(\,|R_w\cdot(d_2^{-\frac{1}{2}}\xi)^*|^2\right)$$

and

$$\|\xi\|^2 = w_2\left(|R_{w'}(d_2^{-\frac{1}{2}}\xi)^+|^2 \right),$$

and the first part of our statement is a consequence of the weak operator density in M of the linear span of

$$\left\{ |R_{w'}(d_2^{-\frac{1}{2}}\xi)^+|^2 = \xi \in D(H,w_2) \right\}.$$

In order to prove the second part of our statement, if we take into account propositions 2.3 and 2.4, we need only to prove that $d_{1,2}(z)$ is bounded for $z \in S$ and that the mapping $z \to d_{1,2}(z)$ is bounded on S. We get, in the above hypothesis, as in the proof of prop. 2.3, for $\xi \in D(M,w_2)$,

$\eta \in \mathcal{D}(d_1^{\frac{1}{2}})$ and $\varphi(z) = \langle \eta, d_{1,2}(z)\xi \rangle$:

$$|\varphi(it)| < \|\eta\| \ \|\xi\|$$

$$(1+it) < \|\eta\| \, w_1\left(\sigma_{w_2}^{\frac{t}{2}}(\zeta_{w_2}(\xi))\right) \le \alpha \|\eta\| \, w_2\left(\zeta_{w_2}(\xi)\right) = \alpha \|\eta\| \ \|\xi\|$$

So by the three lines theorem

$$|\varphi(z)| < \alpha^{\mathrm{Re}\frac{z}{2}} \|\eta\| \ \|\xi\|, \text{ which implies } \|d_{1,2}(z)\| \le \alpha^{\mathrm{Re}\frac{z}{2}}.$$

2.13. Remark. In order to have $w_1 \le \alpha w_2$, by prop. 2.3. it is enough to assume the existence of a mapping $z \to a(z)$ from S to M, bounded and (weakly) continuous on S and (weakly) analytic on $\mathcal{S}$ such that $a(it) = (Dw_1 : Dw_2)_t$.

2.14. Proposition. If the hypothesis of th. 2.12 are satisfied, then, for $a \in M$, $p \in [1,+\infty)$, we have

$$D_{2,1}^p[q(a,w_1)] = q\left(d_{1,2}^+(\tfrac{1}{p})\, a\, d_{1,2}(\tfrac{1}{p}), w_2\right).$$

Proof. By the results proved in [1], as remarked in the proof of lemma 2.6:

$$D_{2,1}^p\, q(a,w_1) = D_{2,1}^p\left(d_1^{\frac{1}{2p}}\, a\, d^{\frac{1}{2p}}, p, w_1\right) = q\left(d_2^{\frac{1}{2p}}\, d_{1,2}(\tfrac{1}{p})^+\, a\, d_{1,2}(\tfrac{1}{p})\, d^{\frac{1}{2p}}, p, w_2\right) =$$

$$= q\left(d_{1,2}(\tfrac{1}{p})^+\, a\, d_{1,2}(\tfrac{1}{p}), \infty\right).$$

2.15. Theorem (Connes, [4]). In the above hypothesis, $w_2(d_{1,2}(1)^+ a\, d_{1,2}(1)) = w_1(a)$ for $a \in M$.

Proof. We have:

$$w_2\left(d_{1,2}(1)^+ a\, d_{1,2}(1) \right) = \int d_2\, d_{1,2}(1)^+ a\, d_{1,2}(1)\, dw' =$$

$$\int d_2^{\frac{1}{2}}\, d_{1,2}(1)^+ a\, d_{1,2}(1)\, d_2^{\frac{1}{2}}\, dw' = \int d_1^{\frac{1}{2}}\, a\, d_1^{\frac{1}{2}}\, dw' = w_1(a).$$

In the following theorem for a closed positive operator a with spectral decomposition $a = \int_0^{+} \mu\, de_\mu$ we shall set $p_\lambda(a) = \int_0^{\lambda} de_\mu$ and $a_\lambda = \int_0^{\lambda} \mu\, de_\mu$.

2.16. Theorem (cfr. Pedersen-Takesaki, [7]). The three following conditions are equivalent:

a) the closure of $d_{1,2}$ exists and is a positive operator with spectral projections p_λ in M invariant under $\sigma^t_{w_1}$ ($t \in \mathbf{R}$).

b) $w_1(a) = w_1(\sigma^t_{w_2}(a))$ $\quad \forall\, a \in M, \quad \forall\, t \in \mathbf{R}$.

c) there is a closed positive operator b affiliated with M with spectral projections invariant under $\sigma^t_{w_2}$ ($t \in \mathbf{R}$) such that

$$w_1(a) = \lim_{\lambda} w_2(b_\lambda\, a\, b_\lambda).$$

Proof a) $\Rightarrow$ b). Let $\xi \in D(H,w_2)$; then $w_1(\zeta_{w_2}(\xi)) =$

$$= \| d_{1,2}(1)\, \xi \|^2 = \| d_2^{-it}\, d_{1,2}(1)\, d_2^{it}\, \xi \|^2 =$$

$$= \| d_{1,2}(1)\, d_2^{it}\, \xi \| = w_1\left(\sigma^t_{w_2}(\zeta_{w_2}(\xi)) \right) \text{ for } t \in \mathbf{R}$$

and the conclusion follows by the usual density argument.

b) $\Rightarrow$ a). For all $t \in \mathbf{R}$ d_2^{it} is a unitary on H whose action leaves invariant M, M', w_1, w'; so it commutes with d_1. Therefore $d_1^{\frac{1}{2}}$ and $d_2^{-\frac{1}{2}}$ are positive operators with commuting spectral projections; then the closure of their product, $d_{1,2}(1)$ exists and is a positive operator with spectral projections p_i in M by 2.4. We also have:

$$\sigma_{w_1}^{it}\left(p_\lambda\left(d_{1,2}(1)\right)\right) = d_1^{it}\, p_\lambda\left(d_{1,2}(1)\right) d_1^{-it} = p_\lambda\left(d_{1,2}(1)\right).$$

a) $\Rightarrow$ c) By setting $b = d_{1,2}(1)$ our requirements are satisfied, as

$$d_1^{\frac{1}{2}} = \sup_\lambda \; d_{1,2}(1)_\lambda\, d_2^{\frac{1}{2}}, \qquad d_1 = \sup_\lambda \left[d_{1,2}(1)\, d_2^{\frac{1}{2}} \right]^2$$

and, for $a \in M^+$ (see [5])

$$w_1(a) = \int d_1\, a\, dw_1 = \sup_\lambda \int \left[d_{1,2}(1)_\lambda\, d_2^{\frac{1}{2}} \right]^2 a\, dw' =$$

$$= \sup_\lambda \int d_2\, d_{1,2}(1)_\lambda\, a\, d_{1,2}(1)_\lambda\, dw' =$$

$$= \lim_\lambda \; w_2\left(d_{1,2}(1)_\lambda\, a\, d_{1,2}(1_\lambda)\right).$$

c $\Rightarrow$ a) . For all a M_+, using the fact that b commutes with d_2, we get:

$$\int d_1\, a\, dw' = w_1(a) = \sup_\lambda \; w_2(b_\lambda\, a\, b_\lambda) =$$

$$= \sup_\lambda \int d_2\, b_\lambda\, a\, b_\lambda\, dw' = \sup_\lambda \int \left[b_\lambda\, d_2^{\frac{1}{2}} \right]^2 a\, dw'.$$

Therefore $d_1 = \sup \left[b_\lambda\, d_2^{\frac{1}{2}} \right]^2$, for $\xi' \in D(H,w')$, $d_1^{\frac{1}{2}}\, \xi' = \lim_\lambda b_\lambda\, d_2^{\frac{1}{2}}\, \xi'$ and for $\xi \in D(H,w_2)$,

$d_1^{\frac{1}{2}}\, d_2^{-\frac{1}{2}}\, \xi = b\, \xi$, which implies $d_1^{\frac{1}{2}}\, d_2^{\frac{1}{2}} = b$.

3. Let now M, M_o be von Neumann algebras with $M \supset M_o$, w a normal faithful state on M, $w_o = w|_{M_o}$ and φ_o a normal faithful state on M_o. If E_w the w-conditional expectation from M to M_o preserving w is a

norm one projection, the $\varphi = \varphi_o \circ E_w$ is a normal faithful state such that $E_\varphi = E_w$ and $\varphi|_{M_o} = \varphi_o$. For the general E_w, a φ with the above properties does not always exist, as it is proved in [10]. However in [4] the following result was proved.

3.1. Theorem. In the above situation there is a unique canonical extension $\varphi(\varphi_o, E_w)$ of φ_o to M such that

$$E_\varphi(a) = E_w (v^+ a v)$$

for a suitable partial isometry $v = v(\varphi_o, E_w)$.

For the full proof and details see [4]. Here it will only be mentioned that if we work in the standard representation of M with respect to w, with cyclic and separating vector Ω, the $\varphi(\varphi_o, E_w)(a) =$

$= \langle (D\varphi_o : Dw_o)_{\frac{i}{2}}\Omega, a(D\varphi_o : Dw_o)_{\frac{i}{2}}\Omega \rangle$ for $a \in M$, and that the partial isometry $v = v(\varphi_o\ w)$ is

given explicitely. In this way it is possible to prove that if E_w is a norm one projection, then $\varphi(\varphi_o, E_w) = \varphi_o \circ E_w$. It is also useful to note that, as φ is not necessarily faithful on M, E_φ is a natural generalization of the definition of w-conditional expectation given in [1]. Theorem 3.1 gives already a Radon-Nikoym type theorem for w and $\varphi(\varphi_o, E_w)$. It can be combined with the following

3.2. Theorem. Let M, M_o be von Neumann algebras with $M \supset M_o$, w_1, w_2 normal faithful states on M such that $w_1|_{M_o} = w_2|_{M_o} = w_o$ and $w_1 \leq \alpha w_2$ for some $\alpha > 0$. Then for $a \in M$,

$$E_{w_1}(a) = E_{w_2}\left((Dw_1 : Dw_2)_{\frac{i}{2}}\ a(Dw_1 : Dw_2)^+_{\frac{i}{2}} \right)$$

which is a straightforward generalization of theorem 1.1 both in the statement and in the proof, to yield the following result:

3.3. Theorem. Let M, M_o be von Neumann algebras with $M \supset M_o$, w_1, w_2 normal faithful states on M and $\overset{\circ}{w}_2 = w_2|_{M_o}$. If (*) $w_1 \leq \alpha\, \varphi(\overset{\circ}{w}_1, E_{w_2})$, for some positive α, then for $a \in M$

$$E_{w_1}(a) = E_{w_2}\left(\left(Dw_1 : D\varphi(\overset{\circ}{w}_1, E_{w_2}) \right)_{\frac{i}{2}} v(\overset{\circ}{w}_2, w_1)\, a\, v(\overset{\circ}{w}_2, w_1)^+ (Dw_1 : D\varphi(\overset{\circ}{w}_1, E_{w_2}))\right)$$

As it can be proved that (*) holds whenever for w_1, w_2 we substitute two different normal faithful states whose w-conditional expectations coincide respectively with E_{w_1} and E_{w_2}, theorem 3.3 depends in fact only on E_{w_1} and E_{w_2} and not on the particular choice of w_1 and w_2, and is a generalization of a) $\Rightarrow$ b) of theorem 11 for w conditional expectations with a) replaced by condition (*). For full proofs and additional material see [4].

We shall now generalize theorem 3.3 by dropping the majorization condition. If M, M_o are von Neumann algebras with $M \supset M_o$, w a normal faithful state on M and $w_o = w|_{M_o}$, we shall denote by E^1_w the mapping from L(1;M,w) to L(1;M,w_o) obtained by first lifting the w conditional expectation from M to M_o defined in [1] to a mapping from L(∞; M,w) to L(∞;M_o,w_o) and then extending it continuously to L(1; M,w).

3.4. Theorem. Let be M, M_o von Neumann algebras with $M \supset M_o$, w_1, w_2 normal faithful states on M sucht that $w_1|_{M_o} = w_o = w_2|_{M_o}$.

Let $q \in L(1; M,w_1)$. Then

$$E_{w_2}(q) = E_{w_1}(D^1_{1,2}\, q),$$

with $D_{1,2}$ as in 2.5

Proof. Recall that $D^1_{1,2}(q)\left((Dw_1 : Dw_2)_{\frac{i}{2}}\, \xi\right) = q(\xi)$ for $\xi \in D(H,w_2)$, $q \in L(1;M,w_2)$ and that we proved (remark 2.13), that if we choose any normal faithful state w' on M', $(Dw_1 : Dw_2)_{\frac{i}{2}} =$

$$= \left(\frac{dw_1}{dw'}\right)^{1/2} \left(\frac{dw_2}{dw'}\right)^{-1/2} \xi \text{ for all } \xi \in D(H, w_2).$$

If for $q \in L(1;m,w)$, $\varphi \in M_*$, $\xi \in D(H,w)$ we have:

$$q(\xi) = \varphi\left(R_{w'}\left(\left|\left(\frac{dw}{dw'}\right)^{-1/2} \xi\right)^{+}\right|^2\right).$$

then we set $q = q(\varphi)$ and $\varphi = \varphi(q)$. With this notation, we get, for $\xi \in D(H,w_2)$:

$$D^1_{1,2}(q)\ \big((Dw_1 : Dw_2)_{\frac{i}{2}}\, \xi \big) = D^1_{1,2}(q)\ \Big(\big(\frac{dw_1}{dw'}\big)^{1/2} \ \big(\frac{dw_2}{dw'}\big)^{-1/2}\, \xi \Big) =$$

$$[\, \varphi(D^1_{1,2}(q))\,]\Big(| R_{w'}\, \big(\big(\frac{dw_2}{dw'}\big)^{-1/2} \xi \big)^{+} |^2 \Big).$$

On the other hand,

$$q(\xi) = \varphi(q)\ \Big(| R_{w'}\, \big(\big(\frac{dw_2}{dw'}\big)^{-1/2} \xi \big)^{+} |^2 \Big).$$

By the weak operator density of the linear span of the set $\Big\{ | R_{w'}\, \big(\big(\frac{dw_2}{dw'}\big)^{-1/2} \big)^{+} |^2 ,\ \xi \in D(H,w_2)$ in M, this implies $\varphi(q) = \varphi(D^1_{1,2}(q))$. That is, q and $D^1_{1,2}(q)$ are the images, respectively in $L(1;M,w_1)$ and in $L(2;M, w_2)$, of the same $\varphi \in M_*$.

Now, as shown in [3], $E_{w_2}(q)$ is the representative in $L(1;M_\circ,w_\circ)$ of $\varphi(q)|_{M_\circ}$ and $E_{w_1}(D^1_{1,2}(q))$ the representative in $L(1;M_\circ, w_\circ)$ of $\alpha\, D^1_{1,2}(q)\,)\,|_{M_\circ}$. Therefore the established equality $\varphi(q) = \varphi(D^1_{1,2}(q))$ implies our statement. Combining theorems 3.4 and 3.1, we get immediately the following

3.5. <u>Theorem.</u> Let M, $M_\circ$ as before, w_1 and w_2 normal faithful states on M, $w_1|_{M_\circ} = \overset{\circ}{w}_1$, $w_2|_{M_\circ} = \overset{\circ}{w}_2$. Then, for $a \in M$, we get $E_{w_2}(q_a) = E_{w_1}\Big(D_{w_1',\, \varphi(\overset{\circ}{w}_1,\, E_{w_2})}\ q_{v^+ av} \Big)$

(with $v = v(\overset{\circ}{w}_1, E_{w_2})$ as usually).

Bibliography

1. L. Accardi - C. Cecchini: "Conditional expectations in von Neumann algebras and a theorem of Takesaki", J. Funct. Anal. 45 (1982) 245 - 273.

2. C. Cecchini: "Noncommutative L^p spaces and KMS functions", on Quantum Probability and Applications (Proceedings, Heidelberg 1984) Springer Verlag 1985, 136 - 142.

3. C. Cecchini:"Noncommutative integration for states on von Neumannn algebras", J. Op. Th. 15 (1986) 217 - 237.

4. C. Cecchini - D. Petz: "State extensions and a Radon Nikodym theorem for conditional expectations on von Neumann algebras", preprint.

5. A.Connes: "Sur le theoreme de Radon - Nikodym pour les poids normaux fideles semifinis", Bull. Sci. Math. Sec II, 97 (1973) 253 - 258.

6. A. Connes: "Une classification des facteurs de type III", Ann. Sci. Ecole Norm. Sup. 6 (1973) 133 – 252.

7. A. Connes: "On the spatial theory of von Neumann algebras", J. Funct. Anal. 35 (1980) 153 - 164.

8. N. Hilsum: "Les espaces L^p d' une algèbre de von Neumann defines par la derivée spatiale", J. Funct. Anal. 40 (1981) 151 - 169.

9. G. K. Pedersen - M. Takesaki: "The Radon-Nikodym theorem for von Neumann algebras", Acta Math. 130 (1973) 53 - 87.

10. D. Petz: "Sufficient subalgebras and the relative entropy of states of a von Neumann algebra", Comm. Math. Phys. 105 (1986) 123 - 131.

11. M. Terp: "Interpolation spaces between a von Neumann algebra and its predual", J. Op. Th. 8 (1982) 327 - 360.

MULTIDIMENSIONAL QUANTUM DIFFUSIONS

by

M.Evans and R.L. Hudson
Mathematics Department, Nottingham University
Nottingham NG7 2RD,UK

Acknowledgements : Part of this work was completed when R.L. Hudson was Visiting Professor at the Indian Statistical Institute, New Delhi. M. Evans acknowledges an SERC Research Studentship. We have had useful conversations with D.B.Applebaum, W. Bradshaw , P. Robinson and M. Schürmann.

§ 1. Introduction

This paper is about quantum generalisations of the concept of a diffusion on a manifold [6,14]. There are two aspects to this generalisation. Firstly we shall replace the integrator of classical diffusion theory, which is N-dimensional Brownian motion, by the gauge, creation and annihilation processes of quantum stochastic calculus satisfying the quantum Ito formula [13]

	$d\Lambda$	$dA^\dagger$	dA	dt
$d\Lambda$	$d\Lambda$	$dA^\dagger$	0	0
$dA^\dagger$	0	0	0	0
dA	dA	dt	0	0
dt	0	0	0	0

(1.1)

or an appropriate N-dimensional generalisation thereof. Secondly we must follow Connes [4] in finding an appropriate non-commutative generalisation of the notion of manifold. More specificly we need to have an appropriate quantum generalisation of the algebra $C^\infty(V)$ of smooth functions on a compact manifold V in accordance with the scheme [4]

Commutative	Noncommutative generalisation
$L^\infty(V)$	von Neumann algebra
$C(V)$	C^*-algebra
$C^\omega(V)$	?

In fact we shall say rather little about this second aspect, which is considered further in the article by D.B. Applebaum in this volume, contenting ourselves with some examples whose relation to classical manifolds is evident. Suffice it to say that the role of $C^{\infty}(V)$ is to be played by a noncommutative unital *-algebra of operators in a Hilbert space which will in general not be closed in the usual operator topologies leading to C*- and von Neumann algebras.

In § 2, by way of motivation, we consider classical diffusions from the view point of the Accardi-Frigerio-Lewis theory of quantum stochastic processes. In particular we show that a commutative process of the latter type over the algebra $C^{\infty}(V)$ in which the time evolution can be described by a system of stochastic differential equations driven by classical Brownian motion reduces to a diffusion on V. In § 3 we review for present purposes the Fock space quantum stochastic calculus. Included here is a coordinate-free tensorial description of the multidimensional quantum stochastic calculus which is appropriate for our purpose. In § 4 we describe some aspects of the Hochschild algebra cohomology theory which are relevant to quantum diffusions. Quantum diffusions are defined in § 5 as Accardi-Frigerio-Lewis type processes in which time evolution is governed by systems of quantum stochastic differential equations. It is shown that these equations, combined with the quantum Ito formula, generate identities of a cohomological character, and that, in certain circumstances, topological obstructions related to anomalies present themselves to the construction of quantum diffusions. In § 6 we show that cohomologically equivalent quantum diffusions are related by a natural perturbation procedure. It follows that for algebras such as $B(\mathcal{H}^{o})$ for which the cohomology is entirely trivial, quantum diffusions are completely described by multidimensional generalisations of the unitary processes constructed in [13]. In § 7 we consider examples, in particular involving multi-dimensional quantum Poisson processes.

Much of the present theory is described in [9,10] in the one-dimensional case, N = 1. For some earlier work on quantum diffusions see [2,11,12].

§ 2. Classical diffusions as quantum stochastic processes in the sense of Accardi, Frigerio and Lewis

Accardi, Frigerio and Lewis [1] define a C^* or W^* - quantum stochastic process on A as a quadruple $(A,\tilde{A},\tilde{\omega},j)$ comprising C^*- or W^*-algebras A and $\tilde{A}$, a state $\tilde{\omega}$ of $\tilde{A}$ and a family $j = (j_t : t \geq 0)$ of injective homomorphisms from A into $\tilde{A}$. In the von Neumann algebra or W^*-case, the homomorphisms j_t and the state $\tilde{\omega}$ are naturally required to be normal. For our purposes, as explained in the introduction, A will in general be neither a C^*-nor a von Neumann algebra; however we shall always assume that A is a unital normed $*$-algebra, that $\tilde{A}$ is a subalgebra of the unital $*$-algebra $B(\tilde{H})$ of bounded operators on a Hilbert space $\tilde{H}$ and that the homomorphisms j_t are norm-contractive.

Suppose now that $A = C^{\infty}(V)$, the algebra of complex valued smooth functions on a compact Riemann manifold V. Let $(\Omega,F,\mathbb{P})$ be a probability space supporting the N-dimensional Brownian motion $B = (B_1,\ldots,B_N)$. We equip V with its natural Borel σ-field G, and let μ be a probability measure on (V,G). Let A be the algebra $L^{\infty}(V \times \Omega, G \times F, \mu \times \mathbb{P})$, equipped with the natural state

$$\omega(f) = \int fd(\mu \times \mathbb{P}).$$

We consider an Accardi-Lewis-Frigerio quantum stochastic process $(A,\tilde{A},\tilde{\omega},j)$ in which, for each $f \in A$ and $p \in V$, the complex valued stochastic process $j_t f(p,.)$ is adapted to the filtration $(F_t : t \geq 0)$ of the Brownian motion B, and we have a system of stochastic differential equations

$$dj_t f(p,.) = \sum_{i=1}^{N} j_t \beta^i (f)(p,.)dB_i + j_t \tau(F)(p,.)dt, \; j_o f(p,.) = f(p);$$

equivalently we may write

$$d(j_t f) = \sum_{i=1}^{N} j_t \beta^i (f)dB_i + j_t \tau(f)dt, \quad j_o f = f . \tag{2.1}$$

Here $\beta^1,\ldots,\beta^N$ and τ are maps from A to itself. Because each j_t is a linear unital $*$-map, the β^i and τ are linear $*$-maps which vanish on the identity function 1.

Let us investigate the consequences for the β^i and τ of the multiplicativity of the maps j_t,

$$j_t(fg) = j_t(f)j_t(g).$$

Differentiating, and using (2.1) together with the Ito product formula for N-dimensional Brownian motion, we find that

$$\sum_{i=1}^{N} j_t\beta^i(fg)dB_i + j_t\tau(fg)dt = \sum_{i=1}^{N} j_t(\beta^i(f)g + f\beta^i(g))dB_i + (\sum_{i=1}^{N} j_t\, \beta^i(f)\beta^i(g)$$

$$+ j_t(\tau(f)g + f\tau(g)))dt.$$

Comparing coefficients of the differentials dB_i and dt, and cancelling j_t (which is possible since j_t is injective) we find that the maps β^i and τ must satisfy

$$\beta^i(fg) = \beta^i(f)g + f\beta^i(g) \tag{2.2}$$

$$\tau(fg) - \tau(f)g - f\tau(g) = \sum_{i=1}^{N} \beta^i(f)\beta^i(g) . \tag{2.3}$$

(2.2) tells us that each β^i is a derivation of $C^\infty(V)$ or vector field. For given vector fields β^i, the general solution for τ of (2.3) is the sum of a particular solution and another vector field. A particular solution is

$$\tau(f) = \frac{1}{2} \sum_{i=1}^{N} \beta^i(\beta^i(f))$$

thus the general solution is

$$\tau(F) = \beta^o(f) + \frac{1}{2} \sum_{i=1}^{N} \beta^i(\beta^i(f))$$

where β^o is another vector field.

In terms of a local coordinate system $(x_1,\ldots,x_m)$ on V we may write

$$\beta^i(f) = \sum_{j=1}^{m} \sigma^i_j(x) \frac{\partial f}{\partial x_j} , \quad i = 0,\ldots,N .$$

Consequently

$$\tau(f) = \sum_{j=1}^{m} \sigma^o_j(x) \frac{\partial f}{\partial x_j} + \frac{1}{2} \sum_{i=1}^{N} \sum_{j,k=1}^{m} \sigma^i_k(x) \frac{\partial}{\partial x_k} (\sigma^i_j(x)) \frac{\partial f}{\partial x_j}$$

$$+ \frac{1}{2} \sum_{i=1}^{N} \sum_{j,k=1}^{m} \sigma^i_k(x)\sigma^i_j(x) \frac{\partial^2 f}{\partial x_j \partial x_k} .$$

Substituting these forms for the β^i and τ into (2.1) and choosing $F \equiv x_\ell$, locally, we finally arrive at the local diffusion equation in the form

$$dX_\ell = \sum_{k=1}^{N} \sigma_\ell^k(X)dB_k + m_\ell(X)dt, \quad \ell = 1,\dots,m \tag{2.4}$$

where

$$m_\ell(x) = \sigma_\ell^o(x) + \frac{1}{2}\sum_{i=1}^{N}\sum_{k=1}^{m} \sigma_k^i(x)\frac{\partial}{\partial x_k}(\sigma_\ell^i(x)).$$

Even within the context of classical stochastic processes, the advantage of the global, algebraic description (2.1) of a diffusion over the local description (2.4) are apparent [14].

We shall see that introduction of the quantum integrator processes produces a rich generalisation of the cohomological identities (2.2) and (2.3), the analysis of which is the main prupose of this work.

§ 3. Quantum stochastic calculus

Let K be a finite dimensional complex Hilbert space, fixed once and for all. We call K the circumambient space and denote its dimension by N. We denote by h the test-function space $h = L^2(\mathbb{R}_+,K)$ consisting of square-integrable K-valued vector functions on the half line $\mathbb{R}_+$, and by H the Fock space over h. H may be conveniently characterised, to within unitary isomorphism exchanging the exponential vectors, as a Hilbert space equipped with a total family $\psi(f)$, $f \in h$ of exponential vectors satisfying

$$\langle\psi(f), \psi(g)\rangle = \exp\langle f,g\rangle, \quad f,g \in h.$$

The vacuum vector is $\psi_o = \psi(0)$. We denote by E the dense linear manifold in H spanned by the exponential vectors.

Let $f \in h$ and let C and T be respectively a contraction and a bounded operator in h. The creation and annihilation operators corresponding to f, and the differential second quantisation of T are the operators in H defined on the domain E by the actions

$$a^\dagger(f)\psi(g) = \frac{d}{d\varepsilon}\psi(g+\varepsilon f)\Big|_{\varepsilon=o}$$

$$a(f)\psi(g) = \langle f,g\rangle\psi(g)$$

$$\lambda(T)\psi(g) = \frac{d}{d\varepsilon}\psi(e^{\varepsilon T}g)\Big|_{\varepsilon=o}.$$

We assume that there is given, once and for all, an <u>initial</u> Hilbert space $\mathcal{H}^o$ equipped with a dense linear manifold $\mathcal{E}^o$ called the <u>initial domain</u>. We write $\tilde{\mathcal{H}} = \mathcal{H}^o \otimes \mathcal{H}$, $\tilde{\mathcal{E}} = \mathcal{E}^o \underline{\otimes} \mathcal{E}$ where $\otimes$ and $\underline{\otimes}$ denote respectively the Hilbert space and algebraic tensor products.

Corresponding to each $t \in \mathbb{R}_+$ we decompose h as

$$h = h^t \oplus h^{(t}, \quad h^t = L^2([o,t];K), \quad h^{(t} = L^2((t,\infty);K) .$$

There is a corresponding tensor product decomposition

$$\mathcal{H} = \mathcal{H}^t \otimes \mathcal{H}^{(t}$$

of $\mathcal{H}$ into the Hilbert space tensor product of the Fock spaces over h^t and $h^{(t}$ respectively, in which each $\psi(f)$ is a product vector,

$$\psi(f) = \psi(f^t) \otimes \psi(f^{(t})$$

where f^t and $f^{(t}$ are the components of $f \in h$ in h^t and $h^{(t}$. We can write

$$\mathcal{E} = \mathcal{E}^t \underline{\otimes} \mathcal{E}^{(t}$$

where $\mathcal{E}^t$ and $\mathcal{E}^{(t}$ are the spans of the exponential vectors in $\mathcal{H}^t$ and $\mathcal{H}^{(t}$. Also

$$\tilde{\mathcal{H}} = \tilde{\mathcal{H}}^t \otimes \mathcal{H}^{(t}, \quad \tilde{\mathcal{E}} = \tilde{\mathcal{E}}^t \underline{\otimes} \mathcal{E}^{(t}$$

where

$$\tilde{\mathcal{H}}^t = \mathcal{H}^o \otimes \mathcal{H}^t, \quad \tilde{\mathcal{E}}^t = \mathcal{E}^o \underline{\otimes} \mathcal{E}^{(t} .$$

Let L be a complex vector space and let r and s be nonnegative integers. An L-valued tensor of type (r,s) over K is a multilinear map E from the Cartesian product of r copies of K^* and s copies of K, where K^* is the dual space of K. Let E be a tensor of type (r,r). Choosing an orthonormal basis $(\xi_1,\dots,\xi_N)$ of K and denoting by ξ^* the image of ξ under the natural conjugate - isomorphism from K to K^* induced by the inner product, we may form the <u>trace</u> of E, that is the L - valued scalar or tensor of type $(0,0)$

$$\operatorname{tr} E = \sum_{j_1,\dots,j_r=1}^{N} E(\xi^*_{j_1},\dots,\xi^*_{j_N}, \xi_{j_1},\dots,\xi_{j_N}) .$$

There are natural vector space isomorphisms, which we shall use to identify the corresponding spaces, between the vector space tensor products $L \underline{\otimes} B(K)$, $L \underline{\otimes} K$ and $L \underline{\otimes} K^*$, and the spaces of L - valued tensors of types $(1,1)$, $(1,0)$

and (0,1) respectively. These are given, respectively, by linear extensions of the identifications $x \otimes |\xi\rangle\langle\eta| (\xi'^*, \eta') = \langle\xi', \xi\rangle\langle\eta, \eta'\rangle x$, $x \in L, \xi' \eta' \in K$, where $|\xi\rangle\langle\eta|$ is the dyad $|\xi\rangle\langle\eta|\zeta = \langle\eta, \zeta\rangle\xi, \zeta \in K$, and $x \otimes |\xi\rangle(\xi'^*) = \langle\xi', \xi\rangle x, x \otimes \xi^*(\xi') = \langle\xi, \xi'\rangle x$.

When L is an algebra, arbitrary L-valued tensors may be multiplied on the left or right by L-valued scalars. In addition, for $m,n,r \in \{0,1\}$, a tensor of type (m,n) may be multiplied on the right by a tensor of type (n,r) to yield a tensor of type (m,r), by using the identifications above together with the multiplications in L and $B(K)$, the natural left action of $B(K)$ on K and the dual right action given by $\xi^* T = (T^*\xi)^*$, $T \in B(K), \xi \in K$ of $B(K)$ on K^*.

Now let $\tilde{L}$ be the space of operators in $\tilde{H}$, defined on the domain $\tilde{E}$, whose adjoints have domains which include $\tilde{E}$. For $T \in \tilde{L}$ let $T^\dagger = T^*|_{\tilde{E}}$. Let $\tilde{L}^t$ and $L^{(t}$ be the analogous spaces of operators in $\tilde{H}^t$ and $H^{(t}$ respectively. Let E be an $\tilde{L}^t$-valued tensor of type (r,s) and J be an $L^{(t}$-valued tensor of type (s,r). Then we may form the $\tilde{L}$-valued tensor of type $(r+s, r+s)$

$$E \otimes J(\xi_1^*, \ldots, \xi_{r+s}^*, \eta_1, \ldots, \eta_{s+r}) = E(\xi_1^*, \ldots, \xi_r^*, \eta_1, \ldots, \eta_s) J(\xi_{r+1}^*, \ldots, \xi_{r+s}^*, \eta_{s+1}, \ldots, \eta_{s+r})$$

and hence the $\tilde{L}$-valued scalar $\operatorname{tr} E \otimes J$, in which the r contravariant entries of E are paired with the r covariant entries of J, and vice versa. Then

$$(\operatorname{tr} E \otimes J)^\dagger = \operatorname{tr} E^\dagger \otimes J^\dagger$$

where $E^\dagger$ is the $\tilde{L}^t$-valued tensor of type (s,r) given by

$$E^\dagger(\xi_1^*, \ldots, \xi_s^*, \eta_1, \ldots, \eta_r) = [E(\eta_1^*, \ldots, \eta_r^*, \xi_1, \ldots, \xi_s)]^\dagger$$

and $J^\dagger$ is defined analogously.

If $E, \tilde{E}$ are $\tilde{L}$-valued tensors of type (r,s) and $\phi, \tilde{\phi} \in \tilde{E}$, then we may define the $\mathbb{C}$-valued tensor of type $(r+s, r+s)$ $\langle E\phi, \tilde{E}\tilde{\phi}\rangle$ by

$$\langle E\phi, \tilde{E}\tilde{\phi}\rangle(\xi_1^*, \ldots, \xi_{r+s}^*, \eta_1, \ldots, \eta_{s+r})$$

$$= \langle E(\eta_1^*, \ldots, \eta_r^*, \xi_1, \ldots, \xi_s)\phi, \tilde{E}(\xi_{r+1}^*, \ldots, \xi_{r+s}^*, \eta_{s+1}, \ldots, \eta_{s+r})\tilde{\phi}\rangle,$$

and hence the $\mathbb{C}$-valued scalar $\operatorname{tr}\langle E\phi, \tilde{E}\tilde{\phi}\rangle$, in which the r contravariant entries of E become covariant on the left hand side of the inner product and match the r contravariant entries of $\tilde{E}$, and vice versa.

If E is a tensor of type $(r,1)$ over K and $\eta \in K$ is fixed, then $E(.,\eta)$ a tensor of type $(r,0)$ over K.

An adapted process of type (r,s) is a family $(E(t), t \geq 0)$ of $\check{\mathcal{L}}$-valued tensors of type (r,s) such that, for each $t \geq 0$ and $\xi_1,\ldots,\xi_{r+s} \in K$

$$E(t,\xi_1^*,\ldots,\xi_{r+s}) = E^t(\xi_1^*,\ldots,\xi_{r+s}) \underline{\otimes}\, 1^{(t}$$

where E^t is an $\tilde{\mathcal{L}}^t$-valued tensor of type (r,s).

The gauge, creation and annihilation processes are the adapted processes of types $(1,1)$, $(0,1)$ and $(1,0)$ respectively given by

$$\left.\begin{aligned} \Lambda(t,\eta^*,\xi) &= 1 \underline{\otimes}\, \lambda(P_{[o,t]} \otimes |\xi\rangle\langle\eta|) \\ A^\dagger(t,\xi) &= 1 \underline{\otimes}\, a^\dagger(\chi_{[o,t]}\xi) \\ A(t,\eta^*) &= 1 \underline{\otimes}\, a(\chi_{[o,t]}\eta) \end{aligned}\right\} \quad t \geq 0,\ \xi,\eta \in K$$

Here $P_{[o,t]}$ is the projector in $L^2(\mathbb{R}_+)$ of multiplication by the indicator function $\chi_{[o,t]}$ of $[o,t]$, and we use the natural identification $L^2(\mathbb{R}_+;K) = L^2(\mathbb{R}_+) \otimes K$. Together with the time t, there are the basic processes of quantum stochastic calculus; they are the integrators of the theory.

We note that, if J is one of the four basic processes, of type (r,s) and $0 \leq t_1 \leq t_2$ then we may write

$$J(t_2) - J(t_1) = 1 \underline{\otimes}\, J(t_1,t_2)$$

where $J(t_1,t_2)$ is an $\mathcal{L}^{(t_1}$ valued tensor of type (r,s).

Now let J be one of the four basic processes, of type (r,s). We define the stochastic integral against J of an adapted process E of type (s,r) initially in the case when E is elementary, that is, of the form

$$E(t) = \begin{cases} 0, & 0 \leq t < t_1 \\ E(t_1), & t_1 \leq t < t_2 \\ 0 & t_2 \leq t, \end{cases}$$

to be the scalar adapted process $M = \int \operatorname{tr} E \,\underline{\otimes}\, dJ$ given by

$$M(s) = \begin{cases} 0, & 0 \leq s < t_1 \\ \operatorname{tr} E(t_1) \underline{\otimes} J(t_1, s \wedge t_2), & t_1 \leq s \end{cases} .$$

An obvious extension defines the stochastic integral against J of simple process, that is of finite sums of elementary processes. Thus if E, F, G and H are simple processes of types (1,1), (1,0), (0,1) and (0,0) respectively, we may define the scalar process

$$M(s) = \int_0^s \operatorname{tr}(E \underline{\otimes} d\Lambda + F \underline{\otimes} dA^{\dagger} + G \underline{\otimes} dA + H \underline{\otimes} dt).$$

The following is essentially a restatement in tensorial language of Theorems 4.2 and 4.4 of [13].

Theorem 2.1 : Let $M = \int \operatorname{tr}(E \underline{\otimes} d\Lambda + F \underline{\otimes} dA^{\dagger} + G \underline{\otimes} dA + H \underline{\otimes} dt)$ and $\tilde{M} = \int \operatorname{tr}(\tilde{E} \underline{\otimes} d\Lambda + \tilde{F} \underline{\otimes} dA + \tilde{G} \underline{\otimes} dA + \tilde{H} \underline{\otimes} dt)$. Then, for arbitrary $u, v \in \mathcal{E}^{0}$, $f, g \in h$ and $s \geq 0$,

(a) $\langle u \otimes \psi(f), M(s) v \otimes \psi(g) \rangle$

$$= \int_0^s \langle u \otimes \psi(f), \{E(t, f(t)^*, g(t)) + F(t, f^*(t)) + G(t, g(t)) + H(t)\} v \otimes \psi(g) \rangle dt$$

(b) $\langle \tilde{M}(s) u \otimes \psi(f), M(s) v \otimes \psi(g) \rangle$

$$= \int_0^s [\langle \tilde{M}(t) u \otimes \psi(f), \{E(t, f(t)^*, g(t)) + F(t, f^*(t)) + G(t, g(t)) + H(t)\} v \otimes \psi(g) \rangle$$

$$+ \langle \{\tilde{E}(t, g(t)^*, f(t)) + \tilde{f}(t, g^*(t)) + \tilde{G}(t, f(t)) + \tilde{H}(t)\} u \otimes \psi(f), M(t) v \otimes \psi(g) \rangle$$

$$+ \operatorname{tr} \langle \{\tilde{E}(t, \cdot, f(t)) + \tilde{F}(t)\} u \otimes \psi(f), \{E(t, \cdot, g(t)) + F(t)\} v \otimes \psi(g) \rangle] dt.$$

By arguments similar to those of [13], based on the estimate

$$||M(s) u \otimes \psi(f)||^2$$

$$\leq \int_0^s \beta(s,t) \operatorname{tr}[(N+2)(\langle E(t, \cdot, f(t)) u \otimes \psi(f), E(t, \cdot, f(t)) u \otimes \psi(f) \rangle$$

$$+ F(t) u \otimes \psi(f), \mathbf{f}(t) u \otimes \psi(f) \rangle) + \langle G(t) u \otimes \psi(f), G(t) u \otimes \psi(f) \rangle$$

$$+ \langle H(t) u \otimes \psi(f), H(f) u \otimes \psi(f) \rangle] dt \qquad (3.1)$$

where

$$\beta(s,t) = \exp(s - t + (N+2) \int_t^s ||f||^2 ,$$

stochastic integration is extended from simple integrands to measurable processes E, F, G and H which, together with their adjoints, satisfy the local square integrability condition

$$\int_0^s \mathrm{tr}[\langle E(t,..,f(t)u \otimes \psi(f), E(t.,f(t))u \otimes \psi(f)\rangle + \langle f(t)u \otimes \psi(f), f(t)u \otimes \psi(f)\rangle$$
$$+ \langle G(t)u \otimes \psi(f), G(t)u \otimes \psi(f)\rangle + \langle H(t)u \otimes \psi(f), H(t)u \otimes \psi(f)\rangle]dt < \infty$$

for arbitrary $t > 0$, $u \in E^0$ and $f \in h$.

The multidimensional quantum Ito formula may now be expressed as follows. Let M_j, $j = 1,2$ have stochastic differentials $dM_j = \mathrm{tr}(E_j \underline{\otimes} d\Lambda + F_j \underline{\otimes} dA^\dagger + G_j \underline{\otimes} dA + H_j \underline{\otimes} dt)$.

Then, provided that M_1 and M_2, as well as the component processes of the E_j, F_j, G_j and H_j, $j = 1,2$ comprise bounded operators which are locally uniformly bounded in operator bound norm , the product M_1M_2 has stochastic differential

$$d(M_1,M_2) = M_1dM_2 + dM_1M_2 + dM_1 \cdot dM_2$$
$$= \mathrm{tr}[(M_1E_2+E_1M_2+E_1E_2) \underline{\otimes} d\Lambda + (M_1F_2+F_1M_2 + E_1F_2) \underline{\otimes} dA^\dagger$$
$$(M_1G_2+G_1M_2+G_1F_2) \underline{\otimes} dA + (M_1H_2+H_1M_2 + GF_2) \underline{\otimes} dt] \quad . \tag{3.2}$$

The "correction terms" E_1E_2, F_1F_2, G_1E_2 and G_1F_2 occuring in (3.2) are products of $B(\tilde{H})$ - valued tensors as defined above. In this sense, the multidimensional Ito formula takes precisely the same form (1.1) as the one dimensional case.

§ 4. Cohomology

Let M be a (two sided) module for the unital $*$-algebra A, so that M is a complex vector space and there are bilinear maps, called the left and right actions of A on M, $A \times M \ni (x,m) \mapsto x.m \in M$, $M \times A \ni (m,x) \mapsto m.x \in M$ satisfying

$$x_1.(x_2.m) = x_1x_2.m, \quad (m.x_1).x_2 = m.(x_1x_2)$$
$$x_1.(m.x_2) = (x_1.m).x_2$$
$$1.m = m.1 = m$$

for arbitrary x_1, $x_2 \in A$ and $m \in M$.

In the Hochschild algebra cohomology theory for A taking values in M, the space C_n of <u>n-cochains</u> consists of all n-linear maps $\eta : \overset{n}{\underset{j=1}{\times}} A \to M$. The <u>coboundary operator</u>, δ maps each C_n into C_{n+1} by

$$\delta\eta(x_0,x_1,\dots,x_n) = x_0.\eta(x_1,\dots,x_n) + \sum_{j=1}^{n} (-1)^j \eta(x_0,x_1,\dots,x_{j-1}x_j,\dots,x_n)$$
$$+ (-1)^{n+1}\eta(x_0,x_1,\dots,x_{n-1}).x_n .$$

The spaces Z_n of <u>n-cocycles</u> and B_n of <u>n-coboundaries</u> are the kernel of $\delta|_{C_n}$ and the range of $\delta|_{C_{n-1}}$ respectively. Since $\delta^2 = 0$, $B_n \subseteq Z_n$; the quotient space $H_n = Z_n/B_n$ is the nth <u>cohomology space</u> of the theory [8].

The simplest example of the theory is to take M to be A itself, with left and right actions to be ordinary multiplication,

$$x.m = \lambda(x)(m), \quad m.x = \rho(x)(m)$$

where

$$\lambda(x)(m) = xm, \quad \rho(x)(m) = xm, \quad x,m \in A \tag{4.1}$$

We call this the <u>standard</u> cohomology theory over A. Notice that a 1-cocycle for the standard theory is a derivation of A, that is a linear map α from A to itself satisfying

$$\alpha(xy) = x\alpha(y) + \alpha(x)y \qquad (x,y \quad A),$$

while a 1-coboundary is an <u>inner</u> derivation, that is a derivation of the form $\mathrm{ad}\,\ell : x \mapsto \ell x - x\ell$,

As a second example, let σ be a unital $*$-homomorphism from A to $A \otimes B(K)$ where K is the N-dimensional circumambient space of § 3, and let M be the algebraic vector space tensor product $A \underline{\otimes} K$ with left and right actions

$$x.m = (\lambda \otimes 1)\sigma(x)m, \quad m.x = \rho(x) \otimes 1\, m,$$

λ and ρ being defined by (4.1). We denote by C_n^σ, Z_n^σ, B_n^σ, H_n^σ and δ^σ the spaces of n-cochains, n-cocycles, and n-coboundaries, the nth cohomology space and the coboundary operator of this theory, henceforth reserving the symbols C_n, Z_n, B_n, H_n and δ for the corresponding objects of the standard theory.

For $\alpha \in C_1^\sigma$, define $\alpha^\dagger : A \to A \underline{\otimes} K^*$, where K^* is the dual space of K, by

$$\alpha^{\dagger}(x) = [\alpha(x^*)]^{\#} \tag{4.2}$$

where the outer involution # is the tensor product map of the involution of the *-algebra A with the canonical conjugate-isomorphism from the Hilbert space K onto its dual. The tensor product of the multiplication map $A \times A \to A$ and the natural bilinear map $(x^*,y) \to x^*(y)$ from $K^* \times K$ to $\mathbb{C}$ gives a bilinear map ι from $(A \underline{\otimes} K^*) \times (A \underline{\otimes} K)$ to $A \underline{\otimes} \mathbb{C} = A$. We define $\eta_\alpha \in C_2$ by

$$\eta_\alpha(x,y) = - \iota(\alpha^{\dagger}(x),\alpha(y)).$$

<u>Theorem 4.1</u> (a) If $\alpha \in Z_1^\sigma$ then $\eta_\alpha \in Z_2$. (b) If $\alpha \in B_1^\sigma$ then $\eta_\alpha \in B_2$; indeed if $\alpha = \delta^\sigma(\ell)$ then $\eta_\alpha = -\delta(\tau_\ell^\sigma)$ where

$$\tau_\ell^\sigma(x) = -\frac{1}{2}[\iota(\ell^{\#},\ell)x - 2\iota(\ell^{\#},(\lambda \otimes 1)\sigma(x)\ell) + x\,\iota(\ell^{\#},\ell)]$$

<u>Proof</u> : If $\alpha \in Z_1^\sigma$ then $\delta^\sigma\alpha = 0$ and so, for $x,y \in A$

$$x.\alpha(y) - \alpha(xy) + \alpha(x).y = 0 ,$$

that is,

$$\alpha(xy) = \lambda \otimes 1\sigma(x)\alpha(y) + \rho(y) \otimes 1\alpha(x) . \tag{4.3}$$

From this it follows that $\alpha^{\dagger}$ satisfies the identity

$$\alpha^{\dagger}(xy) = \rho \otimes 1\sigma^{\dagger}(y)\alpha^{\dagger}(x) + \lambda(x) \otimes 1\,\alpha^{\dagger}(y) \tag{4.4}$$

where $\sigma^{\dagger}$ is the homomorphism from A to $A \underline{\otimes} B(K^*)$ given by

$$\sigma^{\dagger}(y) = [\sigma(y^*)]^{\flat} ,$$

the outer involution $\flat$ being the tensor product map of the involution in A with the natural conjugate isomorphism from $B(K)$ to $B(K^*)$. Now consider

$$\delta\eta_\alpha(x,y,z) = x.\eta_\alpha(y,z) - \eta_\alpha(xy,z) + \eta_\alpha(x,y,z) - \eta_\alpha(x,y).z$$

From (4.3),

$$\eta_\alpha(x,yz) = -\iota(\alpha^{\dagger}(x),\alpha(yz)) = -\iota(\alpha^{\dagger}(x), \lambda \otimes 1\sigma(y)\alpha(z)+\rho(z)\otimes 1\alpha(y))$$

$$= -\iota(\alpha^{\dagger}(x), \lambda \otimes 1\sigma(y)\alpha(z)) + \eta_\alpha(x,y).z. \tag{4.5}$$

Similarly, from (4.4) we have

$$\eta_\alpha(xy,z) = -\iota(\rho\otimes 1\,\sigma^\dagger(y)\alpha^\dagger(x),\ \alpha(z)) + x.\eta_\alpha(y,z) \tag{4.6}$$

But

$$\iota\ (\alpha^\dagger(x),\ \lambda\otimes 1\,\sigma(y)\alpha(z))$$

$$= \iota\,(\rho\otimes 1\ [\sigma(y^*)]^\flat\ \alpha^\dagger(x),\ \alpha(z))$$

$$= (\rho\otimes 1\,\sigma^\dagger(y)\alpha^\dagger(x),\ \alpha(z)).$$

Thus, subtracting (4.6) from (4.5) we obtain $\delta\eta_\alpha = 0$ as required.

(b). If $\alpha = \delta^\sigma(\ell)$ then, for $x \in A$,

$$\alpha(x) = -\ell.x + x.\ell = -(\rho(x)\otimes 1)\ell + (\lambda\otimes 1)\sigma(x)\ell\ .$$

From this it follows that

$$\alpha^\dagger(x) = -\lambda(x)\otimes 1\ \ell^\# + (\rho\otimes 1)\sigma^\dagger(x)\ell^\#$$

and hence that

$$-\eta_\alpha(x,y) = \iota(\lambda(x)\otimes 1\ \ell^\# - (\rho\otimes 1)\sigma^\dagger(x)\ell^\#,\ \rho(x)\otimes 1\ell - (\lambda\otimes 1)\sigma(x)\ell)$$

$$= x\ \iota(\ell^\#,\ell)y - x\ \iota(\ell^\#,\ (\lambda\otimes 1)\sigma(x)\ell)$$

$$- \iota((\rho\otimes 1)\sigma^\dagger(x)\ell^\#,\ell)y + \iota((\rho\otimes 1)\sigma^\dagger(x)\ell^\#,(\lambda\otimes 1)\sigma(x)\ell)$$

$$= x\ \iota(\ell^\#,\ell)y - x\ \iota(\ell^\#,\ (\lambda\otimes 1)\sigma(y)\ell)$$

$$- \iota(\ell^\#,\ (\lambda\otimes 1)\sigma(x)\ell)y + \iota(\ell^\#,(\lambda\otimes 1)\sigma(xy)\ell).$$

Hence

$$\delta\tau^\sigma_\ell(x,y) = x\tau^\sigma_\ell(y) - \tau^\sigma_\ell(xy) + \tau^\sigma_\ell(x)y$$

$$= -\frac{1}{2}\,[x\ \iota(\ell^\#,\ell)y - 2x\ \iota(\ell^\#,(\lambda\otimes 1)\sigma(y)\ell) + xy\ \iota(\ell^\#,\ell)$$

$$- \iota(\ell^\#,\ell)xy + 2\ \iota(\ell^\#,(\lambda\otimes 1)\sigma(xy)\ell) - \iota(\ell^\#,\ell)xy$$

$$+ (\ell^\#,\ell)xy - 2\,\iota(\ell^\#,(\lambda\otimes 1))\sigma(x)\ )y + x\ \iota(\ell^\#,\ell)y]$$

$$= x\ \iota(\ell^\#(\lambda\otimes 1)\sigma(y)\ell) - \iota(\ell^\#,\lambda\otimes 1\sigma(x)\sigma(y)\ell)$$

$$+ \iota(\ell^\#,(\lambda\otimes 1)\sigma(x)\ell)y - x\ \iota(\ell^\#,\ell)y$$

$$= -\eta_\alpha(x,y)$$

□

§ 5. Quantum diffusions

We use the notations of § 3, but choose the initial domain $\mathcal{E}^{o}$ to be all of $\mathcal{H}^{o}$. Let $\mathcal{A}$ be a unital *-subalgebra of $B(\mathcal{H}^{o})$, having the property that homomorphisms from $\mathcal{A}$ are norm contractive (as will be the case if $\mathcal{A}$ is a C*-algebra, or, more generally, is closed under formation of holomorphic functions of self-adjoint elements).

Definition A quantum diffusion on $\mathcal{A}$ with circumambient space K is a family $j = (j_t : t \geq 0)$ of injective homomorphism from $\mathcal{A}$ into $B(\mathcal{H})$ with the following property. For each $x \in \mathcal{A}$, $j_o(x) = x \otimes 1$, and $x(t) \equiv j_t(x)$ is a scalar adapted process. Furthermore, for each $x \in \mathcal{A}$, there exist $\mathcal{A}$-valued tensors $\gamma(x)$, $\alpha(x)$, $\alpha^{\dagger}(x)$ and $\tau(x)$, of types (1,1), (1,0), (0,1) and (0,0) respectively, such that

$$dj_t(x) = tr[j_t o\gamma(x) \underline{\otimes} d\Lambda + j_t o\, \alpha(x) \underline{\otimes} dA^{\dagger} + j_t o\, \alpha^{\dagger}(x) \underline{\otimes} dA + j_t o\, \tau(x) \underline{\otimes} dt] \tag{5.1}$$

We abbreviate (5.1) as

$$dx = tr[\gamma(x) \underline{\otimes} d\Lambda + \alpha(x) \underline{\otimes} dA^{\dagger} + \alpha^{\dagger}(x) \underline{\otimes} dA + \tau(x) \underline{\otimes} dt]. \tag{5.2}$$

From the linearity, unitality and *-map properties of the homomorphisms j_t, together with the independence of the basic differentials [2], it is clear that $\gamma(x)$, $\alpha(x)$, $\alpha^{\dagger}(x)$ and $\tau(x)$ must be linear in $x \in \mathcal{A}$, vanish at $1 \in \mathcal{A}$, and that

$$\gamma(\xi^*,\eta,x^*) = \gamma(\eta^*,\xi,x)^*, \quad \alpha(\xi^*,x^*) = \alpha^{\dagger}(\xi,x)^*$$

$$\tau(x^*) = \tau(x)^* \tag{5.3}$$

for $\xi,\eta \in K$ and $x \in \mathcal{A}$.

We now investigate the consequences of the multiplicativity of the homomorphisms j_t. Compairing coefficients of the basic differentials between

$$d(xy) = dx\cdot y + x.dy + dx\, dy ,$$

where the left hand side is evaluated by substituting xy for x in (5.2) while the right hand side is evaluated by two uses of (5.2) and (3.2), we find that, for arbitrary $x,y \in \mathcal{A}$,

$$\gamma(xy) = x\gamma(y) + \gamma(x)y + \gamma(x)\gamma(y) \tag{5.4}$$

$$\alpha(xy) = \alpha(x)y + x\alpha(y) + \lambda(x)\alpha(y) \tag{5.5}$$

$$\alpha^{\dagger}(xy) = \alpha^{\dagger}(x)y + x\alpha^{\dagger}(y) + \alpha^{\dagger}(x)\lambda(y) \tag{5.6}$$

$$\tau(xy) = \tau(x)y + x\tau(y) + \alpha^{\dagger}(x)\alpha(y) \tag{5.7}$$

where tensors are multiplied as in § 3.

We add $xy\ 1$ to both sides of (5.4), where 1 is the A-valued tensor $1\,(\xi^*,\eta) = \langle\xi,\eta\rangle\, 1_A$ which corresponds to the identity in $A \underline{\otimes} B(H)$ when type (1,1) tensors are identified with elements of the algebra. Then (5.3) becomes

$$\gamma(xy) + xy\ 1 = (\tau(x) + x\ 1)(\tau(y) + y_1)\ .$$

Thus we can set

$$\gamma(x) = \sigma(x) - x \otimes 1 \tag{5.8}$$

where σ is a homomorphism from A to $A \otimes B(K)$.

Substituting from (5.8) into (5.5), we find that

$$\alpha(xy) = \rho(y) \otimes 1\ \alpha(x) + \lambda \otimes 1\ \sigma(x)\alpha(y).$$

In other words, $\alpha \in Z_1^{\sigma}$.

Finally we note that, in view of (5.3), $\alpha^{\dagger}$ is related to α by (4.2), and (5.6) becomes (4.4) on substituting from (5.8). The multiplication of type (0,1) and (1,0) tensor over A becomes the map ι from $A \underline{\otimes} K^* \underline{\otimes} A \otimes K$ to A. Thus (5.7) can be reformulated in the language of § 4 as

$$-\delta\tau = \eta_{\alpha}\ . \tag{5.9}$$

In particular, we must have $\eta_{\alpha} \in B_2$. (5.9)

For given $\sigma \in \mathrm{Hom}(A, A \underline{\otimes} B(K))$ and $\alpha \in Z_1^{\sigma}$, Theorem 4.1 guarantees that $\eta_{\alpha} \in Z_2$ but not that $\eta_{\alpha} \in B_2$. However if either of the cohomology spaces H_1^{σ} or H_2 is trivial then the stronger conclusion $\eta_{\alpha} \in B_2$ follows. Thus cohomological obstructions to the existence of quantum diffusions arise only in the case when <u>both</u> the cohomology spaces H_1^{σ} and H_2 are non-trivial. An example where this is the case is the so called noncommutative torus [4,5].

§ 6. Perturbations and unitary processes

Given $B(\mathcal{H})$-valued adapted processes L_o, L_1, L_2 and L_3 of types (1,1), (1,0), (0,1) and (0,0) respectively, we consider the stochastic differential equation for a scalar process U

$$dU = \operatorname{tr} U(L_o \underline{\otimes}\, d\Lambda + L_1 \underline{\otimes}\, dA^{\dagger} + L_2 \underline{\otimes}\, dA + L_3 \underline{\otimes}\, dt),\ U(0) = 1 \tag{6.1}$$

If the L_j are ampliations of fixed $B(\mathcal{H}^o)$-valued tensors then existence and uniqueness of the solution of (6.1) can be established iteratively [7] using the estimate (3.1), generalising the one dimensional case [13].

Necessary conditions for the solution of (6.1) to be unitary are found from

$$0 = d(U^{\dagger}U) = dU^{\dagger}U + U^{\dagger}dU + dU^{\dagger}dU\ ;$$

they are formally identical to the one-dimensional case:

$$\left.\begin{aligned} L_o + L_o^{\dagger} + L_o^{\dagger} L_o &= 0 \\ L_1 + L_2^{\dagger} + L_o^{\dagger} L_1 &= 0 \\ L_2 + L_1^{\dagger} + L_1^{\dagger} L_o &= 0 \\ L_3 + L_3^{\dagger} + L_1^{\dagger} L_1 &= 0 \end{aligned}\right\} \tag{6.2}$$

Here $B(\tilde{\mathcal{H}})$-valued tensors are multiplied as in § 3. We make the identifications between the spaces of such tensors of types (1,1), (1,0), (0,1) and $B(\tilde{\mathcal{H}}) \underline{\otimes} B(\mathcal{K}) = B(\tilde{\mathcal{H}} \otimes \mathcal{K})$, $B(\tilde{\mathcal{H}}) \underline{\otimes} \mathcal{K}$, and $B(\tilde{\mathcal{H}}) \underline{\otimes} \mathcal{K}^*$ respectively; then (6.1) and the corresponding set of conditions derived from $d(UU^{\dagger}) = 0$ are equivalent to $(L_o, L_1, L_2, L_3) = (W - 1,\ L,\ -L^{\dagger}W,\ iH - \frac{1}{2} L^{\dagger}L)$ where W is a unitary element of $B(\tilde{\mathcal{H}} \otimes \mathcal{K})$, L an element of $B(\tilde{\mathcal{H}}) \underline{\otimes} \mathcal{K}$, $L^{\dagger}$ is the image of L under the conjugate linear map which extends $T \otimes \xi \to T^* \otimes \xi^*$, $L^{\dagger}W = (W^*L)^{\dagger}$ where $B(\tilde{\mathcal{H}}) \underline{\otimes} B(\mathcal{K})$ acts on $B(\tilde{\mathcal{H}}) \underline{\otimes} \mathcal{K}$ as the tensor product of left multiplication in $B(\tilde{\mathcal{H}})$ with the natural action of $B(\mathcal{K})$ on $\mathcal{K}$, H is a self-adjoint element of $B(\tilde{\mathcal{H}})$ and $L^{\dagger}L$ is $\iota(L^{\dagger}, L)$. Thus (6.1) may be written

$$dU = \operatorname{tr} U((W-1) \underline{\otimes}\, d\Lambda + L \underline{\otimes}\, dA^{\dagger} - L^{\dagger}W \underline{\otimes}\, dA + (iH - \tfrac{1}{2} L^{\dagger}L) \underline{\otimes}\, dt) \tag{6.3}$$

In the case when W, L and H are ampliations to $B(\mathcal{H}) \otimes B(\mathcal{K})$, $B(\mathcal{H}) \otimes \mathcal{K}$ and $B(\mathcal{H})$ respectively of elements of $B(\mathcal{H}^o) \otimes B(\mathcal{K})$, $B(\mathcal{H}^o) \otimes \mathcal{K}$ and $B(\mathcal{H}^o)$ respectively, it can be shown [7], by an extension of the one-dimensional argument [13], that the solution of (6.3) is indeed unitary.

Now consider a quantum diffusion j over the $*$-algebra $\mathcal{A} \subseteq B(\mathcal{H}^o)$, given by

$$dx = \mathrm{tr}[\lambda(x) \underline{\otimes} d\Lambda + \alpha(x) \underline{\otimes} dA^\dagger + \alpha(x) \underline{\otimes} dA^\dagger + \tau(x) \underline{\otimes} dt] \; .$$

Let W_o, L_o and H_o be, respectively, a unitary element of $\mathcal{A} \underline{\otimes} B(\mathcal{K})$, an element of $\mathcal{A} \otimes \mathcal{K}$, and a self-adjoint element of $\mathcal{A}$. Denote by W, L and H the tensor adapted processes of types (1,1), (1,0) and (0,0) given by

$$W(t) = j_t \underline{\otimes} 1(W_o), \; L(t) = j_t \underline{\otimes} 1 \;\; (L_o), \; H(t) = j_t(H_o) \; .$$

Assume that, for these processes, a unitary process U satisfying (6.3) exists, noting that this will certainly be the case when j is the trivial diffusion for which $\lambda = \alpha = \tau = 0$, since then W, L and H retain their initial values. Consider the family of injective homomorphism from $\mathcal{A}$ into $B(\mathcal{H})$

$$\tilde{j}_t(x) \;=\; U(t) j_t(x) \, U(t)^* \; ;$$

Clearly $\tilde{j}_o(x) = j_o(x) = x \otimes 1$. A straightforward but tedious computation based on (1.1) gives

$$d\tilde{j}_t(x) = \mathrm{tr}[\tilde{j}_t \circ \tilde\gamma(x) \underline{\otimes} d\Lambda + \tilde{j}_t \circ \tilde\alpha(x) \underline{\otimes} dA^\dagger + \tilde{j}_t \circ \tilde\alpha^\dagger(x) \underline{\otimes} dA + \tilde{j}_t \circ \tilde\tau(x) \underline{\otimes} dt] \tag{6.4}$$

where

$$\tilde\gamma(x) = \tilde\sigma(x) - x \otimes 1 \; , \; \gamma(x) = \sigma(x) - x \otimes 1$$

and

$$\tilde\sigma(x) \;=\; W_o \, \sigma(x) W_o^{-1} \tag{6.5}$$

$$\tilde\alpha(x) \;=\; W_o \, \alpha(x) - (\lambda \otimes 1) W_o \, \sigma(x) W_o^* L_o + \rho(x) \otimes 1 \, L_o \tag{6.6}$$

$$\tilde\tau(x) \;=\; \tau(x) + i\,\mathrm{ad}H_o(x) \;+\; \tau^\sigma_{L_o}(x) - \mathcal{L}((W_o^* L_o)^\dagger, \alpha(x)) - \mathcal{L}(\alpha^\dagger(x), W_o^* L_o) \tag{6.7}$$

When $\tilde\sigma$ is related to σ by (6.5), the bijective maps $T_n : C_n^\sigma \ni \eta \mapsto W_o \eta$ map C_n^σ to $C_n^{\tilde\sigma} = C_n^\sigma$ in such a way that $\delta^{\tilde\sigma} T_{n-1} = T_n \delta^\sigma$. It follows that T_n maps Z_n^σ onto $Z_n^{\tilde\sigma}$ and B_n^σ onto $B_n^{\tilde\sigma}$ and we may pass to the quotient spaces to obtain an isomorphism between the cohomology spaces H_n^σ and $H_n^{\tilde\sigma}$, $n = 1,2,\ldots$. From (6.6) we see that $\tilde\alpha$ differs from the image of α under the isomorphism T_1 by a $\tilde\sigma$-coboundary. Similarly, $\tilde\tau$ differs by the coboundary ad H_o from the "natural" solution

$$\tilde\tau_o(x) = \tau(x) + \tau_{L_o}^{\tilde\sigma}(x) - \iota((W_o^* L_o)^\dagger, \alpha(x)) - \iota(\alpha^\dagger(x), W_o^* L_o) \text{ of } \delta\tilde\tau = -\eta_{\tilde\alpha}.$$

Thus the coefficient maps driving the quantum diffusion j differ in a cohomologically trivial way from those driving the original diffusion j.

§ 7. Examples

1. Diffusions on $B(H^o)$

Every endomorphism σ from $B(H^o)$ to $B(H^o \otimes K)$ is of from $x \to W(x \otimes 1) W^*$ where W is a unitary element of $B(H^o \otimes K)$. The cohomology spaces H_n^σ and H_n are all trivial. Thus every quantum diffusion on $B(H^o)$ is a perturbation, in the sense of § 6, of the trivial constant diffusion, and is thus given by

$$j_t(x) = U(t)(x \otimes 1)U(t)^\dagger$$

where U is a unitary process of the form (6.3) where W, L and H are ampliations of elements of $B(H^o \otimes K)$, $B(H^o) \otimes K$ and $B(H^o)$ respectively.

2. Multidimensional quantum poisson processes

For an arbitrary initial unital $*$-algebra $A \subsetneq B(H^o)$, let σ be an endomorphism from A to $A \otimes B(K)$, and let L be an element of $A \otimes K$, which we identify as in § 3 with an A-valued tensor of type (1,0). Suppose furthermore that, for all $\xi^* \in K^*$, $L(\xi^*) \in Z(A)$, the centre of A, and that

$$\sigma(L(\xi^*)) = L(\xi^*) \otimes 1 \quad .$$

These conditions are satisfied for example if $L = 1 \otimes \eta$ where η is a fixed element of K.

We regard L as an element of C_o^σ and construct the corresponding 1-coboundary $\alpha = \delta^\sigma(L)$, together with the associated element τ_L^σ of C_1 given by Theorem 4.1B. Then the quantum diffusion equation (5.2) becomes in this case

$$dx = tr(\sigma(x) - x \otimes 1) \underline{\otimes} d\Pi_L \tag{7.1}$$

where Π_L is the type (1,1) adapted process given by

$$\Pi_L(\xi^*,\eta) = \Lambda(\xi^*,\eta) - L(\xi^*) \otimes 1\, A^\dagger(\eta) - L^\dagger(\eta) \otimes 1\, A(\xi^*) + L(\xi^*)L(\eta) \otimes 1\, t\ .$$

In [9] and [10] the Fock space realisation of the Poisson process [13] was generalised to allow an "intensity" which is determined by an element of the initial algebra, and which can therefore itself be subject to a probability distribution. The process Π_L is the multidimensional analog of this generalisation. $tr\,\Pi_L$ is a Poisson process in the sense of [9] and [10] of intensity $\mathcal{L}(L^\dagger,L)$.

In the case when the endomorphism σ can be "diagonalised", in the sense that there exists an orthonormal basis $(\xi_1,\ldots,\xi_n)$ of K such that

$$\sigma(x) = \sum_{j=1}^{N} \sigma_j(x) \otimes |\xi_j><\xi_j|,\ \sigma_j : A \longrightarrow A,$$

the solution to the system (7.1) can be formed as follows.
Introduce the N commuting Poisson processes (in the sense of [9],[10]) $\Pi_L(\xi_j^*,\xi_j)$, $j = 1,\ldots,N$. Define $\{0,1\}$-valued observables $n_{1r},\ldots,n_{1N}, n_{21},\ldots,n_{2N},\ldots$ by

$$n_{jk} = \begin{cases} 1 \text{ if the } j^{th} \text{ jump of } tr\,\Pi_L = \sum_\ell \Pi_L(\xi_\ell^*,\xi_\ell) \text{ is in } \Pi_L(\xi_j^*\,;\,\xi_j) \\ 0 \quad \text{otherwise}\ . \end{cases}$$

Then

$$x(t) = \prod_{j=1}^{tr\,\Pi_L(t)} \sigma_1^{n_{1j}} \cdots \sigma_N^{n_{Nj}}(x)\ ,$$

where the product increases to the left with increasing j, solves (7.1).

References

[1] Accardi, L., Frigerio, A. and Lewis, J.T., Quantum stochastic processes, Proc. Res. Inst. Math. Sci. Kyoto 18, 94-133 (1982).

[2] Applebaum, D.B. and Hudson, R.L., Fermion Ito's formula and stochastic evolutions, Commun. Math. Phys. 96, 473-496 (1984).

[3] Applebaum, D.B. and Hudson, R.L., Fermion diffusions, J. Math. Phys. 25, 858-61 (1984).

[4] Connes, A, Noncommutative differential geometry, Publ. IHES 62, 41-144 (1985).

[5] Connes, A and M.Rieffel, Yang Mills for noncommutative two tori (preprint).

[6] Elworthy, K.D., Stochastic differential equations on manifolds, CUP (1982).

[7] M. Evans, Nottingham thesis in preparation.

[8] Hochschild, G., On the cohomology groups of an associative algebra, Ann. of Math. 48, 326-341 (1945).

[9] Hudson, R.L., Quantum diffusions and cohomology of algebras, to appear in Proceedings of 1st World Congress of Bernoulli Society, Tashkent 1986.

[10] Hudson, R.L., Algebraic theory of quantum diffusions, to appear in Stochastic differential equations, Proceedings, Swansed 1986, ed. A. Truman

[11] Hudson, R.L. and Parthasarathy, K.R., Quantum Diffusions, in Theory and Applications of Random Fields, Proceedings, Bangalore 1982, ed. Kallianpur, Springer LN Control Theory and Inf. Sci. 49 (1983).

[12] Hudson, R.L. and Parthasarathy, K.R., Construction of quantum diffusions, in Quantum probability and applications to quantum theory of irreversible processes, proceedings Rome 1982. ed. Accardi L, Frigerio, A. and Gorini V, Springer LNM 1055 (1984).

[13] Hudson, R.L. and Parthasarathy, K.R., Quantum Ito's formula and stochastic evolutions, Commun. Math. Phys. 93, 301-323 (1984).

[14] Ikeda, N. and Watanabe, S., Stochastic differential equations and diffusion processes, North Holland (1981).

AN APPLICATION OF DE FINETTI'S THEOREM

M. Fannes
Instituut voor Theoretische Fysica, K.U.Leuven
Celestijnenlaan 200 D, B-3030 Heverlee

I. INTRODUCTION

It is well known that symmetric measures (or states) on an infinite product of copies of a measure space (or of a C^*-algebra) have a unique decomposition in terms of symmetric product measures (or states). Recall that a measure (or a state) is called symmetric if it is invariant under the (local) permutations of the factors in the infinite product of measure spaces (or of algebras). There exist various extensions and refinements of the original result due to de Finetti [1] among which [2,3,4]. Evidently this type of result is due to the very high invariance of the symmetric measures (or states) which limits their number to such an extent that they can be parametrized by 'few' variables (the measures on the measures or states on a single factor in the infinite product).

In mathematical physics symmetric states are not completely uninteresting. Although their symmetry is too high to give a realistic description they occur as equilibrium states of the crudest kind of mean-field models. In fact these models are the simplest ones to exhibit a phase transition and they are often used as a first approximation to more realistic models. A formal procedure for computing the equilibrium states of such mean-field models (Haag-Bogoliubov approximation [5,6] consists in reducing the problem to solving the 'gap'-equation. This gap-equation is not anymore an equilibrium equation for a state of the infinite system but rather a non-linear equation for a state of a single site (factor of the infinite product). It is now precisely de Finetti's theorem, combined with a characterization of equilibrium states in terms of correlation inequalities [7], that allows us to prove the validity of the formal computation described above [8].

There are however many interesting models of mean-field type for which rigorous proofs of the validity of the gap-equation method of computing the equilibrium state don't exist yet [9,10]. Typical examples of such models are the discrete B.C.S. model [11], the continuous

B.C.S. model, the H.Y.L. model [12], ... A common feature of these models is that their Hamiltonians are no longer permutation invariant but rather that they split as a sum of a permutation invariant 'mean-field' part and a free part that transforms in a simple way under permutations. Evidently de Finetti's theorem has to be modified to cope with such systems.

This contribution will consist of two main parts. In the first we will prove the validity of the gap-equation approach to symmetric mean-field models and in the second we will extend de Finetti's theorem to non-permutation invariant measures.

II. EQUILIBRIUM STATES FOR PERMUTATION-INVARIANT MEAN FIELD MODELS

In order to fix the ideas and to avoid technical complications, we discuss in this section a discrete quantum spin system. More precisely we will consider the C^*-algebra

$$\mathcal{A}_{\mathbb{N}} = \bigcup_{\Lambda\subset\mathbb{N}} \mathcal{A}_\Lambda$$

which is the inductive limit of the $\{\mathcal{A}_\Lambda \mid \Lambda \subset \mathbb{N}\}$ where $\mathcal{A}_\Lambda = \bigotimes_{i\in\Lambda} \mathcal{A}_i$ and where each $\mathcal{A}_i$ is a copy of say M_N, the $N \times N$ complex matrices. P will denote the group of local permutations of $\mathbb{N}$ and for any $p \in P$ the induced automorphism of $\mathcal{A}_{\mathbb{N}}$ will again be denoted by p. A state ω of $\mathcal{A}_{\mathbb{N}}$ is called symmetric if $\omega \circ p = \omega$ for all $p \in P$. Let ω be a state of M_N, we use the notation ω_ρ for the state $\bigotimes_{i\in\mathbb{N}} \rho_i$ of $\mathcal{A}_{\mathbb{N}}$ where each ρ_i is a copy of ρ. Such states ω_ρ are called symmetric product states. The following theorem holds

<u>Theorem II.1</u> [3,4]
The symmetric states of $\mathcal{A}_{\mathbb{N}}$ form a simplex whose extreme points are the symmetric product states.

■

We now consider Hamiltonians

$$H_\Lambda = \sum_{i\in\Lambda} A_i + \frac{1}{2\#(\Lambda)} \sum_{\substack{i,j\in\Lambda\\ i\neq j}} B_{i,j} \in \mathcal{A}_\Lambda$$

where $A_i \in \mathcal{A}_i$ and $B_{i,j} \in \mathcal{A}_{\{i,j\}}$ are copies of a self-adjoint $A \in M_N$ and of a self-adjoint $B \in M_N \otimes M_N$ which is invariant under the permutation induced by $\phi \otimes \psi \to \psi \otimes \phi$, $\phi,\psi \in \mathbb{C}^N$.
The local Gibbs states $\omega_{\beta,\Lambda}$ of $\mathcal{A}_\Lambda$ are given by

$$\omega_{\beta,\Lambda}(.) = \frac{\mathrm{Tr}\, e^{-\beta H_\Lambda}\,\cdot}{\mathrm{Tr}\, e^{-\rho H_\Lambda}}$$

And we are interested in characterizing the possible limiting Gibbs states as $\Lambda \to \mathbb{N}$ (in the sence that Λ absorbs eventually any finite subset of $\mathbb{N}$). We will now use the following characterization of the canonical Gibbs states [7] :

$$\beta\, \omega_{\beta,\Lambda}(X^\star [H_\Lambda, X]) \geqslant \omega_{\beta,\Lambda}(X^\star X)\ \ln \frac{\omega_{\rho,\Lambda}(X^\star X)}{\omega_{\rho,\Lambda}(XX^\star)} \qquad X \in A_\Lambda \tag{1}$$

Consider any $w^\star$-limit point ω_β of the $\omega_{\beta,\Lambda}$ as $\Lambda \to \mathbb{N}$, which exists by $w^\star$-compactness. Due to the symmetry of the Hamiltonians H_Λ, ω_β will be symmetric and hence there exists by theorem II.1 a unique probability measure $\mu_\beta(d\rho)$ on the states on M_N such that

$$\omega_\beta = \int \mu_\beta(d\rho)\ \omega_\rho \tag{2}$$

We now have :

<u>Lemma II.2</u> For any $X \in A_\Lambda$

$$\beta \int \mu_\beta(d\rho)\ \omega_\rho(X^\star [H^\rho_\Lambda, X]) \geqslant \omega_\beta(X^\star X)\ \ln \frac{\omega_\beta(X^\star X)}{\omega_\beta(XX^\star)} \tag{3}$$

where $H^\rho_\Lambda = \sum_{i\in\Lambda} H^\rho_i$ and $H^\rho_i \in A_i$

is a copy of $H^\rho \in M_N$ where $H^\rho = A + B_\rho$ and B_ρ is determined by $(\sigma \otimes \rho)(B) = \sigma(B_\rho)$, σ any state of M_N.

<u>Proof</u> : Choose $n \in \mathbb{N}\setminus\Lambda$ and let $\{\Lambda_\alpha\}$ be a net tending to $\mathbb{N}$ such that $\omega_\beta = w^\star - \lim_\alpha \omega_{\beta,\Lambda_\alpha}$. For all $\Lambda_\alpha \supset \Lambda \cup \{n\}$ one has using the symmetry of the $\omega_{\beta,\Lambda_\alpha}$:

$$\omega_{\beta,\Lambda_\alpha}(X^\star [H_{\Lambda_\alpha}, X]) =$$

$$\omega_{\beta,\Lambda_\alpha}(X^\star [\sum_{i\in\Lambda} A_i, X]) + \frac{\#(\Lambda_\alpha) - \#(\Lambda)}{\#(\Lambda_\alpha)}\ \omega_{\beta,\Lambda_\alpha}(X^\star [\sum_{i\in\Lambda} B_{i,n}, X])$$

$$+ \frac{1}{2\#(\Lambda_\alpha)} \omega_{\beta,\Lambda_\alpha}(X^\star[\sum_{\substack{i,j\in\Lambda \\ i\neq j}} B_{i,j},X])$$

Taking the limit on α and using (2) one finds

$$\lim_\alpha \omega_{\beta,\Lambda_\alpha}(X^\star[H_{\Lambda_\alpha},X]) = \int \mu_\rho(d\rho)\ \omega_\rho(X^\star[H^\rho_\Lambda,X]) \ .$$

The lemma now immediately follows from (1).

∎

<u>Lemma II.3</u>

For μ_β almost all ρ :

$$[\rho,H^\rho] = 0 \tag{4}$$

<u>Proof</u> :

Choosing in (3) $x = \mathbb{1} + \lambda y$ with $\lambda \in \mathbb{R}$ and $y = y^\star \in A_\Lambda$ one gets

$$\beta \int \mu_\beta(d\rho)\ \omega_\rho((\mathbb{1} + \lambda y)[H^\rho_\Lambda,\lambda y]) \geqslant 0$$

which can only hold if for all $\Lambda \subset \mathbb{N}$ and $X \in A_\Lambda$

$$\int \mu_\beta(d\rho)\ \omega_\rho([H^\rho_\Lambda,X]) = 0 \tag{5}$$

Choose now for $k \in \mathbb{N}$, $y, z_1, \ldots , z_k \in M_N$ and put

$x = y \otimes z_1 \otimes \ldots \otimes z_k$ in (5) to get

$$\int \mu_\beta(d\rho)\{\rho([H^\rho,y])\rho(z_1)\ \ldots\ \rho(z_k) + \rho(y) \sum_{i=1}^k \rho(z_1)\ldots$$

$$\rho([H^\rho,z_i])\ \ldots\ \rho(zk)\} = 0$$

Replacing if necessary y by $y + \lambda$ where $\lambda \in \mathbb{C}$ is suitably chosen it follows that

$$\int \mu_\beta(d\rho)\rho(\ [H^\rho,y])\rho(z_1)\ \ldots\ \rho(z_k) = 0$$

It is now sufficient to observe that the algebra of functions on the states of M_N generated by

$$\rho \to \rho(z) \quad , \quad z \in M_N$$

contains the constant function and separates the states of M_N. Using the Stone-Weierstrass theorem the lemma follows. ∎

<u>Theorem II.4</u>

For μ_β almost all ρ

$$\rho = \frac{e^{-\beta H^\rho}}{\mathrm{Tr}\, e^{-\beta H^\rho}} \tag{6}$$

<u>Proof</u> :

Consider any local $z \in A_\Lambda$ and choose for $N = 1,2,\ldots$ copies $z_1,\ldots,z_N$ of z where $z_i \in A_{\Lambda_i}$, $1 \notin \Lambda_i$ and $\Lambda_i \cap \Lambda_j = \emptyset$ if $i \neq j$. Using (4) we compute that for

$$X^N = \sum_{k=1}^{N} y \otimes \mathbb{1}_{A_{\Lambda_1}} \otimes \ldots \otimes z_k \otimes \ldots \otimes \mathbb{1}_{A_{\Lambda_N}} \quad , \; y \in A_{\{1\}}$$

$$\int \mu_\beta(d\rho)\omega_\rho((X^N)^\star [H^\rho_{\{1\}\cup\Lambda_1 \ldots\cup\Lambda_N}, X^N]) =$$

$$N^2 \int \mu_\beta(d\rho)\rho(y^\star[H^\rho,y]) \; |\bigotimes_{j\in\Lambda} \rho(z)|^2 + O(N)$$

$$\omega_\beta\; ((X^N)^\star X^N) = N^2 \int \mu_\rho(d\rho)\rho(y^\star y) |\bigotimes_{j\in\Lambda} \rho(z)|^2 + O(N)$$

$$\omega_\rho(X^N(X^N)^\star) = N^2 \int \mu_\beta(d\rho)\rho(yy^\star) |\bigotimes_{j\in\Lambda} \rho(z)|^2 + O(N)$$

Therefore inequality (3) becomes on taking the limit $N \to \infty$

$$\beta \int \mu_\beta(d\rho)\rho(y^\star[H^\rho,y]) |\bigotimes_{j\in\Lambda} \rho(z)|^2 \geq$$

$$\int \mu_\rho(d\rho)\rho(y^\star y) |\times_{j\in\Lambda} \rho(z)|^2 \ln \frac{\int\mu_\rho(d\rho)\rho(y^\star y) |\bigotimes_{j\in\Lambda} \rho(z)|^2}{\int\mu_\rho(d\rho)\rho(yy^\star) |\bigotimes_{j\in\Lambda} \rho(z)|^2}$$

Using again the Stone-Weierstrass theorem we conclude that for every non-negative continuous function f on the states of M_N :

$$\beta \int \mu_\beta(d\rho)\rho(y^\star[H^\rho,y])\, f(\rho) \geq$$

$$\int \mu_\rho(d\rho)\rho(y^\star y) f(\rho) \ln \frac{\int\mu_\rho(d\rho)\rho(y^\star y)\, f(\rho)}{\int\mu_\rho(d\rho)\rho(yy^\star)\, f(\rho)}$$

which can only hold if for μ_β almost all ρ

$$\beta\ \rho(y^{\star}[H^{\rho},y]) \geqslant \rho(y^{\star}y)\ \ln \frac{\rho(y^{\star}y)}{\rho(yy^{\star})} \qquad y \in M_N$$

Repeating the argument of [7] yields then the gap-equation

$$\rho = \frac{e^{-\beta H^{\rho}}}{\mathrm{Tr}\ e^{-\beta H^{\rho}}}$$

■

In general the gap-equation (6) will admit several solutions. A further selection of the solutions occuring in limiting Gibbs states can be made on the basis of the variational principle of thermodynamics : only those solutions which minimize the free energy density should be taken into account. For more details on this matter and an explicit example we refer to [8].

III. NON-PERMUTATION INVARIANT PRODUCT MEASURES

In this section we will deal with a classical lattice system and for simplicity we will only consider here a configuration space of the type $\{+,-\}^{\mathbb{N}}$, the more general case being treated in [13]. Therefore the algebra $A_{\mathbb{N}}$ of real-valued continuous functions on the configuration space is generated by the classical spin functions σ_i, $i \in \mathbb{N}$ where $\sigma_i(x) = \pm 1$ according to whether the configuration x has a + or a - at site i. As in section II A_Λ denotes the observables of the region Λ i.e. $C(\{+,-\}^{\Lambda})$ and P the group of local permutations of $\mathbb{N}$ which induces by transposition automorphisms of $A_{\mathbb{N}}$:

$$p(f)(x) = f(\tilde{p}(x)) \quad \text{and} \quad \tilde{p}(x)(i) = x(p^{-1}(i)) \ , \ f \in A_{\mathbb{N}}$$

$$x \in \{+,-\}^{\mathbb{N}} .$$

As explained in the introduction it is our aim to characterize equilibrium measures for models with hamiltonians of the type

$$H_\Lambda = \sum_{i\in\Lambda} \phi_i \sigma_i + H_\Lambda^{\mathrm{int}} \in A_\Lambda \tag{7}$$

where $i \to \phi_i \in \mathbb{R}$ is a given 1-body potential (external field) and where the interaction part H_Λ^{int} is supposed to be permutation invariant i.e. $p(H_\Lambda^{\mathrm{int}}) = H_\Lambda^{\mathrm{int}}$ for all $\Lambda \subset \mathbb{N}$ and $p \in P$ such that $p(\Lambda) = \Lambda$.

Let ω now be an equilibrium measure at inverse temperature $\beta = 1$ then ω satisfies the D.L.R. equations

$$\omega(\tau^{-1}(f)) = \omega(f\ e^{-\Delta(\tau)}) \qquad f \in A_{\mathbb{N}}$$

$$\Delta(\tau) = \lim_{\Lambda} \tau(H_\Lambda) - H_\Lambda$$

and where τ is any local automorphism of $A_{\mathbb{N}}$ [14].
Restricting in the D.L.R. equations the local automorphisms to local permutations and taking into account the structure of H_Λ in (7) we get :

$$\omega(p^{-1}(f)) = \omega(f\ e^{-\Delta\phi(p)}) \qquad p \in P\ , \quad \in A_{\mathbb{N}} \tag{8}$$

where $\quad \Delta\phi(p) = \sum_{i\in\mathbb{N}} \phi_i(\sigma_{p(i)} - \sigma_i)$

$$= \sum_{i\in\mathbb{N}} (\phi_{p^{-1}(i)} - \phi_i)\sigma_i$$

The conditions (8) are now replacing the notion of symmetry and a measure ω which satisfies (8) will be called ϕ-symmetric. We now study the decomposition of ϕ-symmetric measures in external ones. Evidently some kind of growth condition on the ϕ_i has to be imposed. Indeed in the extreme situation where $\phi_i = 0$ for $i \leq i_0$ and $\phi_i = +\infty$ for $i > i_0$ equation (8) forces the measure to become the product of a symmetric measure on $A_{\{1,\ldots,i_0\}}$ and the point measure concentrated on the configuration $\{-\}^{\mathbb{N}\setminus\{1,\ldots,i_0\}}$. It is however well known that de Finetti's theorem does not hold for finite products of copies of a measure space. We will assume here that the rather restrictive condition

$$\sup_{i\in\mathbb{N}} |\phi_i| < \infty \tag{9}$$

is satisfied.

<u>Proposition III.1</u>

i) Let $\underset{i\in\mathbb{N}}{\times} \mu_i$ be a ϕ-symmetric product measure.

Then there exists a probability measure μ on $\{+,-\}$ such that

$$\mu_i(\pm) = \frac{e^{\mp\phi_i}\mu(\pm)}{\mu(e^{-\phi_i\sigma})} \tag{10}$$

ii) ϕ-symmetric product measures are two by two disjoint if ϕ is uniformly bounded (9).

<u>Proof</u> :

i) Choose $i \in \{2,3,\ldots\}$ and denote by $p \in P$ the transformation which permutes 1 and i. By ϕ-symmetry one gets

$$\mu_i(\pm) = (\mathop{X}_{j\in\mathbb{N}} \mu_j)\left(\frac{1 \pm \sigma_i}{2}\right)$$

$$= (\mathop{X}_{j\in\mathbb{N}} \mu_j)\left(p^{-1}\left(\frac{1 \pm \sigma_1}{2}\right)\right)$$

$$= (\mathop{X}_{j\in\mathbb{N}} \mu_j)\left(\left(\frac{1 \pm \sigma_1}{2}\right)e^{(\phi_1-\phi_i)\sigma_1+(\phi_i-\phi_1)\sigma_i}\right)$$

$$= e^{\pm(\phi_1-\phi_i)}\mu_1(\pm)\mu_i(e^{(\phi_i-\phi_1)\sigma})$$

It follows that

$$\mu_1(e^{(\phi_1-\phi_i)\sigma})\mu_i(e^{(\phi_i-\phi_1)\sigma}) = 1$$

and that (10) holds if we put

$$\mu(\pm) = \frac{\mu_1(\pm)\, e^{\pm\phi_1}}{\mu_1(e^{\phi_1\sigma})}$$

ii) Consider two probability measures μ and ν on $\{+,-\}$ and the ϕ-symmetric product measures $(\mathop{X}_{j\in\mathbb{N}} \mu_j)$ and $(\mathop{X}_{j\in\mathbb{N}} \nu_j)$ constructed with μ and ν using (10). By a general result of Kakutani [15] these measures will be disjoint iff

$$\sum_{j\in\mathbb{N}} \left(1 - \nu_j\left(\left(\frac{d\mu_j}{d\nu_j}\right)^{1/2}\right)\right) = +\infty$$

or by (10) iff

$$\sum_{j\in\mathbb{N}} \left(1 - \frac{\{e^{-\phi_j}\mu^{1/2}(+)\nu^{1/2}(+) + e^{\phi_j}\mu^{1/2}(-)\nu^{1/2}(-)\}}{\mu(e^{-\phi_j\sigma})^{1/2}\ \nu(e^{-\phi_j\sigma})^{1/2}}\right) = +\infty$$

Putting $\gamma = \sup_{j\in\mathbb{N}} |\phi_j| < \infty$ we can make the following estimates :

$$\sum_{j\in\mathbb{N}} \left(1 - \frac{\{e^{-\phi_j}\mu^{1/2}(+)\nu^{1/2}(+) + e^{\phi_j}\mu^{1/2}(-)\nu^{1/2}(-)\}}{\mu(e^{-\phi_j\sigma})^{1/2}\ \nu(e^{-\phi_j\sigma})^{1/2}}\right)$$

$$\geq \frac{1}{2} \sum_{j\in\mathbb{N}} \left(1 - \frac{\{e^{-\phi_j}\mu^{1/2}(+)\nu^{1/2}(+) + e^{\phi_j}\mu^{1/2}(-)\nu^{1/2}(-)\}^2}{\mu(e^{-\phi_j\sigma})\nu(e^{-\phi_j\sigma})}\right)$$

$$\geq \frac{1}{2} e^{-2\gamma} \sum_{j\in\mathbb{N}} \left(\mu(e^{-\phi_j\sigma})\nu(e^{-\phi_j\sigma}) - \{e^{-\phi_j}\mu^{1/2}(+)\nu^{1/2}(+) + e^{\phi_j}\mu^{1/2}(-)\nu^{1/2}(-)\}^2\right)$$

$$= \frac{1}{2} e^{-2\gamma} \sum_{j\in\mathbb{N}} \left(\mu^{1/2}(+)\nu^{1/2}(-) - \mu^{1/2}(-)\nu^{1/2}(+)\right)^2 = +\infty$$

if $\mu \neq \nu$.

∎

Lemma III.2

Let $a = (a_i)_{i=1,\dots,N}$ and $b = (b_i)_{i=1,\dots,N}$ belong to $(\mathbb{R}^+)^N$, $N = 1,2,\dots$

$$A_{ij} = a_i b_j \qquad i,j = 1,\dots,N$$

and $\gamma = \|a\|\,\|b\| - \langle a|b\rangle$

then $[A_{ij}] + \frac{1}{2}\gamma$

is a positive matrix on $\mathbb{R}^N$.

Proof :

If $\gamma = 0$ and $a \neq 0$ then there is a $\lambda \geq 0$ such that $b = \lambda a$. $[A_{ij}]$, considered as a linear operator on $\mathbb{C}^N$ is $\lambda\|a\|^2$ times the orthogonal projection on $\mathbb{C}a$ and therefore positive on $\mathbb{C}^N$ and hence on $\mathbb{R}^N$.

If $\gamma \neq 0$ then a and b are linearly independent in $\mathbb{C}^N$ and it is sufficient to show that

$$a \otimes b + b \otimes a + \gamma$$

is a positive operator on $\mathbb{C}^N$ where

$$(\phi \otimes \psi)\chi = \langle\psi|\chi\rangle\phi \qquad \phi,\psi,\chi \in \mathbb{C}^N$$

By an explicit computation however it is easy to check that the lowest eigenvalue of the self-adjoint rank 2 operator $a \otimes b + b \otimes a$ is precisely $-\gamma$.

∎

We now prove the main technical lemma :

<u>Lemma III.3</u>

Let μ be a ϕ-symmetric measure where ϕ is uniformly bounded. Suppose that

$$g \in A_{\Lambda_0} \ , \ g \geq 0$$

$$f \in A_{\Lambda} \ , \ \Lambda \cap \Lambda_0 = \emptyset$$

$$p \in P \text{ is such that } p(\Lambda) \cap \Lambda_0 = \emptyset \text{ and } p(\Lambda) \cap \Lambda = \emptyset$$

then one has

$$\mu(g\ f\ p(f)\ e^{\sum_{i\in\Lambda}\phi_i\sigma_i + \sum_{i\in p(\Lambda)}\phi_i\sigma_i}) \geq 0$$

<u>Proof</u> :

Let $\Lambda = \{i_1,\ldots,i_r\}$ and $p(\Lambda) = \{j_1,\ldots,j_r\}$ where $p(i_k) = j_k$, $k = 1,\ldots,r$.

As $\gamma = \sup_{j\in\mathbb{N}} |\phi_j| < \infty$ we can always find a strictly increasing sequence $n_1,n_2,\ldots$ in $\mathbb{N}$ not intersecting $\Lambda_0 \cup \Lambda \cup p(\Lambda)$ such that

$$\lim_{k\to\infty} \phi_{n_k} = \phi_\infty \quad \text{and} \quad \sum_{k\in\mathbb{N}} (\phi_{n_k} - \phi_\infty)^2 < \infty \tag{11}$$

Let now $\Lambda^k = \{n_{(k-1)r+1}, n_{(k-1)r+2}, \ldots, n_{kr}\}$, $k = 1,2,\ldots$ and define local permutations $p_k, q_k \in P$ by

$$p_k(i_\ell) = n_{(k-1)r+\ell} \ , \ q_k(j_\ell) = n_{(k-1)r+\ell} \ , \ p_k^2 = \text{id} \ ,$$

$$q_k^2 = \text{id} \ , \ k = 1,2,\ldots, \quad \ell = 1,\ldots,r \ .$$

For a finite $M \subset \mathbb{N}$ we will use the shorthand notation

$$\phi(M) = \sum_{i \in M} \phi_i \sigma_i .$$

By the ϕ-symmetry of μ one has for $k \neq \ell$

$$\mu(g\ f\ p(f)\ e^{\phi(\Lambda) + \phi(p(\Lambda))}) =$$

$$\mu(g\ e^{\phi(\Lambda) + \phi(p(\Lambda))} p_k(e^{-\phi(\Lambda^k)}) e^{\phi(\Lambda^k)} p_k(f) q_\ell(e^{-\phi(\Lambda^\ell)}) \times$$

$$e^{\phi(\Lambda^\ell)} p\ (f)) \tag{12}$$

and we apply now lemma III.2 with the choices

$$a_k = p_k(e^{-\phi(\Lambda^k)}) \ , \ b_k = q_k(e^{-\phi(\Lambda^k)}) \ , \ k = 1,\dots,N$$

to obtain

$$\sum_{k,\ell=1}^{N} e^{\phi(\Lambda^k)} p_k(f) p_k(^{-\phi(\Lambda^k)}) q_\ell(e^{-\phi(\Lambda^\ell)}) e^{\phi(\Lambda^\ell)} p_\ell(f)$$

$$\geqslant -\frac{1}{2}\gamma(N) \sum_{k=1}^{N} e^{2\phi(\Lambda^k)} p_k(f)^2 \tag{13}$$

where

$$\gamma(N) = \{\sum_{k=1}^{N} p_k(e^{-2\phi(\Lambda^k)})\}^{1/2} \{\sum_{k=1}^{N} q_k(e^{-2\phi(\Lambda^k)})\}^{1/2} -$$

$$\sum_{k=1}^{N} p_k(e^{-\phi(\Lambda^k)}) q_k(e^{-\phi(\Lambda^k)})$$

We now multiply both sides of (13) with the positive function $g\ e^{\phi(\Lambda) + \phi(p(\Lambda))}$ and we take the expectation with respect to μ. We then find using (12) :

$$N(N-1)\mu(g\ f\ p(f) e^{\phi(\Lambda) + \phi(p(\Lambda))}) +$$

$$\sum_{k=1}^{N} \mu(g\ e^{\phi(\Lambda) + \phi(p(\Lambda))} e^{2\phi(\Lambda^k)} p_k(f)^2 p_k(^{-\phi(\Lambda^k)}) q_k(e^{-\phi(\Lambda^k)})$$

$$-\frac{1}{2}\sum_{k=1}^{N}\mu(g\ e^{\phi(\Lambda)}+\phi(p(\Lambda))e^{2\phi(\Lambda^k)}p_k(f)^2\gamma(N)) \qquad (14)$$

But

$$\sum_{k=1}^{N}\mu(g\ e^{\phi(\Lambda)}+\phi(p(\Lambda))e^{2\phi(\Lambda^k)}p_k(f)^2p_k(e^{-\phi(\Lambda^k)})q_k(e^{-\phi(\Lambda^k)}))$$

$$\leqslant N\ \|g\|\ \|f\|^2\ e^{6\gamma r}$$

and

$$\sum_{k=1}^{N}\mu(g\ e^{\phi(\Lambda)}+\phi(p(\Lambda))e^{2\phi(\Lambda^k)}p_k(f)^2\gamma(N))$$

$$\leqslant N\ \|g\|\ \|f\|^2\ e^{4\ r}\ \|\gamma(N)\|$$

We now estimate $\|\gamma(N)\|$. Denote by ε and δ two 'configurations' of length r : $\varepsilon=(\varepsilon_1,\dots,\varepsilon_r)$ and $\delta=(\delta_1,\dots,\delta_r)$ with $\varepsilon_i,\delta_i\in\{-1,+1\}$.

$$\|\gamma(N)\| = \sup_{\varepsilon,\delta}\ \{\sum_{k=1}^{N}\prod_{i=1}^{r} e^{-2\phi_{n_{(k-1)r+i}}\varepsilon_i}\}^{1/2}\ \times$$

$$\{\sum_{k=1}^{N}\prod_{i=1}^{r} e^{-2\phi_{n_{(k-1)r+i}}\sigma_i}\}^{1/2}\ -\sum_{k=1}^{N}\prod_{i=1}^{r} e^{-2\phi_{n_{(k-1)r+i}}(\varepsilon_i+\delta_i)}$$

$$=\sup_{\varepsilon,\delta}(\prod_{i=1}^{r} e^{-\phi_\infty(\varepsilon_i+\delta_i)})\times(\sum_{k=1}^{N}\prod_{i=1}^{r} e^{2(\phi_\infty-\phi_{n_{(k-1)r+i}})\varepsilon_i}\}^{1/2}\ \times$$

$$\{\sum_{k=1}^{N}\prod_{i=1}^{r} e^{2(\phi_\infty-\phi_{n_{(k-1)r+i}})\delta_i}\}^{1/2}\ -\sum_{k=1}^{N}\prod_{i=1}^{r} e^{(\phi_\infty-\phi_{n_{(k-1)r+i}})(\varepsilon_i+\delta_i)})$$

$$\leqslant\frac{1}{2}e^{2\gamma r}\sup_{\varepsilon,\delta}\sum_{k=1}^{N}(\prod_{i=1}^{r} e^{(\phi_\infty-\phi_{(k-1)r+i})\varepsilon_i}-\prod_{i=1}^{r} e^{(\phi_\infty-\phi_{(k-1)r+i})\delta_i})^2$$

$$\leqslant 2e^{2\gamma r}\sup_{\varepsilon}\sum_{k=1}^{N}(\prod_{i=1}^{r} e^{(\phi_\infty-\phi_{n_{(k-1)r+i}})\varepsilon_i}-1)^2$$

$$\leqslant 2e^{6\gamma r}\sum_{k=1}^{N}(\sum_{i=1}^{r}|\phi_\infty-\phi_{n_{(k-1)r+i}}|)^2$$

$$\leq 2\, r\, e^{6\gamma r} \sum_{k=1}^{Nr} (\phi_\infty - \phi_{n_k})^2$$

which is uniformly bounded in N by (11).
Collecting now these estimates and taking in (14) the limit $N \to \infty$ the lemma follows.

■

We now arrive at the decomposition theorem for ϕ-symmetric states :

Theorem III.4

Let ϕ be uniformly bounded, then the set of ϕ-symmetric probability measures on $\{+,-\}^{\mathbb{N}}$ is a simplex whose extreme points are the ϕ-symmetric product measures.

Proof :

i) Suppose that μ is a ϕ-symmetric product measure which can be decomposed as

$$\mu = \lambda\mu_1 + (1-\lambda)\mu_2 \qquad 0 < \lambda < 1$$

where μ_1 and μ_2 are ϕ-symmetric. We have to show that $\mu_1 = \mu_2 = \mu$. As μ is product it follows from proposition III.1 that for any $f \in A_\Lambda$, $\Lambda \subset \mathbb{N}$ finite and for any $p \in P$ such that $\Lambda \cap p(\Lambda) = \emptyset$

$$\mu((f - \nu)(p(f) - \nu)e^{\phi(\Lambda \cup p(\Lambda))}) = 0$$

where $\nu = \dfrac{\mu(f\; e^{\phi(\Lambda)})}{\mu(e^{\phi(\Lambda)})}$.

It then follows from lemma III.3 that

$$\mu_i((f - \nu)(p(f) - \nu)e^{\phi(\Lambda \cup p(\Lambda))}) = 0 \qquad i = 1,2$$

which implies that both μ_1 and μ_2 are also product states and must therefore coincide with μ.

ii) The converse argument is quite standard and consists in constructing for a non-product ϕ-symmetric measure a non-trivial decomposition in ϕ-symmetric measures.

iii) The uniqueness of the decomposition in ϕ-symmetric product states is an immediate consequence of the Stone-Weierstrass theorem.

■

In order to conclude I would like to make the following comment. The condition (9) of uniform boundedness of ϕ can certainly be weakened. A reasonable decomposition of the type of theorem III.4 can be expected to hold as long as the ϕ-symmetric product states are two by two disjoint. In fact, from the point of view of applications, one is rather interested in potentials ϕ such that for all $\beta > 0$ the $\beta\phi$-symmetric product states are two by two disjoint. This condition turns out to be equivalent to the existence of a strictly increasing sequence n_r in $\mathbb{N}$ such that

$$\lim_{N\to\infty} \frac{1}{N \ln N} \sum_{k=1}^{N} |\phi_{n_r}| = 0$$

and should replace the uniform boundedness of ϕ. One can however not expect to allow for still more divergent potentials as this could induce phase transitions due to the non-translation invariance of the potential [16].

IV. REFERENCES

[1] B. de Finetti; Ann. Inst. H. Poincaré, 1, 1 (1937)
[2] E. Hewitt; L.S. Savage, Trans. Am. Math. Soc., 80, 470 (1955)
[3] E. Størmer; J. Funct. Anal., 3, 48 (1969)
[4] R.L. Hudson, G.R. Moody; Z. Wahrsch. verw. Geb., 33, 343 (1976)
[5] N.N. Bogoliubov; J. Phys. Moscow, 11, 23 (1947)
[6] R. Haag; Nuovo Cimento, 25, 287 (1962)
[7] M. Fannes, A. Verbeure; Commun. Math. Phys. 57, 165 (1977)
[8] M. Fannes, H. Spohn, A. Verbeure; J. Math. Phys., 21, 355 (1980)
[9] F. Jelinek; Commun. Math. Phys., 9, 169 (1968)
[10] M. van den Berg, J.T. Lewis, J.V. Pulè; The large deviation principle and some models of an interacting Bose gas; preprint Dublin Institute for Advanced Studies.
[11] W. Thirring; Commun. Math. Phys., 7, 181 (1968)
[12] K. Huang, C.N. Yang, J.M. Luttinger; Phys. Rev., 105, 776 (1957)
[13] M. Fannes; Non-translation invariant product states of classical lattice systems; preprint K.U.Leuven
[14] M. Fannes, P. Vanheuverzwijn, A. Verbeure; J. Stat. Phys., 29, 547 (1982)
[15] S. Kakutani; Ann. Math., 49, 214 (1948)
[16] W.G. Sullivan; Commun. Math. Phys., 40, 249 (1975)

THE QUANTUM LANGEVIN EQUATION FROM THE INDEPENDENT-OSCILLATOR MODEL

G.W. Ford
Department of Physics
The University of Michigan
Ann Arbor, MI 48109-1120/USA

The quantum Langevin equation for a Brownian particle in an external potential V(x) has the form:[1]

$$m\ddot{x} + \int_{-\infty}^{t} dt' \mu(t - t')\dot{x}(t') + V'(x) = F(t) \ . \tag{1}$$

This is a Heisenberg equation of motion for the position operator x(t). The coupling to the heat bath corresponds to two terms: a friction or radiation-reaction force characterized by the memory function $\mu(t)$ and a random operator force F(t). The random operator force is Gaussian with mean zero and with (symmetric) correlation:

$$\begin{aligned} &\tfrac{1}{2}\langle F(t)F(t') + F(t')F(t)\rangle \\ &\quad = \frac{1}{\pi}\int_{0}^{\infty} d\omega \, \mathrm{Re}\{\tilde{\mu}(\omega + i0^{+})\} \, \hbar\omega \coth(\hbar\omega/2kT) \cos\omega(t - t') \ , \end{aligned} \tag{2}$$

It has nonequal-time commutator:

$$[F(t),F(t')] = \frac{2}{i\pi}\int_{0}^{\infty} d\omega \, \mathrm{Re}\{\tilde{\mu}(\omega + i0^{+})\} \, \hbar\omega \sin\omega(t - t') \ . \tag{3}$$

Here $\tilde{\mu}$ is the Fourier transform of the memory function,

$$\tilde{\mu}(z) = \int_{0}^{\infty} dt e^{izt}\mu(t) \ , \ \mathrm{Im}z > 0 \ . \tag{4}$$

The function $\tilde{\mu}(z)$ is obviously analytic in the upper half plane, Imz > 0. This is a consequence of the requirement of causality; only the past history of the particle motion appears in the friction force. It is a consequence of the second law or thermodynamics that the boundary value of $\tilde{\mu}(z)$ on the real axis must be a positive distribution,

$$\mathrm{Re}\{\tilde{\mu}(\omega + i0^{+})\} \geq 0 \qquad , \ -\infty < \omega < \infty \ . \tag{5}$$

Finally, the fact that $\mu(t)$ is real leads to the reality condition:

$$\tilde{\mu}(-\omega + 0^{+}) = \tilde{\mu}(\omega + 0^{+})^{*} \ . \tag{6}$$

These three properties, analytic in the upper half plane, real part positive on the real axis, and reality, characterize $\tilde{\mu}(z)$ as what is termed a positive real function.[2,3] Positive real functions have many important properties, here I will only stress the fact that they can always be represented by the Stieltjes inversion theorem:

$$\tilde{\mu}(z) = -icz + \frac{2iz}{\pi} \int_0^\infty d\omega \, \frac{\mathrm{Re}\{\tilde{\mu}(\omega + i0^+)\}}{z^2 - \omega^2} \,, \tag{7}$$

where c is a positive constant.

With this representation we see that the quantum Langevin equation, like its classical counterpart, is completely characterized by the single real, even distribution $\mathrm{Re}\{\tilde{\mu}(\omega + i0^+)\}$. I have given here but a brief review of the remarkable properties of this equation. For a more thorough discussion I refer to a forthcoming publication,[4] where also more complete references are given. I should, however, point out for the benefit of this audience that, except in the classical limit, the random force is *never* δ-function correlated.

My aim here is to show that these properties can be derived on the basis of the independent-oscillator (IO) model of the heat bath. In this model the Brownian particle is surrounded by a large number of bath particles, to each of which it is coupled with a spring. The Hamiltonian for the IO model is

$$H = \frac{p^2}{2m} + V(x) + \sum_j \left[\frac{p_j^2}{2m_j} + \frac{1}{2}k_j(q_j - x)^2 \right] \tag{8}$$

We recall the canonical commutation rules:

$$[x,p] = i\hbar \;, \quad [q_j, p_k] = i\hbar\delta_{jk} \;. \tag{9}$$

The Heisenberg equations of motion for the coupled motion of the system of particle plus bath are then

$$\begin{aligned} \dot{x} &= [x,H]/i\hbar = p/m \\ \dot{p} &= [p,H]/i\hbar = -V'(x) + \sum_j m_j\omega_j^2(q_j - x) \\ \dot{q}_j &= [q_j,H]/i\hbar = p_j/m_j \\ \dot{p}_j &= [p_j,H]/i\hbar = -m_j\omega_j^2(q_j - x) \quad , \end{aligned} \tag{10}$$

where dot and prime represent, respectively, differentiation with respect to t and x. Eliminating the the momentum variables, we can write these in the form:

$$m\ddot{x} + V'(x) = \sum_j m_j\omega_j^2(q_j - x) \;, \tag{11}$$

$$\ddot{q}_j + \omega_j^2 q_j = \omega_j^2 x \;. \tag{12}$$

The equations (12) are inhomogeneous differential equations for the

q_j, whose general solution is

$$q_j(t) = q_j^h(t) + x(t) - \int_{-\infty}^{t} dt' \ \cos\omega_j(t - t') \ \dot{x}(t') \ , \tag{13}$$

where $q_j^h(t)$ is the general solution of the homogeneous equation ($x \equiv 0$). This is given by

$$q_j^h(t) = q_j \cos w_j t + p_j \frac{\sin\omega_j t}{m_j \omega_j} \ , \tag{14}$$

where q_j and p_j are time independent operators satisfying the same commutation rules (9). Putting (13) in (11), we can write the result in the form of the quantum Langevin equation (1) withthe memory function given by:

$$\mu(t) = \sum_j m_j \omega_j^2 \cos\omega_j t \ , \tag{15}$$

and with the random operator force given by:

$$F(t) = \sum_j m_j \omega_j^2 q_j^h(t) \ . \tag{16}$$

To calculate the correlation and commutator of the random force F(t), we can imagine, for example, that in the distant past the Brownian particle is fastened to a very large mass at $x = 0$. This was already implied by our choice of the solution (13). The oscillators are then at equilibrium with respect to the bath Hamiltonian

$$H_B = \sum_j \left(\frac{1}{2m_j} p_j^2 + \frac{1}{2} m_j \omega_j^2 q_j^2 \right) \ . \tag{17}$$

Then, with the aid of the well known results:

$$\begin{aligned}
\langle q_j q_k \rangle &\equiv \mathrm{Tr}\left\{ q_j q_k \exp\left(- H_B/kT \right) \right\} \Big/ \mathrm{Tr}\left\{ \exp\left(- H_B/kT \right) \right\} \\
&= \frac{\hbar}{2m_j\omega_j} \coth(\hbar\omega_j/2kT) \ \delta_{jk} \ , \\
\langle p_j p_k \rangle &= \frac{\hbar m_j \omega_j}{2} \coth(\hbar\omega_j/2kT) \ \delta_{jk} \ , \\
\langle q_j p_k \rangle &= - \langle p_j q_k \rangle = \frac{1}{2} i\hbar \ \delta_{jk} \ ,
\end{aligned} \tag{18}$$

we see that the correlation of the random force is

$$\begin{aligned}
&\frac{1}{2}\langle F(t)F(t') + F(t')F(t) \rangle \\
&\qquad = \frac{1}{2}\sum_j \hbar m_j \omega_j^3 \coth(\hbar\omega_j/2kT) \cos\omega_j(t - t') \ .
\end{aligned} \tag{19}$$

Finally, using the canonical commutations (9) we see that the commutator of the random force is

$$[F(t), F(t')] = -i\sum_j \hbar m_j \omega_j^3 \sin\omega_j(t-t') \ . \tag{20}$$

To see that these results are equivalent to the general expressions (2) and (3) we form the Fourier transform of the memory function (15) to get

$$\tilde{\mu}(z) = \int_0^\infty dt\; e^{izt} \sum_j m_j \omega_j^2 \cos\omega_j t$$
$$= iz \sum_j m_j \omega_j^2 \frac{1}{z^2 - \omega_j^2} \; . \tag{21}$$

Using the well known result: $1/(x + i0^+) = P(1/x) - i\pi\delta(x)$, we see that

$$\mathrm{Re}\{\tilde{\mu}(\omega + i0^+) = \frac{\pi}{2} \sum_j m_j \omega_j^2 [\,\delta(\omega - \omega_j) + \delta(\omega + \omega_j)] \; . \tag{22}$$

With this one can readily verify the expressions (19) and (20) can be written, respectively, in the forms (2) and (3). Finally, as a check, one can verify the Stieltjes inversion theorem (7) with, however, the constant $c = 0$. This is to be expected in our application since the constant c can always be absorbed into the mass m. (Beware, this is not mass renormalization.)

The expression (22) for $\mathrm{Re}\{\tilde{\mu}(\omega + i0^+)$ makes clear that, by an appropriate choice of the distribution of the frequencies and masses of the independent oscillators, the most general positive real function $\tilde{\mu}(z)$ and therefore the most general quantum Langevin equation can be realized by an IO model of the heat bath. I hasten to stress that I do not say that in any given physical application the bath *is* a set of independent oscillators, rather that the bath cannot be distinguished from such an oscillator bath. It is remarkable that such a simple, almost trivial, model has such generality.

REFERENCES

[1] G.W. Ford, M. Kac and P. Mazur, J. Math Phys. 6, 504 (1965)

[2] E.A. Guillemin, *Synthesis of passive networks* (Wiley, New York 1957)

[3] J. Meixner, in *Statistical mechanics of equilibrium and nonequilibrium*, ed. J. Meixner (North-Holland, Amsterdam 1965)

[4] G.W. Ford, R.F.O'Connell, and J.T. Lewis, to be published.

QUANTUM POISSON PROCESSES : PHYSICAL MOTIVATIONS AND APPLICATIONS .

Alberto Frigerio
Istituto di Matematica Informatica e Sistemistica
Università di Udine, 33100 Udine, Italy .

(Based on joint work with Hans Maassen)

1. Introduction.

According to conventional wisdom, the reduced time evolution of a system S coupled to a reservoir R displays a Markovian irreversible behaviour when the characteristic relaxation time τ_R of correlations in R is much shorter than the characteristic time τ_S for appreciable effects on S of its interaction with R . Supposing, for the sake of definiteness, that S is a spatially confined system located in a fixed region of space, and that R is a gas whose molecules can collide with S , one may conceive of at least two physical situations in which the condition $\tau_S >> \tau_R$ is satisfied:

(1) Weak coupling: The interaction between S and the molecules of R is feeble, so that an appreciable influence of R on S can only be observed as the long-time cumulative effect of very many weak collisions.

(2) Low density: R is a very dilute gas, so that collisions of molecules of R with S only happen at infrequent time instants.

It is reasonable to expect that, in the extreme weak-coupling or low-density limit, the system S will appear to be driven by a stochastic process, which is Brownian motion in the former case and a Poisson process in the latter case. We refer to the review article by Spohn [1] for rigorous results substantiating these ideas for the case of classical systems and reservoirs.

In quantum mechanics, the weak coupling limit has been studied by Davies [2] , and the low density limit by Dümcke [3]. In both cases, it has been possible to prove that, under suitable technical assumptions, the reduced dynamics of the observables of S converges to a quantum dynamical semigroup $T_t = \exp L t$. Although the distinction between diffusions and jump processes becomes blurred [1, 3], nevertheless a quantum Brownian motion [4, 5, 6] and a quantum Poisson process [7, 8, 9, 10, 11] have emerged or are emerging in the literature. It has become clear that the closest quantum analogue of (classical) Brownian motion is the so-called "finite temperature quantum Brownian motion" of Hudson and Lindsay [6]cf.[5]. The notion of a quantum Poisson process, generalizing the representation of the classical Poisson process in terms of quantum martingales given by Hudson and Parthasarathy [4], has come out recently from independent work of several people, including Frigerio and Maassen [7], Kümmerer [8], Evans and Hudson [9], Parthasarathy and Sinha [10], Accardi Journé and Lindsay [11]. It should be noted that several aspects of the construction of a quantum Poisson process had been actually considered around twenty years ago, long before the birth of quantum stochastic calculus, in connection with the problem of constructing infinitely divisible representations of the current algebra (Streater and Wulfsohn [12], Araki [13], Streater [14]).

In the present work we shall give an informal review of the weak-coupling and low-density limits, together with heuristic arguments indicating the relationship of the weak coupling limit with quantum Brownian motion (in this connection, see also Dümcke [15], Frigerio and Gorini [16], Gardiner and Collett [17], who consider the essentially equivalent singular coupling limit) and of the low density limit with the quantum Poisson process. The quantum Poisson process itself is described in the version of [7], to which we refer for details and proofs. Additional possible applications of the quantum Poisson process are briefly mentioned in the end.

2. Physical model.

We consider a spatially confined quantum system S interacting with a reservoir R which is a gas of free (Bose or Fermi) particles. In order to eliminate Poincaré recurrences and obtain a well-defined asymptotic behaviour in the limit as $t \to \infty$, R must be taken to be infinitely extended.

The Hilbert spaces for the system S and for the reservoir R will be denoted by $\mathcal{H}_0$ and by $\mathcal{H}'$ respectively. $\mathcal{H}'$ is the space of a cyclic representation of the canonical commutation or anticommutation relations over the test function space (one-particle space) $\mathcal{H}_1 = L^2(\mathbb{R}^3)$, determined by the state $\nu^{(z)}$ which is the grand-canonical equilibrium state corresponding to inverse temperature β and fugacity z ; it is the quasi-free gauge-invariant state with two-point function given by

$$\nu^{(z)}(a^\dagger(f)a(g)) = \int \hat{f}(\underline{k})\,\overline{\hat{g}(\underline{k})}\,\{z^{-1}\exp(\beta k^2/2) \pm 1\}^{-1}(2\pi)^{-3}\,d^3\underline{k} \quad (2.1)$$

(minus sign for bosons, plus sign for fermions). The total Hamiltonian of the composite system $S + R$ is

$$H^{(\lambda)} = H_S \otimes \mathbb{1} + \mathbb{1} \otimes H_R + \lambda V , \quad (2.2)$$

where H_R is the second quantization of the one-particle Hamiltonian $H_1 = -\frac{1}{2}\Delta$. The initial state will be taken to be of product form $\varphi \otimes \nu^{(z)}$, where $\varphi = \mathrm{Tr}\{\rho_S \,\cdot\}$ is an arbitrary normal state on $\mathcal{B}(\mathcal{H}_0)$ and $\nu^{(z)}$ is given by (2.1).

Since the present work has only an illustrative purpose and does not aim at the greatest generality, we shall assume

$$V = \frac{1}{i}[B \otimes C^\dagger - B^\dagger \otimes C] , \quad (2.3)$$

where the system operator B satisfies

$$\exp[iH_S t]\, B \exp[-iH_S t] = e^{-i\omega_0 t}\, B \quad (\omega_0 > 0) \quad (2.4)$$

and the reservoir operator C and its adjoint $C^\dagger$ satisfy

$$\begin{aligned} &\nu^{(z)}(C) = \nu^{(z)}(C^\dagger) = 0 \\ &\nu^{(z)}(C\,C(t)) = \nu^{(z)}(C^\dagger C^\dagger(t)) = 0 \quad \text{for all } t \end{aligned} \quad (2.5)$$

where

$$C(t) = \exp[iH_R t]\, C \exp[-iH_R t] . \quad (2.6)$$

In order to consider the low density limit, we shall assume, more specifically, that C is a one-particle operator, i.e.

$$C = a^{\dagger}(f)\, a(g) \tag{2.7}$$

for some normalized test functions f and g in $\mathcal{H}_1$. Then the Hamiltonian $H^{(\lambda)}$ corresponds to an interaction of S with one reservoir particle given by

$$H_1^{(\lambda)} = H_S \otimes \mathbb{1} + \mathbb{1} \otimes H_1 + \lambda V_1 , \tag{2.8}$$

where

$$V_1 = \frac{1}{i} [B \otimes |g\rangle\langle f| - B^{\dagger} \otimes |f\rangle\langle g|] . \tag{2.9}$$

Both in the weak coupling limit $\lambda \to 0$ and in the low density limit $z \to 0$ ("low density" is the same as "small fugacity"), we shall have to consider the time evolution of observables up to very large microscopic times, in order to see an appreciable macroscopic effect. During a large microscopic time interval, very many oscillations given by the free Hamiltonian $H^{(0)}$ would occur; so we have to eliminate them by going first to the interaction picture. Then we shall consider the time evolution of an observable Y to be given by

$$Y \mapsto U^{(\lambda)\dagger}(t) Y U^{(\lambda)}(t) , \tag{2.10}$$

where

$$U^{(\lambda)}(t) = \exp[\, i H^{(0)} t] \exp[-i H^{(\lambda)} t] . \tag{2.11}$$

The reduced dynamics of the system S in the weak coupling limit is now given by

$$\varphi(T_t(X)) = \lim_{\lambda \to 0} (\varphi \otimes \nu)(U^{(\lambda)\dagger}(t/\lambda^2)(X \otimes \mathbb{1})U^{(\lambda)}(t/\lambda^2)) , \tag{2.12}$$

for all initial states φ and observables X for S, and for all $t \geq 0$; the unscaled parameter z is omitted from notation. Similarly, the reduced dynamics S_t in the low density limit is given by

$$\varphi(S_t(X)) = \lim_{z\to 0} (\varphi \otimes \nu^{(z)})(U^{\dagger}(t/z)(X \otimes \mathbb{1})U(t/z)) \quad , \tag{2.13}$$

the unscaled parameter λ being omitted from notation.

Here we do not wish to go into the (natural or technical) assumptions and the methods of proof under which and by which it can be shown that T_t and S_t are quantum dynamical semigroups (semigroups of completely positive identity-preserving normal linear maps of $\mathcal{B}(H_0)$ into itself) whose infinitesimal generators L and K can be explicitly computed, and we refer to the original papers by Davies [2] and by Dümcke [3]. Keeping in with the informal style of the present work, we shall just assume that T_t and S_t are semigroups, and try to compute their generators L and K through some heuristic arguments which will relate the weak coupling limit with quantum Brownian motion and the low density limit with the quantum Poisson process.

3. Weak coupling.

In order to compute the generator L, we need the expression of $U^{(\lambda)}(\delta t/\lambda^2)$ for small δt. Due to the presence of λ^2 in the denominator, we must consider (at least) the explicit form of the terms in the Dyson series for $U^{(\lambda)}(t)$ up to second order in λ included. Recalling the explicit form (2.3) of V and introducing the notation

$$\left.\begin{aligned} A_t^{(\lambda)} &= \lambda \int_0^{t/\lambda^2} e^{+i\omega_0 s}\, C(s)\, ds \,, \\ A_t^{(\lambda)\dagger} &= \lambda \int_0^{t/\lambda^2} e^{-i\omega_0 s}\, C^{\dagger}(s)\, ds \,, \end{aligned}\right\} \tag{3.1}$$

we find

$$U^{(\lambda)}(t/\lambda^2) = \mathbb{1} - \left[B \otimes A_t^{(\lambda)\dagger} - B^{\dagger} \otimes A_t^{(\lambda)} \right]$$

$$- \; t \; r^{-1} \int\int_{0 \leq t_1 \leq t_2 \leq r} V(t_2)V(t_1)\, dt_1\, dt_2 + \ldots \tag{3.2}$$

where $r = t/\lambda^2 \to \infty$ in the weak coupling limit, and where, explicitly,

$$V(t) = \frac{1}{i}[B \otimes e^{-i\omega_0 t} C^\dagger(t) - B^\dagger \otimes e^{i\omega_0 t} C(t)] \quad . \tag{3.3}$$

In (3.2) there appears a time average of $V(t_2) \int_0^{t_1} V(t_1)\, dt_1$ which, under the kind of ergodicity conditions that can be expected to hold for infinite systems, should become $\int_0^\infty (id \otimes \nu)(V(t_1 + u)V(t_1))\, du$, which is actually independent of t_1 and is just

$$\int_0^\infty e^{i\omega_0 t} \nu(C(t)C^\dagger)\, dt\, B^\dagger B + \int_0^\infty e^{-i\omega_0 t} \nu(C^\dagger(t)C)\, dt\, B\, B^\dagger \quad . \tag{3.4}$$

Now we turn our attention on the operators $A_t^{(\lambda)}$ and $A_t^{(\lambda)\dagger}$. We must suppose that the functions $\nu(C\, C^\dagger(t))$ and $\nu(C^\dagger C(t))$ are in $L^1(\mathbb{R})$. For all real ω , we have

$$\int_{-\infty}^\infty e^{i\omega t} \nu(C^\dagger C(t))\, dt = e^{-\beta\omega} \int_{-\infty}^\infty e^{-i\omega t} \nu(C\, C^\dagger(t))\, dt \tag{3.5}$$

(KMS condition). Upon multiplying B and C by mutually inverse numerical factors, we may assume without loss of generality that

$$\int_{-\infty}^\infty e^{-i\omega_0 t} \nu(C\, C^\dagger(t))\, dt = (1 - e^{-\beta\omega_0})^{-1} \quad , \tag{3.6}$$

$$\int_{-\infty}^\infty e^{i\omega_0 t} \nu(C^\dagger C(t))\, dt = (e^{\beta\omega_0} - 1)^{-1} \quad . \tag{3.7}$$

Under the above assumptions, we have the following

<u>Lemma</u>. The operators $A_t^{(\lambda)}$ and $A_t^{(\lambda)\dagger}$ "converge" to the finite-temperature quantum Brownian motion A_t and $A_t^\dagger$ of Hudson and Lindsay [6] (*) in the sense that

$$\lim_{\lambda\to 0} \nu(A_s^{(\lambda)} A_t^{(\lambda)\dagger}) = (s \wedge t)(1 - e^{-\beta\omega_0})^{-1} \quad , \tag{3.8}$$

$$\lim_{\lambda\to 0} \nu(A_t^{(\lambda)\dagger} A_s^{(\lambda)}) = (s \wedge t)(e^{\beta\omega_0} - 1)^{-1} \quad , \tag{3.9}$$

which are the covariances of the finite temperature quantum Brownian motion.

(*) With a different language, the finite temperature quantum Brownian motion had been introduced earlier by Barnett Streater and Wilde [5].

<u>Proof</u>. Let $f(u) = e^{i\omega_0 u} \nu(C(t)\, C^{\dagger}(t-u)) \in L^1(\mathbb{R})$. Then

$$\nu(A_s^{(\lambda)} A_t^{(\lambda)\dagger}) = \lambda^2 \int_{s'=0}^{s/\lambda^2} \int_{t'=0}^{t/\lambda^2} f(s'-t')\, ds'\, dt'$$

$$= \int_{\tau=0}^{t} \{ \int_{u=-\tau/\lambda^2}^{(s-\tau)/\lambda^2} f(u)\, du\} \quad d\tau \quad .$$

In the limit as $\lambda \to 0$, the expression within braces becomes 0 if $s < \tau$, and $\int f(u)\, du$ if $s > \tau$. This proves (3.8); (3.9) is proved by a similar argument.

Taking into account the Lemma, it is easy to realize that the weak-coupling limit generator L is given by

$$L(X) = i\,[\,H\,,\,X\,] + (1 - e^{-\beta\omega_0})^{-1}(B^{\dagger} X B - \tfrac{1}{2}\{B^{\dagger}B\,,\,X\}) + (e^{\beta\omega_0} - 1)^{-1}(B\, X\, B^{\dagger} - \tfrac{1}{2}\{B\, B^{\dagger},\, X\})\,, \quad (3.10)$$

where

$$i\,H = (\int_0^{\infty} e^{-i\omega_0 t}\, \nu(C\, C^{\dagger}(t))\, dt - \tfrac{1}{2}(1 - e^{-\beta\omega_0})^{-1})\, B^{\dagger}B + (\int_0^{\infty} e^{i\omega_0 t}\, \nu(C^{\dagger}C(t))\, dt - \tfrac{1}{2}(e^{\beta\omega_0} - 1)^{-1})\, B\, B^{\dagger} \quad . \quad (3.11)$$

Moreover, the Lemma suggests the following conjectures:

<u>Conjecture 1.</u> Under suitable assumptions (yet to be described), the process $A_t^{(\lambda)}, A_t^{(\lambda)\dagger}$ "converges in law" as $\lambda \to 0$ to the finite temperature quantum Brownian motion.

<u>Conjecture 2.</u> Under suitable assumptions (same as above or stronger), the family of operators $U^{(\lambda)}(t/\lambda^2)$ converges (in some sense) to the solution of the quantum stochastic differential equation

$$d\,U(t) = [-B\, dA_t^{\dagger} + B^{\dagger}\, dA_t + K\, dt]\, U(t) \quad , \quad (3.12)$$

where

$$K = -\,i\,H - \tfrac{1}{2}(1 - e^{-\beta\omega_0})^{-1}\, B^{\dagger}B - \tfrac{1}{2}(e^{\beta\omega_0} - 1)^{-1}\, B\, B^{\dagger} \quad . \quad (3.13)$$

A few remarks are in order:

(a) The solution $U(t)$ of the quantum stochastic differential equation (3.12) can indeed be used to construct a dilation of the quantum dynamical semigroup with generator (3.10).

(b) We expect that the conjectures should be true under the assumption that the reservoir algebra of observables and time evolution be L^1-asymptotically abelian [18] in some suitable sense, meaning roughly that the integrals $\int [C, C^{\dagger}(t)]\,dt$ and similar ones are convergent.

(c) If we were to assume convergence of the integrals of the form $\int \{C, C^{\dagger}(t)\}\,dt$, we should expect $A_t^{(\lambda)}$, $A_t^{(\lambda)\dagger}$ to converge to the fermionic analogue of the finite temperature quantum Brownian motion (cf.[5] and Applebaum [19]). In particular, if C is an even (odd) monomial in fermionic creation and annihilation operators, we expect to obtain boson (fermion) quantum Brownian motion in the limit as $\lambda \to 0$.

(d) The above conjectures are much more ambitious than the results of [15, 16] (cf. [17] for a more physically oriented discussion), where C is considered to be just an annihilation operator.

4. Low density.

The physical idea of the low density limit is that, in a time of the order of $\delta t/z$, with δt sufficiently small, only one reservoir particle should collide with S, so that $K(X)$ can be computed from the one-particle reduced evolution as

$$K(X) \approx \frac{1}{\delta t}(\mathrm{id}\otimes \nu^{(z)})(U^{\dagger}(\delta t/z)(X\otimes \mathbb{1})U(\delta t/z) - X\otimes\mathbb{1})$$

$$\approx \frac{1}{\delta t}(\mathrm{id}\otimes \mu^{(z)})(U_1^{\dagger}(\delta t/z)(X\otimes\mathbb{1})U_1(\delta t/z) - X\otimes\mathbb{1}), \tag{4.1}$$

where $U_1(t)$ is the unitary evolution (in the interaction picture) corresponding to the interaction of S with one reservoir particle, given by

$$U_1(t) = \exp[\,i(H_S + H_1)t]\,\exp[-i(H_S + H_1 + \lambda V_1)t] \quad , \tag{4.2}$$

and where $\mu^{(z)}$ is the one-particle reduced functional, defined by

$$\mu^{(z)}(|f\rangle\langle g|) = \nu^{(z)}(a^{\dagger}(f)a(g)) \quad , \tag{4.3}$$

which is given by (2.1) and is asymptotic, as $z \sim 0$, to the Maxwell-Boltzmann distribution

$$z\,\mu(|f\rangle\langle g|) = z \int \hat{f}(\underline{k})\,\overline{\hat{g}(\underline{k})}\,\exp(-\beta k^2/2)\,(2\pi)^{-3}\,d^3\underline{k} \quad . \tag{4.4}$$

Contributions of double, triple,... collisions should be of higher order in z, and hence negligible in the low density limit (at least when δt is small). A quantum version of the BBGKY hierarchy [3] is needed to substantiate this idea, moreover, the cluster theorem of multi-particle scattering theory [20] is required to reduce the problem to repeated two-body scattering (the two bodies being S and one reservoir particle). Conditions on V_1 are required to make scattering theory work in the present context, in particular, no problem arises when S is an N-level system, because then V_1 is a finite rank operator.

Note that formula (4.1) cannot be taken literally, since $\mu(\mathbb{1}) = +\infty$ for an infinitely extended reservoir, so that (4.1) gives $\infty - \infty$. However, this is easily remedied by rewriting

$$\begin{aligned} U^{\dagger} X U - X &= \tfrac{1}{2}\{ U^{\dagger}[X, U] + [U^{\dagger}, X] U \} \\ &= \tfrac{1}{2}\{ (U^{\dagger} - \mathbb{1})[X, (U - \mathbb{1})] + [(U^{\dagger} - \mathbb{1}), X](U - \mathbb{1}) + [(U^{\dagger} - \mathbb{1}) - (U - \mathbb{1}), X]\} \end{aligned} \tag{4.5}$$

and observing that $U_1(\delta t/z) - \mathbb{1}$ is trace class, at least when S is an N-level system.

Very roughly, we have from scattering theory that $U_1(t)$ is asymptotic as $t\to\infty$ to the scattering operator $\mathcal{S}$ for the system interacting with one resevoir particle. We have also

$$\langle\alpha|(\mathcal{S} - \mathbb{1})\alpha'\rangle = -2\pi i\,\delta(\text{unperturbed energy difference})\langle\alpha|T\alpha'\rangle \quad , \tag{4.6}$$

where the transition operator T is given by

$$T = \lim_{t\to\infty} V_1 \exp[-i H_1^{(\lambda)} t] \exp[i H_1^{(0)} t] \quad . \tag{4.7}$$

The generator K is obviously related with transition rates among (improper) eigenstates of the unperturbed energy, involving square moduli of the T - matrix elements and a Dirac δ-function of the unperturbed energy difference.

The above considerations make plausible the correct expression of K , which has been rigorously obtained by Dümcke [3] to be given by

$$K(X) = 2\pi \sum_{\omega} \int \int \delta(k'^2/2 - k^2/2 + \omega) . e^{-\beta k^2/2}$$

$$\left(T_{\omega}^{\dagger}(\underline{k}',\underline{k}) \, X \, T_{\omega}(\underline{k}',\underline{k}) - \tfrac{1}{2}\{T_{\omega}^{\dagger}(\underline{k}',\underline{k}) \, T_{\omega}(\underline{k}',\underline{k}) , X\} \right) d^3\underline{k} \, d^3\underline{k}' \tag{4.8}$$

$$+ \, i \sum_{\varepsilon_n = \varepsilon_{n'}} \int e^{-\beta k^2/2} \, 2 \, \mathrm{Re} \, \langle n\underline{k}|T|n'\underline{k}\rangle \, [\,|n\rangle\langle n'|, X] \, d^3\underline{k} \quad ,$$

where

$$H_S|n\underline{k}\rangle = \varepsilon_n|n\underline{k}\rangle \quad , \quad H_1|n\underline{k}\rangle = (k^2/2)|n\underline{k}\rangle \; , \tag{4.9}$$

$$T_{\omega}(\underline{k}',\underline{k}) = \sum_{n,n':\varepsilon_{n'}-\varepsilon_n=\omega} \langle n'\underline{k}'|T|n\underline{k}\rangle \; |n'\rangle\langle n| \quad . \tag{4.10}$$

The quantum Poisson process as will be described in the next Section allows to construct a family $U(t)$ of unitary operators on $\mathcal{H}_0 \otimes \mathcal{H}'$ which are infinitesimally determined by a unitary operator U_1 on $\mathcal{H}_0 \otimes \mathcal{H}_1$, very much like the situation that we have described above.

5. Poisson processes.

Here we shall sketch the construction of the quantum Poisson process (as given in [7]) in a sequence of five steps. Step 1 is well-known, and has been included only for pedagogical clarity. Steps 2 and 4 can be already found in [12 , 13 , 14]. An alternative construction of a quantum Poisson process, based on coupling a quantum process with independent increments in discrete time with a classical Poisson process, can be found in Kümmerer's paper [8], see also his contribution to the present volume.

Step 1. Consider the one-dimensional quantum harmonic oscillator. It can be characterized by saying that there exists a complete orthonormal set $\{\phi_n : n = 0,1,2,\ldots\}$ in the associated Hilbert space such that the Hamiltonian H is of the form $H = \hbar\omega(N + \frac{1}{2})$, where the number operator N is defined by $N\phi_n = n\phi_n$. For any complex number z, consider the exponential vector

$$\psi(z) = \sum_{n=0} z^n (n!)^{-\frac{1}{2}} \phi_n ,$$

with norm $\|\psi(z)\|^2 = \exp(-|z|^2)$. The probability to find the value n for the number N when the oscillator is in the coherent state $\psi(z)/\|\psi(z)\|$ is given by the Poisson distribution with parameter z :

$$\mathbb{P}(N = n|z) = (n!)^{-1} e^{-|z|^2} |z|^{2n} ,$$

and the corresponding characteristic function is

$$\mathbb{E}(e^{i\alpha N}) = \exp\{(e^{i\alpha} - 1)|z|^2\} \quad : \quad \alpha \in \mathbb{R} .$$

Step 2. Let now H_1 be a Hilbert space, whose elements we denote by $\xi, \eta, \zeta, \ldots$, and let H' be the symmetric (boson) Fock space $\Gamma(H_1)$ over H_1, generated by the exponential vectors

$$\psi(\zeta) = (1,\zeta,\ldots,(n!)^{-\frac{1}{2}}\underbrace{\zeta\otimes\ldots\otimes\zeta}_{n\ \text{times}},\ldots) \quad : \zeta \in H_1 , \qquad (5.1)$$

satisfying $\langle\psi(\xi)|\psi(\eta)\rangle = \exp\langle\xi|\eta\rangle$ for all ξ, η in H_1. The Weyl operators $W(\eta)$ are defined by

$$W(\eta)\ \psi(\zeta) = \exp\{ -\tfrac{1}{2}\|\eta\|^2 - \langle\eta|\zeta\rangle\}\, \psi(\eta + \zeta) \ : \ \eta,\zeta \in H_1, \qquad (5.2)$$

and extend to unitary operators on H'. The Fock functor Γ maps the unit ball of $B(H_1)$ into the unit ball of $B(H')$ via

$$\Gamma(X)\ \psi(\zeta) = \psi(X\zeta) \quad : \quad X \in B_1(H_1) , \quad \zeta \in H_1 . \qquad (5.3)$$

For any self-adjoint operator A in H_1, and for all ξ in H_1, we have that $W(\xi)^{-1}\,\Gamma(e^{i\alpha A})\,W(\xi)$ is a strongly continuous group of unitaries on H', so that there exists a self-adjoint operator $N(A;\xi)$ in H' such that

$$W(\xi)^{-1}\,\Gamma(e^{i\alpha A})\,W(\xi) = \exp[i\alpha N(A;\xi)] \quad : \alpha \in \mathbb{R} . \tag{5.4}$$

Note that

$$\begin{aligned} \langle\psi(0)|\exp[i\alpha N(A;\xi)]\psi(0)\rangle &= \|\psi(\xi)\|^{-2}\langle\psi(\xi)|\Gamma(e^{i\alpha A})\psi(\xi)\rangle \\ &= \exp[\langle\xi|e^{i\alpha A}\xi\rangle - \langle\xi|\xi\rangle] . \end{aligned} \tag{5.5}$$

In the special case when A is a projection, this is the characteristic function of a Poisson distribution with parameter $\|A\xi\|^2$.

Any element of $B(H_1)$ may be uniquely expressed as $A + iB$, where A and B are self-adjoint. Then we define

$$N(A + iB;\xi) = N(A;\xi) + i\,N(B;\xi) , \tag{5.6}$$

the sum being well-defined at least on the exponential vectors $\psi(\zeta)$. The linear map $X \mapsto N(X;\xi)$ of $B(H_1)$ into the unbounded operators in H' will be called a quantum Poisson process.

Step 3. Let $M \subset B(H_1)$ be a von Neumann algebra, and let $\xi \in H_1$ be cyclic and separating for M. Then we have

Theorem. [7] The vector $\psi(0)$ is cyclic and separating for the von Neumann algebra N generated by $\{\exp[i\alpha N(A;\xi)]: A = A^\dagger \in M, \alpha \in \mathbb{R}\}$. Moreover, if the finite weight μ on M is defined by

$$\mu(X) = \langle\xi|X\xi\rangle \quad : \quad X \in M , \tag{5.7}$$

and the state ν is defined on N by

$$\nu(Y) = \langle\psi(0)|Y\psi(0)\rangle : Y \in N , \tag{5.8}$$

and if σ_t^μ and σ_t^ν are the respective modular automorphism groups, then

$$\sigma_t^\nu(\exp[i\alpha N(A;\xi)]) = \exp[i\alpha N(\sigma_t^\mu(A);\xi)] : A = A^\dagger \in M . \tag{5.9}$$

From now on, we shall omit the explicit indication of ξ in the notation, unless required for clarity. The triple $(N, \nu, N(.) = N(.;\xi))$ will be called the <u>quantum Poisson process over</u> (M, μ).

<u>Step 4.</u> We wish to introduce a time dependence in the above structure, upon replacing the von Neumann algebra M by $L^{\infty}(\mathbb{R}) \otimes M$ and the weight μ by $\lambda \otimes \mu$, where $\lambda(f) = \int f(t)\,dt$. λ is not a finite weight, but we can give meaning to $N(f \otimes X; g \otimes \xi)$ for f and g of compact support. On $H' = \Gamma(L^2(\mathbb{R}) \otimes H_1)$ we define

$$N_{s,t}(X) = \frac{1}{i} W(1_{[s,t]} \otimes \xi)^{-1} \frac{d}{d\alpha} \Gamma(\exp[i\alpha 1_{[s,t]} \otimes X])\Big|_{\alpha=0} W(1_{[s,t]} \otimes \xi) \quad (5.10)$$

for all $s < t$ in $\mathbb{R}$ and X in M, where $1_{[s,t]}$ denotes the indicator function of the interval $[s,t]$. We shall write just $N_t(X)$ for $N_{0,t}(X)$.

<u>Step 5.</u> We wish to use N_t to perform some quantum stochastic calculus. To this end, we first realize that ([7], cf. [4])

$$dN_t(X) = dA_t(X\xi) + d\Lambda_t(X) + dA_t(X^{\dagger}\xi) + \langle\xi | X\xi\rangle dt \quad (5.11)$$

in terms of the (Fock) creation, preservation (or gauge) and annihilation martingales A_t, Λ_t and A_t of Hudson and Parthasarathy [4]. A quantum Itô's formula then follows, in the form [7]

$$dN_t(X)\, dN_t(Y) = dN_t(XY) \quad : \quad X, Y \in M \quad . \quad (5.12)$$

Then we must add an initial space H_0. Let A be a von Neumann subalgebra of $B(H_0)$. Then, for $X = B \otimes C$, $B \in A$, $C \in M$, we define

$$\hat{N}_t(X) = B \otimes N_t(C) \quad : \quad X = B \otimes C \in A \otimes M \, . \quad (5.13)$$

Then $\hat{N}_t$ extends by linearity to the whole of $A \otimes M$ if either A or M is finite-dimensional. More generally, if both A and M are infinite-dimensional, but there exists a complete orthonormal set in H_1 of the form $\{\xi_j = C_j\xi : C_j \in M ; j = 1,2,\ldots\}$ then $\hat{N}_t$ can be extended by linearity and continuity to those elements X of $A \otimes M$ which can be expressed in the form $X = \sum_{j=1}^{\infty} B_j \otimes C_j$ (strong convergence) such that also $X^\dagger = \sum_{j=1}^{\infty} B_j^\dagger \otimes C_j^\dagger$ converges strongly. This extension is based on the following estimate (for X of the form $B \times C$):

$$\|\hat{N}_t(X)(\phi \otimes \psi(0))\|^2 = \|B\phi\|^2 \,\|[A_t^\dagger(C\xi) + \langle\xi|C\xi\rangle t]\psi(0)\|^2$$

$$= \|B\phi\|^2 (\|C\xi\|^2 + |\langle\xi|C\xi\rangle|^2 t)t$$

$$\leq t(1+t)\|B\phi\|^2 \|C\xi\|^2 = t(1+t)\|X(\phi \otimes \xi)\|^2 ; \quad (5.14)$$

and on the observation that $\{\phi \otimes \psi(0) : \phi \in H_0\}$ is a separating family of vectors for $A \otimes M$.

In analogy to Step 3, if φ is a faithful normal state on A, and σ_t^φ is the associated modular automorphism group, we have (with some abuse of notation, but the meaning is clear)

$$\sigma_s^{\varphi\otimes\nu}(\hat{N}_t(X)) = \hat{N}_t(\sigma_s^{\varphi\otimes\mu}(X)) : \quad s, t \in \mathbb{R}, \quad X \in A \otimes M. \quad (5.15)$$

6. Applications.

We wish to present some infinite-dimensional generalization of the results of [7] concerning quantum stochastic differential equations. The techniques are similar to those of Hudson and Lindsay [6], Applebaum and Frigerio [21].

Suppose, for the sake of definiteness, that M is isomorphic to the algebra of all bounded linear operators on a separable Hilbert space K_1. Then the faithful normal semifinite weight μ can be expressed as

$$\mu(C) = \sum_{j=1}^{\infty} \xi_j^2 \langle v_j | C v_j \rangle \quad : \quad C \in M \quad , \tag{6.1}$$

$\{v_j : j = 1, 2, \ldots\}$ being a complete orthonormal set in K_1 and $\{\xi_j : j = 1, 2, \ldots\}$ being a square summable sequence of positive real numbers. Then M may be regarded as acting by left multiplication on the Hilbert space H_1 of Hilbert-Schmidt operators on K_1, with cyclic and separating vector ξ given by

$$\xi = \sum_{j=1}^{\infty} \xi_j \, |v_j\rangle\langle v_j| \tag{6.2}$$

such that (5.7) holds. Let A be another von Neumann algebra of operators on a Hilbert space H_0, and consider those elements X of $A \otimes M$ that can be expanded as

$$X = \sum_{i,j} X_{ij} \otimes |v_i\rangle\langle v_j| \quad (X_{ij} \in A) \tag{6.3}$$

such that, for all ϕ in H_0, the following series are convergent in the norm of $H_0 \otimes H_1$:

$$\begin{aligned} X \phi \otimes \xi &= \sum_{i,j} X_{ij} \phi \otimes \xi_j |v_i\rangle\langle v_j| \\ X^{\dagger} \phi \otimes \xi &= \sum_{i,j} X_{ij}^{\dagger} \phi \otimes \xi_i |v_j\rangle\langle v_i| \quad . \end{aligned} \tag{6.4}$$

Note that $\{ w_{ij} = \xi_j |v_i\rangle\langle v_j| : i, j = 1, 2, \ldots\}$ is an orthogonal family in H_1, with

$$\langle w_{ij} | w_{k\ell} \rangle = \mathrm{Tr}\{ \xi_j |v_j\rangle\langle v_i | v_k\rangle\langle v_\ell | \xi_\ell \} = \xi_j^2 \, \delta_{ik} \, \delta_{j\ell} \quad . \tag{6.5}$$

By the reasoning sketched in Step 5 of the previous Section, we may define $\hat{N}_t(X)$ as an unbounded operator affiliated with $A \otimes N$ satisfying the inequality

$$\| \hat{N}_t(X)\phi \otimes \psi(0)\|^2 \leq \sum_{i,j} \| X_{ij}\phi \otimes N_t(|v_i\rangle\langle v_j|)\psi(0)\|^2$$

$$= t(1+t) \sum_{i,j} \| X_{ij}\phi\|^2 \; \| w_{ij}\|^2$$

$$= t(1+t) \| X \quad \phi \otimes \xi\|^2 \leq t(1+t)\| X\|^2 \; \|\phi\|^2 \; \|\xi\|^2$$

$$(\leq 2t \quad \| X \|^2 \; \|\phi\|^2 \; \|\xi\|^2 \qquad \text{if} \quad t \leq 1) . \tag{6.6}$$

We now recall from Hudson and Parthasarathy [4] that an adapted process $F(t)$ in $\mathcal{H}_0 \otimes \Gamma(L^2(\mathbb{R}) \otimes \mathcal{H}_1)$ is a family of densely defined operators such that, for all $t \geq 0$, $F(t)$ maps $\mathcal{H}_{t]} = \mathcal{H}_0 \otimes \Gamma(L^2(-\infty, t] \otimes \mathcal{H}_1)$ into itself and leaves $\mathcal{H}_{[t} = \mathcal{H}_0 \otimes \Gamma(L^2(t, +\infty) \otimes \mathcal{H}_1)$ undisturbed in the sense that, if $\phi \in \mathcal{D}(F(t))$ is of the form $\phi_{t]} \otimes \phi_{[t}$, $\phi_{t]} \in \mathcal{H}_{t]}$, $\phi_{[t} \in \mathcal{H}_{[t}$, then

$$F(t)(\phi_{t]} \otimes \phi_{[t}) = (F(t)\phi_{t]}) \otimes \phi_{[t} \quad . \tag{6.7}$$

Using (6.7) and the same reasoning as in (6.6), we can easily prove the following

<u>Lemma 6.1</u>. For any adapted process $F(t)$ and for any vector ϕ in $\mathcal{H}_0$, we have

$$\| d\hat{N}_t(X) \; F(t) \; \phi \otimes \psi(0)\|^2$$

$$\leq \quad 2 \; \| X\|^2 \; \|\xi\|^2 \; \| F(t) \; \phi \otimes \psi(0)\|^2 \; dt \; . \tag{6.8}$$

Then we consider the stochastic differential equation

$$dU(t) = d\hat{N}_t(X) \; U(t) \quad , \tag{6.9}$$

with initial condition $U(0) = \mathbb{1}$. We try to solve it by iteration, letting

$$U^{(n+1)}(t) = \mathbb{1} + \int_0^t d\hat{N}_s(X) \; U^{(n)}(s) \quad . \tag{6.10}$$

It follows from (6.8) that

$$\| [U^{(n+1)}(t) - U^{(n)}(t)]\phi \otimes \psi(0) \|^2$$

$$\leq 2 \| X\|^2 \|\xi\|^2 \int_0^t \| [U^{(n)}(s) - U^{(n-1)}(s)] \phi \otimes \psi(0)\|^2 ds$$

$$\leq (2 \| X\|^2 \|\xi\|^2 t)^n (n!)^{-1} \|\phi\|^2 \quad ; \qquad (6.11)$$

so that the iteration scheme (6.10) converges on vectors of the form $\phi \otimes \psi(0)$.

Take now ϕ to be cyclic and separating for A , and let $\varphi(B) = \langle\phi|B\phi\rangle$ for all B in A . Then $\phi \otimes \psi(0)$ is cyclic and separating for $A \otimes N$. In order to prove that there exists $U(t)$ affiliated with $A \otimes N$ and satisfying (6.9), it suffices to prove that the iteration scheme (6.10) and its adjoint converge on $\phi \otimes \psi(0)$. We have already proved the convergence of (6.10). Supposing that X is invariant under the modular automorphism group $\sigma_t^{\varphi\otimes\mu}$ of $A \otimes M$, then $\hat{N}_t(X)$ and $U^{(n)}(t)$ are invariant under the modular automorphism group $\sigma_t^{\varphi\otimes\nu}$ of $A \otimes N$, so that the action on $\phi \otimes \psi(0)$ of the adjoint of (6.10) is obtained by just multiplying on the left with the modular antiunitary involution J . Then also the adjoint iteration scheme converges. We have thus proved the following

<u>Theorem 6.2</u>. Let X in $A \otimes M$ be of the form (6.3), such that the series (6.4) are convergent and $\sigma_t^{\varphi\otimes\mu}(X) = X$ for all real t . Then the stochastic differential equation (6.9) has a unique adapted solution, which is affiliated with $A \otimes N$ and invariant under $\sigma_t^{\varphi\otimes\nu}$.

By reasoning as in Hudson and Parthasarathy [4], Frigerio and Maassen [7] , Evans and Hudson [9], we can also prove

<u>Theorem 6.3</u>. If X of Theorem 6.2 is of the form $U_1 - 1\!\!1$, U_1 being unitary in $A \otimes M$, then each $U(t)$ is a unitary operator

belonging to $A \otimes N$. Moreover, $U^\dagger(t)$ is a cocycle with respect to the right shift τ_t on $H_0 \otimes \Gamma(L^2(\mathbb{R}) \otimes H_1)$ in the sense that

$$U^\dagger(t+s) = U^\dagger(t)\,\tau_t(U^\dagger(s)) \quad . \qquad (6.12)$$

Adapted unitary cocycles which are invariant under the modular automorphism group can be used to construct Markov dilations of dynamical semigroups with a global stationary state, see Kümmerer ([8] and this volume). Here we confine ourselves to the following observation:

Theorem 6.4. For each $t \geq 0$, let S_t be the map on A defined by

$$S_t(X) = (\mathrm{id} \otimes \nu)(U^\dagger(t)(X \otimes \mathbb{1})U(t)) \quad : \quad X \in A \,. \qquad (6.13)$$

Then S_t is a dynamical semigroup, whose infinitesimal generator K is given by

$$K(X) = (\mathrm{id} \otimes \mu)(U_1^\dagger(X \otimes \mathbb{1})U_1 - (X \otimes \mathbb{1}))\,. \qquad (6.14)$$

Proof (Sketch, see [7]). We have

$$dS_t(X) = (\mathrm{id} \otimes \nu)(U^\dagger(t)\{d\hat{N}_t(U_1^\dagger - \mathbb{1})(X \otimes \mathbb{1}) + (X \otimes \mathbb{1})d\hat{N}_t(U_1 - $$

$$+ d\hat{N}_t(U_1^\dagger - \mathbb{1})(X \otimes \mathbb{1})d\hat{N}_t(U_1 - \mathbb{1})\}U(t))$$

$$= (\mathrm{id} \otimes \nu)(U^\dagger(t)\{d\hat{N}_t(U_1^\dagger(X \otimes \mathbb{1})U_1 - (X \otimes \mathbb{1}))\}U(t))$$

$$= (\mathrm{id} \otimes \nu)(U^\dagger(t)\{(\mathrm{id} \otimes \mu)(dt \cdot (U_1^\dagger(X \otimes \mathbb{1})U_1 - (X \otimes \mathbb{1}))\}U(t))$$

$$= S_t(K(X))\,dt \,. \qquad (6.15)$$

Remark. The form (6.14) of the generator K shows some analogy with the form (4.1), (4.8) obtained in the low density limit.

Finally, we wish to sketch briefly two additional possible applications of the quantum Poisson process, which are currently under investigation.

Approximation of unbounded generators. There has been some effort to construct unitary solutions of quantum stochastic differential equations of the form

$$dU(t) = [- B\, dA_t^{\dagger} + B^{\dagger}\, dA_t + Z\, dt]\, U(t) \quad , \quad U(0) = \mathbb{1} \quad , \tag{6.16}$$

with unbounded operator coefficients B , $B^{\dagger}$ and Z (see Journé [22]). The idea is that, under suitable conditions, (6.16) could be approximated by

$$dU_{\alpha}(t) = d\hat{N}_t(e^{i\alpha H};\alpha^{-1}\xi)\, U_{\alpha}(t) \quad , \qquad U_{\alpha}(0) = \mathbb{1} \quad , \tag{6.17}$$

where $\alpha > 0$ and where

$$H = \frac{1}{i}\, [B \otimes |v_2\rangle\langle v_1| - B^{\dagger} \otimes |v_1\rangle\langle v_2|] \ , \tag{6.18}$$

$$\xi = \xi_1 |v_1\rangle\langle v_1| + \xi_2 |v_2\rangle\langle v_2| \ . \tag{6.19}$$

This works formally if $Z = -\frac{1}{2}(\xi_1^2\, B^{\dagger}B + \xi_2^2\, B\, B^{\dagger})$ and A_t , $A_t^{\dagger}$ are the non-Fock quantum Brownian motion with

$$dA_t\, dA_t^{\dagger} = \xi_1^2\, dt \quad , \quad dA_t^{\dagger}\, dA_t = \xi_2^2\, dt \ . \tag{6.20}$$

It is hoped that the argument can be made rigorous under suitable assumptions (including self-adjointness of H).

Stochastic dilation of the Boltzmann equation. Upon replacing the fixed ξ by a time-dependent $\xi(t)$, it is possible to obtain a time evolution S_t of observables satisfying an equation of the form

$$\frac{d}{dt} S_t(X) = S_t\Big[K_{\mu(t)}(X)\Big] \ : \qquad X \in \mathcal{A} \ , \tag{6.21}$$

where $\mu(t)(X) = \langle\xi(t)|X\xi(t)\rangle$ and where

$$K_{\mu(t)}(X) = (\mathrm{id} \otimes \mu)(U_1^{\dagger}(X \otimes \mathbb{1})U_1 - (X \otimes \mathbb{1})) \ . \tag{6.22}$$

With $\mathcal{A} = \mathcal{M}$, one could try to determine $\mu(t)$ self-consistently from an initial condition $\mu(0)$ so as to have

$$\mu(0)(S_t(X)) = \mu(t)(X) \quad \text{for all} \quad X \quad \text{in} \quad \mathcal{A} \ . \tag{6.23}$$

Note that, if $\mu(t)(X) = \mathrm{Tr}\{\rho(t)\, X\}$, then the trace-class operator $\rho(t)$ satisfies the nonlinear "quantum Boltzmann equation"

$$\frac{d}{dt}\rho(t) = \mathrm{tr}_2 \{U_1\, \rho(t) \otimes \rho(t)\, U_1^{\dagger} - \rho(t) \otimes \rho(t)\} \ , \tag{6.24}$$

where tr_2 denotes the partial trace over the second factor of the tensor product.

The solution $U_\xi(t)$ of the stochastic differential equation

$$dU_\xi(t) = d\hat{N}_t(U_1 - \mathbb{1}; \xi(t))\, U_\xi(t) \quad , \quad U_\xi(0) = \mathbb{1} \ , \tag{6.25}$$

would then provide a unitary dilation of the nonlinear Boltzmann equation (6.24). Of course, a similar construction would not be any progress towards the understanding of the Boltzmann equation from a mechanical model, nevertheless, it might have some interest. Note that no effect of quantum statistics on the occupation of final states has been included, and this is consistent with the interpretation of the quantum Poisson process as describing a situation of low density.

References.

1. H. Spohn: Kinetic equations from Hamiltonian dynamics: Markovian limits. Rev. Mod. Phys. 53, 569-615 (1980)

2. E.B. Davies: Markovian master equations. Commun. Math. Phys. 39, 91-110 (1974)

3. R. Dümcke: The low density limit for an N-level system interacting with a free Bose or Fermi gas. Commun. Math. Phys. 97, 331-359 (1985)

4. R.L. Hudson and K.R. Parthasarathy: Quantum Ito's formula and stochastic evolutions. Commun. Math. Phys. 93, 301-323 (1984)

5. C. Barnett, R.F. Streater and I. Wilde: Quasi-free quantum stochastic integrals for the CAR and CCR. J. Funct. Anal. 52, 19-47 (1983)

6(a). R.L. Hudson and J.M. Lindsay: Stochastic integration and a martingale representation theorem for non-Fock quantum Brownian motion. J. Funct. Anal. 61, 202-221(1985)

6(b). R.L. Hudson and J.M. Lindsay: Uses of non-Fock quantum Brownian motion and a quantum martingale representation theorem. In *Quantum Probability and Applications II*, edited by L. Accardi and W. von Waldenfels. Lecture Notes in Mathematics vol. 1136, pp. 276-305. Springer Verlag, Berlin Heidelberg New York Tokyo (1985)

7. A. Frigerio and H. Maassen: Quantum Poisson processes and dilations of dynamical semigroups. Preprint.

8. B. Kümmerer: Markov dilations and non-commutative Poisson processes. Preprint.

9. M. Evans and R.L. Hudson: Algebraic theory of quantum diffusions II; many-dimensional diffusions. Preprint.

10. K.R. Parthasarathy and K. Sinha: In preparation.

11. L. Accardi, J.L. Journé and J.M. Lindsay: In preparation.

12. R.F. Streater and A. Wulfsohn: Continuous tensor products and generalized random fields. Nuovo Cimento 57 B, 330-339 (1968)

13. H. Araki: Factorizable representations of current algebra. Publ. RIMS Kyoto Univ. 5, 361-442 (1970)

14. R.F. Streater: Infinitely divisible representations of Lie algebras. Z. f. Wahrscheinlichkeitstheorie , 67-80 (1971)

15. R. Dümcke: Convergence of multi-time correlation functions in the weak and singular coupling limits. J. Math. Phys. 24, 311-315(1983)

16. A. Frigerio and V. Gorini: Markov dilations and quantum detailed balance. Commun. Math. Phys. 93, 517-532 (1984)

17. C.W. Gardiner and M.J. Collett: Input and output in damped quantum systems: quantum stochastic differential equations and the master equation. Phys. Rev. A 31, 3761-3774(1985)

18. O. Bratteli and D.W. Robinson: *Operator Algebras and Quantum Statistical Mechanics* (two volumes). Springer-Verlag, New York Heidelberg Berlin (1979/81)

19. D. Applebaum: Quasi-free stochastic evolutions. In *Quantum Probability and Applications II* (see Ref. 6), pp. 46-56 (1985)

20. W. Hunziker: Cluster properties of multiparticle systems. J. Math. Phys. 6, 6-10 (1965)

21. D. Applebaum and A. Frigerio: Stationary dilations of W*-dynamical systems via quantum stochastic differential equations. In *From local times to global geometry, control and physics*, edited by K.D. Elworthy. Pitman Research Notes in Mathematics Series, vol. 150, pp. 1-38 . Longman Scientific & Technical, Harlow, Essex, UK (1987)

22. J.L. Journé: Structure des cocycles markoviens sur l'espace de Fock. Preprint.

A NONCOMMUTATIVE GENERALIZATION OF CONDITIONALLY POSITIVE DEFINITE FUNCTIONS

A.S.Holevo

Steklov Mathematical Institute, Academy of Sciences of the USSR. Vavilova 42, Moscow 117966, USSR

Introduction

Let G be a group, $\mathcal{A}$ a C^*-algebra of operators in a Hilbert space $\mathcal{H}$ and $\mathcal{F}(\mathcal{A})$ the Banach algebra of bounded linear maps of $\mathcal{A}$ into $\mathcal{A}$. We call a function $\Phi(g)$, $g \in G$, with values in $\mathcal{F}(\mathcal{A})$ positive definite (p.d.) if for all finite sets $\{\psi_j\} \subset \mathcal{H}$, $\{g_j\} \subset G$, $\{X_j\} \subset \mathcal{A}$ holds

$$\sum_{j,k} (\psi_j \mid \Phi(g_j^{-1} g_k)[X_j^* X_k]\psi_k) \geq 0. \qquad (*)$$

In the case $\dim \mathcal{A} = 1$ this reduces to the standard notion of positive definite function on G, while in the case $G = \{e\}$ this reduces to the notion of completely positive map of $\mathcal{A}$. A function $\Phi(g)$ is called conditionally positive definite (c.p.d.) if (*) holds under the restriction $\sum_j X_j \psi_j = 0$. The two notions are related by an extension of Schoenberg's theorem proved in §1: the function $\exp t\mathcal{L}(g)$ is p.d. for all $t > 0$ if and only if $\mathcal{L}(g)$ is hermitean and c.p.d.. The p.d. function $\exp \mathcal{L}(g)$ is infinitely divisible in the sense of pointwise composition.

Our main goal here is to describe hermitean c.p.d. functions and hence, to an extent, infinitely divisible p.d. functions with values in $\mathcal{F}(\mathcal{A})$. Using a specific generalization of the GNS construction we obtain in §2 the canonical representations for p.d. functions and for kernels related to c.p.d. functions. Then with some facts from cohomology of groups and algebras we prove in §3 the main representation theorem 1, which enables us to give a complete description of $\mathcal{F}(\mathcal{A})$-valued hermitean c.p.d. functions. The continuity properties of p.d. and c.p.d. functions are treated in §4 where the general form of continuous hermitean c.p.d. function on a compact group is established. In §5 we consider the special case where $\mathcal{A}$ is type I factor.

The initial motivation for this work came from the theory of continuous quantum measurement [1], [2], where examples of p.d. and c.p.d. functions appeared without stating the property (*) explicitly (see § 5). Theorem 1 implies, in particular, a general description of the generator of a continuous measurement process (see [3], [4] for more detail).

§ 1. Positive definite and conditionally positive definite kernels and functions.

Let $\mathcal{H}$ be a Hilbert space, $\mathcal{L}(\mathcal{H})$ the algebra of all bounded operators in $\mathcal{H}$. We denote for $X, Y \in \mathcal{L}(\mathcal{H})$

$$X \circ Y = \frac{1}{2}(XY + YX), \quad [X, Y] = XY - YX.$$

We also put

$$\operatorname{Re} X = \frac{1}{2}(X + X^*), \quad \operatorname{Im} X = \frac{1}{2i}(X - X^*).$$

If $\mathcal{K}$ is another Hilbert space, then the Banach space of all bounded linear operators from $\mathcal{H}$ to $\mathcal{K}$ is denoted $\mathcal{L}(\mathcal{H}, \mathcal{K})$.

Let $\mathcal{A}$ be a C^*-subalgebra of $\mathcal{L}(\mathcal{H})$ containing the unit operator I. Then $\mathcal{A}^h$ denotes the space of hermitean elements of $\mathcal{A}$. By $\mathcal{F}$ we denote the Banach space of bounded linear maps from $\mathcal{A}$ into $\mathcal{L}(\mathcal{H})$. If $\Phi \in \mathcal{F}$, then $\Phi^* \in \mathcal{F}$ is defined by the relation

$$\Phi^*[X] = (\Phi[X^*])^*, \quad X \in \mathcal{A}.$$

By $\mathcal{F}(\mathcal{A})$ we denote the Banach algebra of bounded linear maps from $\mathcal{A}$ to $\mathcal{A}$. The identical map is denoted Id.

Let S be an arbitrary set. Kernel is a function $\Phi(\tau, s)$; $\tau, s \in S$, with values in $\mathcal{F}$. A kernel is called hermitean if $\Phi(s, \tau) = \Phi(\tau, s)^*$, positive definite, if for any finite sets $\{\psi_j\} \subset \mathcal{H}$, $\{s_j\} \subset S$, $\{x_j\} \subset \mathcal{A}$

$$\sum_{j,k} (\psi_j | \Phi(s_j, s_k)[x_j^* x_k] \psi_k) \geq 0, \tag{1.1}$$

and conditionally positive definite, if (1.1) holds for the sets, satisfying

$$\sum_j x_j \psi_j = 0. \tag{1.2}$$

Proposition 1. Let $\mathcal{L}(\tau, s)$; $\tau, s \in S$, be a kernel with values in $\mathcal{F}(\mathcal{A})$. The following conditions are equivalent:

1) the kernels $\exp t\mathcal{L}(\tau, s)$ are positive definite (p.d.) for all $t \geq 0$;

2) the kernel $\mathcal{L}(\tau, s)$ is hermitean conditionally positive definite (h.c.p.d.);

3) the kernel $\mathcal{L}(\tau, s)$ is hermitean and for any $s_0 \in S$ and any finite sets $\{\psi_j\} \subset \mathcal{H}$, $\{s_j\} \subset S$, $\{x_j\} \subset \mathcal{A}$

$$\sum_{j,k} (\psi_j | D_0 \mathcal{L}(s_j, s_k; x_j, x_k) \psi_k) \geq 0, \tag{1.3}$$

where

$$D_0 \mathcal{L}(\tau, s; X, Y) = \mathcal{L}(\tau,s)[X^*Y] - \mathcal{L}(\tau, s_0)[X^*]Y - X^*\mathcal{L}(s_0, s)[Y] + X^*\mathcal{L}(s_0, s_0)[I]Y. \tag{1.4}$$

Proof. We show 1) $\Rightarrow$ 2) $\Rightarrow$ 3) $\Rightarrow$ 1).

1) $\Rightarrow$ 2). By using (1.1) for the case where j assumes only two values, one can show that $\exp t\mathcal{L}(s,\tau) = (\exp t\mathcal{L}(\tau,s))^*$. Differentiating, we obtain $\mathcal{L}(s,\tau) = \mathcal{L}(\tau,s)^*$. Let $\{\psi_j\}$, $\{x_j\}$ satisfy (1.2) then putting

$$f(t) = \sum_{j,k} (\psi_j \mid \exp t\mathcal{L}(s_j, s_k)[x_j^* x_k]\psi_k),$$

we have $f(t) \geq f(0) = 0$ for $t > 0$, whence

$$\frac{d f(0)}{dt} = \sum_{j,k} (\psi_j \mid \mathcal{L}(s_j, s_k)[x_j^* x_k]\psi_k) \geq 0.$$

2) $\Rightarrow$ 3). Fix $\{s_j\} \subset S$, $\{\psi_j\} \subset \mathcal{H}$, $\{x_j\} \subset \mathcal{A}$ and let $j = 1, \dots, n$. Put

$$s_j' = \begin{cases} s_j ; \; j = 1, \dots, n \\ s_0 ; \; j = n+1, \dots, 2n \end{cases} \quad \psi_j' = \begin{cases} \psi_j ; \; j = 1, \dots, n \\ -X_j\psi_j ; \; j = n+1, \dots, 2n \end{cases} \quad X_j' = \begin{cases} X_j ; \; j = 1, \dots, n \\ I ; \; j = n+1, \dots, 2n \end{cases}$$

Then $\sum_{j=1}^{2n} X_j'\psi_j' = 0$. Writing (1.1) for $\Phi(\tau, s) = \mathcal{L}(\tau, s)$ and for the sets $\{s_j'\}$, $\{\psi_j'\}$, $\{X_j'\}$, we obtain (1.3).

3) $\Rightarrow$ 1). The proof is based on the following statement: if $\mathcal{L} \in \mathcal{F}(\mathcal{A})$, $\mathcal{L}^* = \mathcal{L}$, then $\exp t\mathcal{L}$ is positive for all $t \geq 0$ if and only if $X, Y \in \mathcal{A}$, $YX = 0$ implies $X^*\mathcal{L}[Y^*Y]X \geq 0$ [5]. Fix the set $\{s_j\}$ with $j = 1, \dots, n$ and consider the C^*-algebra $\mathcal{A}_n$, consisting of $n \times n$ -matrices $\underline{X} = [X_{jk}]$ where $X_{jk} \in \mathcal{A}$. Define the map $\underline{\mathcal{L}} \in \mathcal{F}(\mathcal{A}_n)$ by putting $\underline{\mathcal{L}}[\underline{X}] = [\mathcal{L}(s_j, s_k)[X_{jk}]]$. Then $\underline{\mathcal{L}}^* = \underline{\mathcal{L}}$ since $\mathcal{L}(\tau, s)$ is hermitean. To establish (1.1) for all $\{\psi_j\} \subset \mathcal{H}$, $\{x_j\} \subset \mathcal{A}$ it is sufficient to prove that the map $\exp t\underline{\mathcal{L}}$ is positive for $t \geq 0$.

Let $\underline{X} = [X_{jk}]$, $\underline{Y} = [Y_{jk}]$ and $\underline{Y}\,\underline{X} = 0$, i.e. $\sum_{k=1}^{n} Y_{ik}X_{kj} = 0$ for all i, j. Then we have

$$(\underline{\psi} \mid \underline{X}^* \underline{\mathcal{L}}[\underline{Y}^*\underline{Y}]\underline{X}\,\underline{\psi}) = \sum_{i=1}^{n} \sum_{j,k} (\tilde{\psi}_j \mid \mathcal{L}(s_j, s_k)[Y_{ij}^* Y_{ik}]\tilde{\psi}_k), \tag{1.5}$$

where $\tilde{\psi}_k = (\underline{X}\,\underline{\psi})_k$, so that $\sum_k Y_{ik}\tilde{\psi}_k = 0$ for all i and (1.5) is nonnegative by 3). Applying the statement we get the result.

If $\mathcal{H} = \mathcal{A} = \mathcal{F}(\mathcal{A}) = \mathbb{C}$, then proposition 1 reduces to the well-known result about scalar kernels [6],[8]. Consider the other extreme case where S is the one-point set. Then we are dealing with a single map $\Phi \in \mathcal{F}(\mathcal{A})$ and (1.1) reduces to the condition that Φ is completely positive (c.p.). Proposition 1 reduces to the results obtained by Lindblad [9] and Evans, Lewis [10]. The map $\mathcal{L}$ satisfying the condition, corresponding to 3) is called completely dissipative (c.d.).

Let $\mathcal{L}(\tau, s)$ be a kernel satisfying the conditions of proposition 1. The family $\Phi_t(\tau, s) = \exp t\mathcal{L}(\tau, s)$; $t \geq 0$, is a semigroup of p.d. kernels satisfying $\Phi_0(\tau, s) \equiv Id$. The kernel $\Phi_t(\tau, s)$ is infinitely divisible in the sense that for any $n = 1, 2, \dots$

$$\Phi_t(\tau, s) = \Phi_{t/n}(\tau, s) \cdot \ldots \cdot \Phi_{t/n}(\tau, s) \equiv (\Phi_{t/n}(\tau, s))^n,$$

where $\Phi_{t/n}(\tau, s)$ is again a p.d. kernel. An example of h.c.p.d. kernel is given by $\mathcal{L}(\tau, s) = \Phi(\tau, s) + c\, Id$, where $\Phi(\tau, s)$ is a p.d. kernel, $c \in \mathbb{R}$. Pointwise norm limit of such kernels is again a h.c.p.d. kernel. Proposition 1 implies the inverse: any h.c.p.d. kernel is a limit of such kernels. Indeed

$$\mathcal{L}(\tau, s) = \lim_{n \to \infty} n\,(\Phi_{t/n}(\tau, s) - Id).$$

Remark. Let $\Phi(\tau, s)$ be a p.d. kernel which is infinitely divisible in the above sense. It is natural to ask whether it can be embedded in a semigroup of p.d. kernels. This question is nontrivial even in the case of a single c.p. map Φ . If $\mathcal{A}$ is finite-dimensional then the positive answer follows from the solution of general embedding problem in a topological semigroup [11], [12]. Then one can infer that a c.p. map Φ of a finite-dimensional C^*-algebra $\mathcal{A}$ is infinitely divisible if and only if it has the form $\Phi = \exp \mathcal{L} \cdot \mathcal{E}$,

where $\mathcal{E}$ is an expectation of $\mathcal{A}$ onto a subalgebra $\mathcal{B}$ of $\mathcal{A}$, and $\mathcal{L}$ is a c.d.map of $\mathcal{B}$. Since the problem is far from general solution, we shall not discuss it further here.

Let G be a group with the neutral element e . A function $\Phi(g)$, $g \in G$, with values in $\mathcal{F}$ is hermitean, positive definite, conditionally positive definite if the kernel $\Phi(g^{-1}h)$; $g, h \in G$ is such. Hermiticity means that $\Phi(g^{-1}) = \Phi(g)^*$. The following corollary is straightforward.

Corollary 1. Let $\mathcal{L}(g)$, $g \in G$, be a function with values in $\mathcal{F}(\mathcal{A})$. The following conditions are equivalent:

1) $\Phi_t(g) = \exp t\mathcal{L}(g)$ are p.d. functions for all $t \geq 0$;
2) $\mathcal{L}(g)$ is h.c.p.d. function;
3) $\mathcal{L}(g)$ is hermitean and for any finite sets $\{\psi_j\} \subset \mathcal{H}$, $\{g_j\} \subset G$, $\{x_j\} \subset \mathcal{A}$

$$\sum_{j,k} (\psi_j | D\mathcal{L}(g_j, g_k; x_j, x_k)\psi_k) \geq 0,$$

where

$$D\mathcal{L}(g,h;X,Y)=\mathcal{L}(g^{-1}h)[X^*Y]-\mathcal{L}(g^{-1})[X^*]Y-$$
$$-X^*\mathcal{L}(h)[Y]+X^*\mathcal{L}(e)[I]Y.$$

P.d. function $\Phi(g)$ (c.p.d. function $\mathcal{L}(g)$) is called <u>normalized</u> if $\Phi(e)[I]=I$ (correspondingly $\mathcal{L}(e)[I]=0$). Note that $\Phi_t(e)[I]=I$ for all $t\geq 0$ if and only if $\mathcal{L}(g)$ is normalized.

Let G be abelian locally compact group. A large class of normalized h.c.p.d. functions can be constructed as follows. Denote $\mathcal{X}=\hat{G}$ the dual group and let $g(x)$ be the value of the character g of the group $\mathcal{X}$ on the element $x\in\mathcal{X}$. Denote by ε the neutral element of $\mathcal{X}$ and let $\mathcal{B}_\varepsilon$ be the σ-ring of Borel sets of $\mathcal{X}$ which do not contain ε . Let $\mathcal{P}(dx)$ be a set function on $\mathcal{B}_\varepsilon$ with the properties:

1) for any $B\in\mathcal{B}_\varepsilon$, $\mathcal{P}(B)$ is a c.p. map from $\mathcal{F}$;
2) the set function $\mathcal{P}(B)$, $B\in\mathcal{B}_\varepsilon$,is norm σ-additive;
3) $\sup\limits_{B\in\mathcal{B}_\varepsilon}\ \sup\limits_{X\in\mathcal{A},\|X\|\leq 1}\|g(x)\mathcal{P}(B)[X]-\mathcal{P}(B)[I]\circ X\|<\infty$

for all $g\in G$.

Under these conditions the Bochner integral

$$\mathcal{L}(g)[X]=\int\limits_{\mathcal{X}\setminus\{\varepsilon\}}(g(x)\mathcal{P}(dx)[X]-\mathcal{P}(dx)[I]\circ X);\ g\in G,\tag{1.6}$$

is well defined and determines a normalized h.c.p.d. function. It is sufficient to establish it for the functions

$$\mathcal{L}_{K_1,K_2}(g)[X]=\int\limits_{K_2\setminus K_1}(g(x)\mathcal{P}(dx)[X]-\mathcal{P}(dx)[I]\circ X),\tag{1.7}$$

where $K_1\subset K_2$ are compact neighbourhoods of ε . Functions (1.7) are obviously hermitean and normalized. Let $\{\psi_j\}\subset\mathcal{H}$, $\{X_j\}\subset\mathcal{A}$ satisfy (1.2), then

$$\sum_{j,k}(\psi_j|\mathcal{L}_{K_1,K_2}(g_j^{-1}g_k)[X_j^*X_k]\psi_k)=$$

$$=\sum_{j,k}\int\limits_{K_2\setminus K_1}\overline{g_j(x)}\,g_k(x)\,(\psi_j|\mathcal{P}(dx)[X_j^*X_k]\psi_k).$$

Approximating continuous functions $g_j(x)$ uniformly on $K_2\setminus K_1$ by simple functions, we see that the expression in the right-hand side is approximated by sums of quantities of the form $\sum_{j,k}\bar{c}_jc_k(\psi_j|\mathcal{P}(B)[X_j^*X_k]\psi_k)$, which are nonnegative by property 1). Thus (1.7) and hence (1.6) are normalized h.c.p.d. functions.

Let, for example, $G=\mathcal{X}=\mathbb{R}$, $\mathcal{L}_0$ be a c.d. map from $\mathcal{F}(\mathcal{A})$

and $\beta > \| \mathcal{L}_0 \|$. Then the set function $\mathcal{P}(dx)$, determined by the density

$$p(x) = \begin{cases} 0\,; & x < 0 \\ x^{-1} \exp x(\mathcal{L}_0 - \beta\, Id)\,; & x \geq 0 \end{cases}$$

with respect to the Lebesgue measure dx, has the properties 1)-3). The integral (1.6) can be calculated explicitly

$$\mathcal{L}(g)[X] = -\ln((\beta - i\lambda)Id - \mathcal{L}_0)[X] + \ln(\beta Id - \mathcal{L}_0)[I] \circ X. \tag{1.8}$$

§2. <u>The canonical representations.</u>

<u>Proposition 2.</u> Let $\Phi(g)$, $g \in G$, be a p.d. function with values in $\mathcal{F}$. Then there exist Hilbert space $\mathcal{K}$, operator $F \in \mathcal{B}(\mathcal{H}, \mathcal{K})$, unitary representation $g \to V_g$ of the group G in $\mathcal{K}$ and $*$--representation $X \to \rho[X]$ of the algebra $\mathcal{A}$ in $\mathcal{K}$ such that $[V_g, \rho[X]] = 0$ for all $g \in G$, $X \in \mathcal{A}$ and

$$\Phi(g)[X] = F^* V_g \rho[X] F\,; \quad g \in G,\ X \in \mathcal{A}. \tag{2.1}$$

Conversely any function of such form is positive definite. It is normalized if and only if F is an isometric operator.

This proposition combines the Gelfand-Raikov representation for scalar p.d. functions and the Stinespring representation for c.p. maps. The proof is rather standard and we just sketch it.

<u>Proof.</u> Let $\tilde{G}$ be the formal linear span of the set G and let $\tilde{\mathcal{K}} = \tilde{G} \otimes \mathcal{H} \otimes \mathcal{A}$ be the algebraic tensor product, generated by the elements $g \otimes \varphi \otimes X$, with $g \in G$, $\varphi \in \mathcal{H}$, $X \in \mathcal{A}$. Defining pre-inner product on $\tilde{\mathcal{K}}$ by

$$(g \otimes \varphi \otimes X, h \otimes \psi \otimes Y) = (\varphi \,|\, \Phi(g^{-1}h)[X^* Y]\psi), \tag{2.2}$$

we construct $\mathcal{K}$ as the corresponding completion of $\tilde{\mathcal{K}}$. The operator F is the canonical extension of the operator $\tilde{F}: \mathcal{H} \to \tilde{\mathcal{K}}$ defined by

$$\tilde{F}\varphi = e \otimes \varphi \otimes I\,; \quad \varphi \in \mathcal{H}, \tag{2.3}$$

and the representations V, ρ are the canonical extensions of maps $\tilde{V}$, $\tilde{\rho}$, defined by

$$\tilde{V}_h\,(g \otimes \varphi \otimes X) = hg \otimes \varphi \otimes X\,; \tag{2.4}$$

$$\tilde{\rho}[Y]\,(g \otimes \varphi \otimes X) = g \otimes \varphi \otimes YX. \tag{2.5}$$

The properties of F, V, ρ and the representation (2.1) follow from the definitions (2.2)-(2.5).

The converse statement follows from the relation

$$\Phi(g^{-1}h)[X^*Y] = (\rho[X]V_g F)^*(\rho[Y]V_h F).$$

Proposition 2 implies useful inequalities for p.d.functions. First, for all $g \in G$, $X \in \mathcal{A}$,

$$\Phi(g)[X]^*\Phi(g)[X] \leq \|\Phi(e)[I]\| \cdot \Phi(e)[X^*X] \leq \leq \|\Phi(e)[I]\| \cdot \|X\|^2 \cdot \Phi(e)[I]. \tag{2.6}$$

Indeed, for all $\psi \in \mathcal{H}$ we have

$$\|\Phi(g)[X]\psi\|^2 = (\psi | F^*\rho[X^*]V_g^* FF^* V_g \rho[X]F\psi) \leq \leq \|FF^*\| \cdot \|\rho[X]F\psi\|^2. \tag{2.7}$$

Using the equality $\|FF^*\| = \|F^*F\| = \|\Phi(e)[I]\|$, we get (2.6). In the same way,

$$\begin{aligned}(\Phi(g)[X] - \Phi(h)[X])^*(\Phi(g)[X] - \Phi(h)[X]) &\leq \\ \leq \|\Phi(e)[I]\| \cdot (2\Phi(e)[X^*X] - \Phi(g^{-1}h)[X^*X] - \Phi(h^{-1}g)[X^*X]) &\leq \\ \leq \|\Phi(e)[I]\| \cdot \|X\|^2 \cdot (2\Phi(e)[I] - \Phi(g^{-1}h)[I] - \Phi(h^{-1}g)[I]),&\end{aligned} \tag{2.8}$$

for all $g, h \in G$, $X \in \mathcal{A}$.

The following corollary is an analog of Schur's lemma concerning product of scalar p.d. functions.

<u>Corollary 2.</u> If $\Phi(g)$, $\Psi(g)$ are p.d. functions with values in $\mathcal{F}(\mathcal{A})$, then $\Phi(g)\cdot\Psi(g)$ is a p.d. function.

<u>Proof.</u> Fix the finite sets $\{\psi_j\} \subset \mathcal{H}$, $\{g_j\} \subset G$, $\{X_j\} \subset \mathcal{A}$ and consider the matrix B with the elements $\Psi(g_j^{-1}g_k)[X_j^* X_k] \in \mathcal{A}$. By proposition 2 B is a positive matrix, hence $B = A^*A$, where A is a matrix with elements $A_{jk} \in \mathcal{A}$. By positive definiteness of $\Phi(g)$,

$$\sum_{j,k} (\psi_j | \Phi(g_j^{-1}g_k)[\Psi(g_j^{-1}g_k)[X_j^* X_k]]\psi_k) = = \sum_{\ell} \sum_{j,k} (\psi_j | \Phi(g_j^{-1}g_k)[A_{\ell j}^* A_{\ell k}]\psi_k) \geq 0.$$

<u>Remark.</u> It is not difficult to prove that for a p.d. kernel $\Phi(\tau, s)$; $\tau, s \in S$, there exist a Hilbert space $\mathcal{K}$, a $*$-representation ρ of $\mathcal{A}$ in $\mathcal{K}$ and a function F_s; $s \in S$, with values in $\mathcal{L}(\mathcal{H}, \mathcal{K})$ such that

$$\Phi(\tau, s)[X] = F_\tau^* \rho[X] F_s; \quad \tau, s \in S; \ X \in \mathcal{A}.$$

Basing on this representation one can prove the analog of corollary 2 for p.d. kernels.

Proposition 3. Let $\mathcal{L}(g)$; $g \in G$, be a h.c.p.d. function with values in $\mathcal{F}$. Then there exist a Hilbert space $\mathcal{K}$, a unitary representation $g \to W_g$ of the group G in $\mathcal{K}$, a $*$ -representation $X \to \pi[X]$ of the algebra $\mathcal{A}$ in $\mathcal{K}$, such that

$$[W_g, \pi[X]] = 0 ; \quad g \in G, X \in \mathcal{A}, \tag{2.9}$$

and the family of bounded linear maps $\mathcal{B}(g)[X]$; $g \in G$, from $\mathcal{A}$ to $\mathcal{L}(\mathcal{H}, \mathcal{K})$, satisfying the cocycle equation

$$\mathcal{B}(hg)[YX] = W_h \pi[Y] \mathcal{B}(g)[X] + \mathcal{B}(h)[Y]X; \quad g, h \in G; \; X, Y \in \mathcal{A}, \tag{2.10}$$

such that

$$\mathcal{D}\mathcal{L}(g,h; X,Y) = \mathcal{B}(g)[X]^* \mathcal{B}(h)[Y]; \quad g,h \in G; \; X,Y \in \mathcal{A}. \tag{2.11}$$

Proof. Consider the algebraic tensor product $\tilde{\mathcal{K}} = \tilde{G} \otimes \mathcal{H} \otimes \mathcal{A}$ and define pre-inner product on $\tilde{\mathcal{K}}$ by the relation

$$(g \otimes \varphi \otimes X, h \otimes \psi \otimes Y) = (\varphi | \mathcal{D}\mathcal{L}(g,h; X,Y) \psi). \tag{2.12}$$

Let $\mathcal{K}$ be the corresponding completion of $\tilde{\mathcal{K}}$. Put

$$\tilde{W}_h (g \otimes \varphi \otimes X) = hg \otimes \varphi \otimes X - h \otimes X\varphi \otimes I, \tag{2.13}$$

then $\tilde{W}_{h_1} \tilde{W}_{h_2} = \tilde{W}_{h_1 h_2}$; $h_1, h_2 \in G$. Using (2.13), (2.12) and the identity

$$\mathcal{D}\mathcal{L}(hg, hg'; X, X') - X^* \mathcal{D}\mathcal{L}(h, hg'; I, X') - \mathcal{D}\mathcal{L}(hg, h; X, I) X' + X^* \mathcal{D}\mathcal{L}(h,h; I, I) X' \equiv \mathcal{D}\mathcal{L}(g, g'; X, X'),$$

we obtain

$$(\tilde{W}_h (g \otimes \varphi \otimes X), \tilde{W}_h (g' \otimes \varphi' \otimes X')) = (g \otimes \varphi \otimes X, g' \otimes \varphi' \otimes X').$$

Therefore $\tilde{W}_h$ canonically extends to a unitary representation W of the group G in $\mathcal{K}$.

Put

$$\tilde{\pi}[Y](g \otimes \varphi \otimes X) = g \otimes \varphi \otimes YX - e \otimes X\varphi \otimes Y, \tag{2.14}$$

then $\tilde{\pi}[Y_1] \tilde{\pi}[Y_2] = \tilde{\pi}[Y_1 Y_2]$. Using (2.14), (2.12) and the identity

$$\mathcal{D}\mathcal{L}(g, g'; X, YX') + X^* \mathcal{D}\mathcal{L}(e, g'; Y^*, X') - \mathcal{D}\mathcal{L}(g, g'; Y^* X, X') - \mathcal{D}\mathcal{L}(g, e; X, Y) X' \equiv 0,$$

we get

$$\left(\tilde{\pi}[Y](\varphi\otimes g\otimes X),(\varphi'\otimes g'\otimes X')\right)=\left((\varphi\otimes g\otimes X),\tilde{\pi}[Y^*](\varphi'\otimes g'\otimes X')\right).$$

Therefore $\tilde{\pi}$ canonically extends to $*$-representation π of the algebra $\mathcal{A}$ in $\mathcal{K}$. From (2.13) and (2.14)

$$\tilde{W}_h\,\tilde{\pi}[Y](g\otimes\varphi\otimes X)=\tilde{\pi}[Y]\,\tilde{W}_h\,(g\otimes\varphi\otimes X)= \\ =hg\otimes\varphi\otimes YX-h\otimes Y\varphi\otimes X, \tag{2.15}$$

in particular, (2.9) follows.

For $g\in G$ and $X\in\mathcal{A}$ define the map $\mathcal{B}(g)[X]\in\mathcal{L}(\mathcal{H},\mathcal{K})$ as the canonical extension of

$$\tilde{\mathcal{B}}(g)[X]\varphi=g\otimes\varphi\otimes X\,;\ \varphi\in\mathcal{H}.$$

Then (2.10) follows from (2.15) and (2.11) from (2.12).

The following result shows that a h.c.p.d. function has "at most quadratic growth".

<u>Corollary 3.</u> Let $\mathcal{L}(g)$, $g\in G$, be a h.c.p.d. function. Then for any $g\in G$ and $n=1,2,\dots$

$$\|\mathcal{L}(g^n)\|\leq(\|\mathcal{L}(e)\|+\|\mathcal{L}(g)\|)n^2-\|\mathcal{L}(e)\|. \tag{2.16}$$

<u>Proof.</u> It follows from (2.10) that $\|\mathcal{B}(gh)\|\leq\|\mathcal{B}(g)\|+\|\mathcal{B}(h)\|$, whence

$$\|\mathcal{B}(g^k)\|\leq k\,\|\mathcal{B}(g)\|. \tag{2.17}$$

Putting $g=h$, $X=Y$ in (2.11) and taking into account that $\|\mathcal{L}(g^{-1})\|=\|\mathcal{L}(g)\|$, we get

$$\|\mathcal{B}(g)[X]\|^2=\|D\mathcal{L}(g,g;X,X)\|\leq 2(\|\mathcal{L}(e)\|+\|\mathcal{L}(g)\|)\cdot\|X\|^2.$$

Therefore

$$\|\mathcal{B}(g)\|\leq\sqrt{2(\|\mathcal{L}(e)\|+\|\mathcal{L}(g)\|)}\,,\quad g\in G. \tag{2.18}$$

Substituting g^{-1} instead of g in (2.11), we get

$$\mathcal{L}(gh)[X^*Y]=\mathcal{B}(g^{-1})[X]^*\mathcal{B}(h)[Y]-X^*\mathcal{L}(e)[I]Y+ \\ +\mathcal{L}(g)[X^*]Y+X^*\mathcal{L}(h)[Y].$$

It follows that

$$\|\mathcal{L}(gh)\|\leq\|\mathcal{B}(g^{-1})\|\cdot\|\mathcal{B}(h)\|+\|\mathcal{L}(e)\|+\|\mathcal{L}(g)\|+\|\mathcal{L}(h)\|.$$

Putting $h=g^k$ and taking into account (2.17), (2.18) we have

$$\|\mathcal{L}(g^{k+1})\|\leq 2(\|\mathcal{L}(e)\|+\|\mathcal{L}(g)\|)\cdot k+(\|\mathcal{L}(e)\|+\|\mathcal{L}(g)\|)+ \\ +\|\mathcal{L}(g^k)\|.$$

Summing over k from 1 to $n-1$, we get (2.16).

§ 3. Representation of hermitean conditionally positive functions.

As usually, $\mathcal{A}'$ denotes the commutant of the algebra $\mathcal{A} \subset \mathcal{L}(\mathcal{H})$, and $\mathcal{A}''$ the von Neumann algebra generated by $\mathcal{A}$. We now state and prove the main result.

Theorem 1. Let $\mathcal{L}(g)$, $g \in G$, be a h.c.p.d. function with values in $\mathcal{F}(\mathcal{A})$. There exist:

a) a Hilbert space $\mathcal{K}$, a unitary representation $g \to W_g$ of the group G in $\mathcal{K}$, and a $*$-representation $X \to \pi[X]$ of the algebra $\mathcal{A}$ in $\mathcal{K}$, satisfying (2.9);

b) an operator $A \in \mathcal{L}(\mathcal{H}, \mathcal{K})$ and a function $B(g)$, $g \in G$,with values in $\mathcal{L}(\mathcal{H}, \mathcal{K})$, such that

$$B(g)\,X = \pi[X]\,B(g); \; g \in G,\; X \in \mathcal{A}, \tag{3.1}$$

and satisfying the cocycle equation for the representation $g \to W_g$

$$B(hg) = W_h\,B(g) + B(h)\; ; \; h, g \in G. \tag{3.2}$$

c) function $z(g)$, $g \in G$, with values in $\mathfrak{Z}^h$, where $\mathfrak{Z} = \mathcal{A}' \cap \mathcal{A}''$ is the center of the von Neumann algebra $\mathcal{A}''$, satisfying the equation

$$z(gh) - z(g) - z(h) = \operatorname{Im} B(g^{-1})^* B(h); \; g, h \in G, \tag{3.3}$$

and the operator $C \in \mathcal{A}''$, such that

$$\begin{aligned} \mathcal{L}(g)[X] &= A^* W_g\,\pi[X]\,A + A^* B(g) X + X B(g^{-1})^* A + \\ &+ X\left[-\tfrac{1}{2} B(g)^* B(g) + i\, z(g)\right] + C^* X + X C\; ; \; g \in G,\; X \in \mathcal{A}. \end{aligned} \tag{3.4}$$

For any set of objects satisfying a)-c), the relation (3.4) defines a h.c.p.d. function with values in $\mathcal{F}$. It is normalized if and only if $\operatorname{Re} C = -\frac{1}{2} A^* A$.

For scalar h.c.p.d. the representation can be reduced to the term in squared brackets; the relation between h.c.p.d. functions and first order cocycles has been noticed by Araki [13] (see also Parthasarathy and Schmidt [8], and especially Guichardet [14]). In the case $G = \{e\}$ theorem 1 and proposition 2 give the general form of a completely dissipative map $\mathcal{L} \in \mathcal{F}(\mathcal{A})$:

$$\mathcal{L}[X] = \Phi[X] + C^* X + X C,$$

where $C \in \mathcal{A}''$ and Φ is a completely positive mapping from $\mathcal{A}$ to $\mathcal{A}''$ (Lindblad [9], Christensen and Evans [15]).

Proof. If $\mathcal{L}(g)$ is defined by (3.4) then

$$D\mathcal{L}(g,h;X,Y) = (W_g \pi[X]A - AX + B(g)X)^* \cdot (W_h \pi[Y]A - AY + B(h)Y), \tag{3.5}$$

whence the converse statement follows. The last statement follows from the fact that $B(e)=0$ and $z(e)=0$.

Let now $\mathcal{L}(g)$ be a h.c.p.d. function with values in $\mathcal{F}(\mathcal{A})$. Let $\mathcal{K}$, W, π and $\mathcal{B}(g)[X]$ be as in proof of proposition 3. Introduce the subspace of $\mathcal{L}(\mathcal{H},\mathcal{K})$ of the form

$$\mathcal{M} = cl\{\mathcal{B}(g)[X]Y;\ g \in G;\ X,Y \in \mathcal{A}\},$$

where $cl\,\mathcal{N}$ means ultraweak closure of linear span of the set $\mathcal{N}$. Note that if $A_1, A_2 \in \mathcal{M}$, then $A_1^* A_2 \in \mathcal{A}''$ because of (2.11).

Lemma 1. Any solution of the cocycle equation (2.10) has the form

$$\mathcal{B}(g)[X] = W_g \pi[X]A - AX + B(g)X, \tag{3.6}$$

where $A \in \mathcal{M}$ and $B(g)$, $g \in G$, is a function with values in $\mathcal{M}$, having the properties (3.1), (3.2).

Proof. Decompose $\mathcal{B}(g)[X]$ as

$$\mathcal{B}(g)[X] - \mathcal{B}_0(g)[X] + V[X], \tag{3.7}$$

where $V[X] = \mathcal{B}(e)[X]$. Putting $g=h=e$ in (2.10) we get

$$V[YX] = \pi[Y]V[X] + V[Y]X;\ X,Y \in \mathcal{A}.$$

Moreover $V[X]^*V[Y] \in \mathcal{A}$ if $X,Y \in \mathcal{A}$ since according to (2.11) $V[X]^*V[Y] = D\mathcal{L}(e,e;X,Y)$. From a result of Christensen and Evans ([15], theorem 2.1) it follows that there is $A \in \mathcal{L}(\mathcal{H},\mathcal{K})$ such that

$$V[X] = \pi[X]A - AX;\ X \in \mathcal{A}. \tag{3.8}$$

Moreover $A \in cl\{V[X]Y;\ X,Y \in \mathcal{A}\} \subset \mathcal{M}$.

For $\mathcal{B}_0(g)[X]$ we get from (2.10), (3.7)

$$\mathcal{B}_0(hg)[YX] = W_h \pi[Y]\mathcal{B}_0(g)[X] + \mathcal{B}_0(h)[Y]X + (W_h - I)\pi[Y]V[X]. \tag{3.9}$$

Putting $h=e$, $X=I$ and taking into account that $\mathcal{B}_0(e)[X]=0$, we get

$$\mathcal{B}_0(g)[Y] = \pi[Y]B_0(g), \tag{3.10}$$

where $B_0(g) = \mathcal{B}_0(g)[I]$. Putting $X=Y=I$ in (3.9) and taking into account that $V[I]=0$, we see that $B_0(g)$ satisfies (3.2).

Substituting (3.10) into (3.9), putting $Y=I$, and taking into account (3.2), we obtain

$$\pi[X]B_0(h) - B_0(h)X = (W_h - I)V[X].$$

Substituting $V[X]$ from (3.7) and denoting

$$B(h) = B_0(h) - (W_h - I)A, \qquad (3.11)$$

we see that $B(h)$ satisfies (3.1); since $(W_h - I)A$ is a cocycle of the representation $h \to W_h$, then $B(h)$ satisfies the cocycle equation (3.2) together with $B_0(h)$.

Putting together (3.7), (3.8), (3.10), (3.11), we obtain the relation (3.6). Since $B_0(h) = \mathcal{B}_0(h)[I] \in \mathcal{M}$ and $W_h A \in \mathcal{M}$ because of (2.10) and the fact that $A \in \mathcal{M}$, then $B(h) \in \mathcal{M}$.

<u>Lemma 2.</u> $B(g)^* B(h) \in \mathcal{Z}$ for all $g, h \in G$.

<u>Proof.</u> The property (3.1) implies that $B(g)^* B(h) X =$ $B(g)^* \pi[X] B(h) = X B(g)^* B(h)$, so that $B(g)^* B(h) \in \mathcal{A}'$. Since $B(g) \in \mathcal{M}$ by lemma 1, then also $B(g)^* B(h) \in \mathcal{A}''$.

Combining (2.11) with (3.6) we obtain (3.5). Consider the hermitean function

$$\mathcal{L}_0(g)[X] = \mathcal{L}(g)[X] - A^* W_g \pi[X] A - A^* B(g) X - X B(g^{-1})^* A. \qquad (3.12)$$

A direct but rather labourious calculation using (3.1), (3.2) and lemma 2 shows that

$$D\mathcal{L}_0(g, h; X, Y) = X^* Y \cdot B(g)^* B(h). \qquad (3.13)$$

Put

$$\begin{aligned} \Delta(g) &= \mathcal{L}_0(g)[I] \equiv \\ &\mathcal{L}(g)[I] - A^* W_g A - A^* B(g) - B(g^{-1})^* A. \end{aligned} \qquad (3.14)$$

<u>Lemma 3.</u> The operator-valued function $\Delta(g)$, $g \in G$, has the properties:

1) $\Delta(g) \in \mathcal{A}''$, $g \in G$;
2) $\Delta(g^{-1}) = \Delta(g)^*$, $g \in G$;
3) $\Delta(gh) - \Delta(g) - \Delta(h) + \Delta(e) = B(g^{-1})^* B(h)$; $g, h \in G$.

<u>Proof.</u> 1) Obviously, $\mathcal{L}(g)[I] \in \mathcal{A}$; the other terms in (3.14) belong to $\mathcal{A}''$ because A , $W_g A$, $B(g) \in \mathcal{M}$ (see the proof of lemma 1). Property 2) follows from the hermiticity of $\mathcal{L}_0(g)$; property 3) follows from (3.13) with $X = Y = I$.

Let $\mathcal{E}$ be a conditional expectation from the von Neumann algebra $\mathcal{A}''$ onto its center $\mathcal{Z}$ (which always exists, see e.g. [16]). Put

$$z_0(g) = \mathcal{E}\,(Im\,\Delta(g)) \qquad (3.15)$$

From the definition (3.15) and lemmas 2,3 we get the following properties of the operator-valued function $z_0(g)$; $g \in G$:

1) $z_0(g) \in \mathfrak{Z}^h$; $g \in G$;

2) $z_0(g^{-1}) = -z_0(g)$; $g \in G$;

3) $z_0(gh) - z_0(g) - z_0(h) = \mathrm{Im}\, B(g^{-1})^* B(h)$; $g, h \in G$.

Consider the operator-valued function

$$\ell_0(g) = -\tfrac{1}{2} B(g)^* B(g) + i z_0(g); \; g \in G.$$

This function has the properties:

1) $\ell_0(g) \in \mathfrak{Z}$, $g \in G$; $\ell_0(e) = 0$;

2) $\ell_0(g^{-1}) = \ell_0(g)^*$, $g \in G$;

3) $\ell_0(gh) - \ell_0(g) - \ell_0(h) = B(g^{-1})^* B(h)$.

The first two follow from the corresponding properties of $z_0(g)$ and from the relation $B(g^{-1}) = -W_g^* B(g)$. The third property is verified by direct calculations using the cocycle equation and the property 3) of $z_0(g)$.

The relation (3.13) implies then that the hermitean function

$$\Lambda(g)[X] = \Upsilon_0(g)[X] - \ell_0(g)\cdot X \qquad (3.16)$$

satisfies the equation

$$D\Lambda(g, h; X, Y) = 0; \; g, h \in G; \; X, Y \in \mathcal{A}. \qquad (3.17)$$

<u>Lemma 4.</u> Let $\Lambda(g)$, $g \in G$, be a hermitean function with values in $\mathcal{F}$ such that $\Lambda(g)[X] \in \mathcal{A}''$ for all $X \in \mathcal{A}$, $g \in G$, and satisfying the equation (3.17). Then

$$\Lambda(g)[X] = C^* X + XC + i\lambda(g) X, \qquad (3.18)$$

where $C \in \mathcal{A}''$, and $\lambda(g)$ is a morphism of G into the additive group of $\mathfrak{Z}^h$.

<u>Proof.</u> Put

$$\Lambda(g)[X] = \Lambda(e)[I] \circ X + \Lambda_0(g)[X], \qquad (3.19)$$

then (3.17) reduces to

$$\Lambda_0(gh)[XY] = \Lambda_0(g)[X]Y + X\Lambda_0(h)[Y]. \qquad (3.20)$$

It follows, in particular, that $\Lambda_0(e)$ is a differentiation from $\mathcal{A}$ to $\mathcal{A}''$ and according to ([15], corollary 2.3), there is $C_1 \in \mathcal{A}''$ such that

$$\Lambda_0(e)[X] = C_1 X - X C_1. \qquad (3.21)$$

Putting in (3.19) $Y=I$, $h=e$ and $X=I$, $g=e$, we obtain, correspondingly

$$\begin{aligned}\Lambda_0(g)[X] &= \Lambda_0(e)[X] + X\Lambda_0(g)[I],\\ \Lambda_0(h)[Y] &= \Lambda_0(h)[I]Y + \Lambda_0(e)[Y].\end{aligned} \tag{3.22}$$

It follows that $\lambda_0(g) \equiv \Lambda_0(g)[I] \in \mathcal{A}'$ and therefore $\lambda_0(g) \in \mathfrak{Z}$. Putting $X=Y=I$ in (3.20) we see that $\lambda_0(g)$ is a morphism of G into the additive group of $\mathfrak{Z}$. From (3.21), (3.22)

$$\Lambda_0(g)[X] = C_1 X - X C_1 + \lambda_0(g) X. \tag{3.23}$$

We now use the hermiticity of $\Lambda(g)$; substituting (3.23) into the relation $\Lambda_0(g^{-1}) = \Lambda_0(g)^*$ we get

$$[\operatorname{Re} C_1, X] = X \operatorname{Re} \lambda_0(g)\ ;\ X \in \mathcal{A}.$$

It follows that $\operatorname{Re}\lambda_0(g) \equiv 0$ and $\operatorname{Re} C_1 = 0$. Therefore $C_1 = iH$, where $H \in \mathcal{A}^h$ and $\lambda_0(g) = i\lambda(g)$, where $\lambda(g)$ is a morphism of G into $\mathfrak{Z}^h$. Substituting this into (3.23) and (3.19) and putting $C = \frac{1}{2}\Lambda(e)[I] - iH$, we get (3.18).

To finish the proof of theorem 1 remark that $\Lambda(g)$ defined by the relation (3.16) satisfies the conditions of lemma 4 and therefore it can be represented in the form (3.18). Denoting $z(g) = z_0(g) + \lambda(g)$, we see that $z(g)$ satisfies the condition c). The relation (3.4) follows from (3.16), (3.18), and the theorem is proved.

Remark. Using elements of the proof of theorem 1, one can show that arbitrary h.c.p.d. kernel $\mathcal{L}(\tau, s)$; $\tau, s \in \mathcal{S}$, with values in $\mathcal{F}(\mathcal{A})$ can be represented as

$$\mathcal{L}(\tau, s)[X] = V_\tau^* \pi[X] V_s + C_\tau^* X + X C_s\ ;\ X \in \mathcal{A},$$

where π is a $*$-representation of $\mathcal{A}$ in a Hilbert space $\mathcal{K}$, V_s; $s \in \mathcal{S}$, is a function with values in $\mathcal{L}(\mathcal{H}, \mathcal{K})$, and C_s ; $s \in \mathcal{S}$ is a function with values in $\mathcal{A}''$.

§ 4. Topological properties.

From a variety of locally-convex topologies which can be introduced in the space $\mathcal{F}$ of bounded linear maps from $\mathcal{A}$ to $\mathcal{L}(\mathcal{H})$ we shall use the following two, defined by the families of seminorms

$$\mathcal{T}_1 : \Phi \to |(\varphi | \Phi[X]\psi)|\ ;\quad \varphi, \psi \in \mathcal{H};\ X \in \mathcal{A},$$

$$\mathcal{T}_2 : \Phi \to \sup_{X \in \mathcal{A},\ \|X\| \le 1} \|\Phi[X]\psi\|\ ;\quad \psi \in \mathcal{H},$$

and also the topology, corresponding to the norm $\|\Phi\| = \sup_{\|X\|\le 1} \|\Phi[X]\|$.

In what follows $\mathcal{A}$ is a von Neumann algebra. Denote by $\mathcal{F}_\sigma$ ($\mathcal{F}(\mathcal{A})_\sigma$) the space of all ultraweakly continuous linear maps from $\mathcal{A}$ to $\mathcal{L}(\mathcal{H})$ (correspondingly, from $\mathcal{A}$ to $\mathcal{A}$); $\mathcal{F}_\sigma$ is a norm-closed subspace of $\mathcal{F}$. The group G will be a topological group.

<u>Proposition 4.</u> Let $\Phi(g)$, $g\in G$, be a p.d. function with values in $\mathcal{F}_\sigma$ such that the operator-valued function $\Phi(g)[I]$, $g\in G$, is weakly continuous at $g=e$. Then $\Phi(g)$ is $\mathcal{T}_2$-continuous uniformly in $g\in G$. Moreover in proposition 2 the representation $g\to V_g$ can be chosen continuous, and the representation $X\to\rho[X]$ normal.

<u>Proof.</u> The first statement follows from (2.8). Proof of the second statement for the representations V, ρ, defined by (2.4), (2.5) is more or less standard and is omitted.

<u>Proposition 5.</u> Let $\mathcal{L}(g)$, $g\in G$, be $\mathcal{T}_1$-continuous h.c.p.d. function with values in $\mathcal{F}(\mathcal{A})_\sigma$. Then in theorem 1 the representation $g\to W_g$ can be chosen continuous, the representation $X\to\pi[X]$ normal, and the operator-valued function $z(g)$, $g\in G$, weakly continuous. If moreover, the center $\mathfrak{Z}$ of $\mathcal{A}$ is finite-dimensional, or if the group G is locally compact and the center $\mathfrak{Z}$ is purely atomic, then $\mathcal{L}(g)$, $g\in G$, is $\mathcal{T}_2$-continuous.

<u>Proof.</u> The proof of the asserted continuity of the representations W and π, defined by(2.13), (2.14) is standard.Consider $B(h)$ defined by the equation (3.11); it follows that it is sufficient to prove strong continuity of $B_0(h) = \mathcal{B}(h)[I]$.
Since $B_0(h)$ is a cocycle of the representation W, it is sufficient to prove its continuily at $h=e$. According to (2.11)

$$\|\mathcal{B}(h)[I]\psi\|^2 = (\psi|D\mathcal{L}(h,h,I,I)\psi),$$

which tends to zero as $g\to e$ because of the $\mathcal{T}_1$-continuity of $\mathcal{L}(g)$.
The weak continuity of $z(g)$ follows from (3.4) with $X=I$.

The first three terms in (3.4) are $\mathcal{T}_2$-continuous. The first term is a p.d. function satisfying the condition of the proposition 4. Using (3.1) we have the estimate for the second term

$$\sup_{\|X\|\le 1}\|A^*(B(g)-B(h))X\psi\| = \sup_{\|X\|\le 1}\|A^*\pi[X](B(g)-B(h))\psi\| \le$$
$$\le \|A^*\|\cdot\|(B(g)-B(h))\psi\|.$$

Analogous argument takes place also for the third term.

Consider the operator-valued function

$$l(g) = -\tfrac{1}{2}B(g)^*B(g) + iz(g);\ g\in G.$$

This function assumes its values in $\mathfrak{Z}$ and is weakly continuous. If $\mathfrak{Z}$ is finite-dimensional then weak and strong topologies of $\mathfrak{Z}$ coincide and $\ell(g)$ is strongly continuous. If G is locally compact and $\mathfrak{Z}$ is purely atomic, then the weak continuity of a function on G with values in $\mathfrak{Z}$ also implies the strong continuity. Indeed, by the uniform boundedness theorem,

$$\sup_{g \in K} \| \ell(g) \| \equiv c_K < \infty$$

for any compact $K \subset G$. Since $\mathfrak{Z}$ is purely atomic,

$$\ell(g) = \sum_\alpha \ell_\alpha(g) P_\alpha,$$

where $\{P_\alpha\}$ is the orthogonal resolution of identity, generating $\mathfrak{Z}$, $\ell_\alpha(g)$ are complex continuous functions, such that $\sup_{g \in K} \sup_\alpha |\ell_\alpha(g)| \leq$ $\leq c_K$. Then

$$\| (\ell(h) - \ell(g)) \psi \|^2 \leq \sum_\alpha | \ell_\alpha(h) - \ell_\alpha(g) |^2 \, \| P_\alpha \psi \|^2,$$

where $\sum_\alpha \| P_\alpha \psi \|^2 < \infty$ and $| \ell_\alpha(h) - \ell_\alpha(g) | \leq 2 c_K$ for $g, h \in K$. It follows that $\ell(h) \to \ell(g)$ strongly if $h \to g$. Therefore the term $X \ell(g)$ in (3.4) is $\mathcal{T}_2$-continuous and so is $\mathcal{L}(g)$.

<u>Theorem 2.</u> Let G be a compact group, $\mathcal{L}(g)$, $g \in G$, be a $\mathcal{T}_1$-continuous h.c.p.d. function with values in $\mathcal{F}(\mathcal{A})_\sigma$. There exist a Hilbert space $\mathcal{K}$, an operator $A \in \mathcal{L}(\mathcal{H}, \mathcal{K})$, a continuous representation $g \to W_g$ of the group G in $\mathcal{K}$, a normal representation $X \to \pi[X]$ of the algebra $\mathcal{A}$ in $\mathcal{K}$, satisfying (2.9), and an operator $C \in \mathcal{A}$ such that

$$\mathcal{L}(g)[X] = A^* W_g \pi[X] A + C^* X + X C. \tag{4.1}$$

In particular, $\mathcal{L}(g)$ is $\mathcal{T}_2$-continuous.

For the proof we need two lemmas.

<u>Lemma 5.</u> Let G be a compact group and $B(g)$, $g \in G$, be a strongly continuous function with values in $\mathfrak{M}$, satisfying the equation (3.2). There exists an operator $B \in \mathfrak{M}$, such that

$$B X = \pi[X] B; \quad X \in \mathcal{A}, \tag{4.2}$$

and

$$B(g) = (W_g - I) B; \quad g \in G. \tag{4.3}$$

<u>Proof.</u> For a fixed $\psi \in \mathcal{H}$ the function $B(g)\psi$, $g \in G$, is a continuous cocycle of the representation $g \to W_g$ in $\mathcal{K}$. According to ([14], App.B) there is $\varphi \in \mathcal{K}$, such that

$$B(g)\psi = (W_g - I)\varphi; \quad g \in G. \tag{4.4}$$

Consider

$$\mathcal{A}_W = cl\{(W_g - I);\ g \in G\}.$$

Since $g \to W_g$ is a unitary representation, $\mathcal{A}_W$ is an ultraweakly closed $*$-algebra. Let P_W be the orthogonal projection onto the closed subspace $\mathcal{K}_W$ of $\mathcal{H}$, generated by the vectors $\{(W_g - I)\psi;\ g \in G,\ \psi \in \mathcal{H}\}$. By the von Neumann's density theorem, P_W is the maximal projection of $\mathcal{A}_W$ and

$$(W_g - I) = P_W(W_g - I) = (W_g - I)P;\ g \in G.$$

Therefore in (4.4) we can take $\varphi \in \mathcal{K}_W$. Such φ is easily seen to be unique. Denote by B the operator which maps ψ into φ. It satisfies (4.3). The operator B is everywhere defined and closed and hence is bounded. From (3.1) and (2.9) it follows for any $g \in G$

$$(W_g - I)BX\psi = B(g)X\psi = \pi[X]B(g)\psi = \pi[X](W_g - I)B\psi =$$
$$= (W_g - I)\pi[X]B\psi,$$

whence $BX = P_W\pi[X]B$. Since $\pi[X] \in \mathcal{A}_W'$ we have $P_W\pi[X] = \pi[X]P_W$, and we get (4.2).

Since $B(g) = (W_g - I)B \in \mathcal{M}$, $g \in G$, and P_W is an ultraweak limit of linear combinations of $(W_g - I)$, $g \in G$, then $B = P_W B \in \mathcal{M}$.

<u>Lemma 6.</u> Let $B(g)$, $g \in G$, be given by (4.3). Then the only weakly continuous solution of the equation (3.3) has the form

$$z(g) = \operatorname{Im} B^* W_g B;\ g \in G. \tag{4.5}$$

<u>Proof.</u> That (4.5) satisfies (3.3) with $B(g)$ given by (4.3) is verified by direct calculation. If $z'(g)$ is another solution then $\delta(g) = z(g) - z'(g)$ satisfies

$$\delta(hg) - \delta(h) - \delta(g) = 0;\ g, h \in G,$$

i.e. is a continuous morphism of the group G intothe additive group of $\mathfrak{Z}$. Since G is compact then $\delta(g) \equiv 0$.

<u>Proof</u> of the theorem 2. Substituting (4.3), (4.5) into (3.4) we get

$$\mathcal{L}(g)[X] = (A+B)^* W_g \pi[X](A+B) +$$
$$+ (C - B^*A - \tfrac{1}{2}B^*B)^* X + X(C - B^*A - \tfrac{1}{2}B^*B).$$

Since $A, B \in \mathcal{M}$, then B^*A, $B^*B \in \mathcal{A}$, and we have the representation (4.1). Since the first term in (4.1) is a p.d. function, then $\mathcal{J}_2$-continuity of $\mathcal{L}(g)$ follows from the proposition 4.

§5. The case $\mathcal{A} = \mathcal{L}(\mathcal{H})$.

Here we shall describe all $\mathcal{T}_1$-continuous h.c.p.d. functions with values in $\mathcal{F}(\mathcal{L}(\mathcal{H}))_\sigma$. Since the center $\mathcal{Z}$ is one-dimensional then $\mathcal{T}_1$-continuity is equivalent to $\mathcal{T}_2$-continuity by proposition 5.

Since any normal representation of $\mathcal{L}(\mathcal{H})$ is unitarily equivalent to a multiple of the identical representation, we can take in theorem 1

$$\mathcal{K} = \mathcal{H} \otimes \mathcal{H}_0 , \quad \pi[X] = X \otimes I_0,$$

where $\mathcal{H}_0$ is a Hilbert space, I_0 is the unit operator in $\mathcal{H}_0$. From (2.9) it follows that

$$W_g = I \otimes W_g^0,$$

where $g \to W_g^0$ is a continuous unitary representation of G in $\mathcal{H}_0$. Then (3.1), (3.2) imply that

$$B(g)\psi = \psi \otimes b_0(g) ; \quad \psi \in \mathcal{H},$$

where $b_0(g)$ is a continuous cocycle of the representation $g \to W_g^0$ in $\mathcal{H}_0$. Then $B(g)X = X \otimes b_0(g)$, where by $X \otimes \psi$ we denote the operator from $\mathcal{H}$ to $\mathcal{H} \otimes \mathcal{H}_0$, mapping the vector ψ into $X\psi \otimes \psi_0$. The function $z(g)$, $g \in G$, is a real function satisfying

$$z(gh) - z(g) - z(h) = \operatorname{Im}\left(b_0(g^{-1}) \mid b_0(h)\right). \tag{5.1}$$

The representation (3.4) takes the form

$$\begin{aligned}\mathcal{L}(g)[X] = A^*(X \otimes W_g^0)A + A^*(X \otimes b_0(g)) + (X \otimes b_0(g^{-1}))^* A + \\ + X\left[-\tfrac{1}{2}\|b_0(g)\|^2 + i z(g)\right] + C^*X + XC,\end{aligned} \tag{5.2}$$

where A is arbitrary bounded operator from $\mathcal{H}$ to $\mathcal{H} \otimes \mathcal{H}_0$, $C \in \mathcal{L}(\mathcal{H})$.

To obtain "coordinate" representation, let

$$\mathcal{H}_0 = \mathcal{H}_1 \oplus \mathcal{H}_2,$$

where $\mathcal{H}_1 = \{\varphi : W_g^0 \varphi = \varphi, \ \forall g \in G\}$ and let $b_0(g) = b_1(g) \oplus b_2(g)$ be the corresponding decomposition of the cocycle $b_0(g)$. Choose an arbitrary unconditional basis $\{\tau_j\}$ in $\mathcal{H}_1$, and let $\{\tau^j\}$ be the conjugate basis. Denote $\tau_j(g) = (\tau_j \mid b_1(g))$, $\gamma_{jk} = (\tau_j \mid \tau_k)$, $\gamma^{jk} = (\tau^j \mid \tau^k)$. Then $\tau_j(g)$ are continuous morphisms of G into $\mathbb{C}$. Let $\{v_j\}$ be an unconditional basis in $\mathcal{H}_2$ and define $R_j, V_j \in \mathcal{L}(\mathcal{H})$ by the relation

$$A\psi = \left(\sum_j R_j \psi \otimes \tau_j\right) \oplus \left(\sum_k V_k \psi \otimes v_k\right), \quad \psi \in \mathcal{H}. \tag{5.3}$$

Then the series

$$\sum_{j,k} \tilde{r}_{jk} R_j^* R_k \;;\; \sum_{j,k} V_j^* V_k (v_j \,|\, W_g^0 v_k), \; g \in G, \tag{5.4}$$

converge ultraweakly. Substituting (5.3) into (5.2) gives

$$\mathcal{L}(g)[X] = \mathcal{L}_1(g)[X] + \mathcal{L}_2(g)[X] + C^*X + XC, \tag{5.5}$$

where

$$\begin{aligned} \mathcal{L}_1(g)[X] = \sum_{j,k} \tilde{r}_{jk} R_j^* X R_k + \sum_j [R_j^* X \tau_j(g) - \overline{\tau_j(g)} X R_j] - \\ - \frac{1}{2} \sum_{j,k} r^{jk} \overline{\tau_j(g)} \tau_k(g) \cdot X \end{aligned} \tag{5.6}$$

is "Gaussian" part and

$$\begin{aligned} \mathcal{L}_2(g)[X] = \sum_{j,k} V_j^* X V_k (v_j \,|\, W_g^0 v_k) + \\ + \sum_j [V_j^* X (v_j \,|\, b_2(g)) + (b_2(g^{-1}) \,|\, v_j) X V_j] + \ell_2(g) \cdot X, \end{aligned} \tag{5.7}$$

with

$$\ell_2(g) = -\frac{1}{2} \| b_2(g) \|^2 + i z(g).$$

More detailed expression for $\ell_2(g)$ may be obtained for concrete classes of groups. The case of abelian locally compact group is treated in [17], [4]. In this case one obtains a noncommutative genegalization of the Levy-Khinchin formula. We shall not repeat it here and give only an example for the case $G = X = \mathbb{R}$. Consider the function (1.8) where $\mathcal{L}_0[X] = V^*X + XV$ with $V \in \mathcal{A} = \mathcal{B}(\mathcal{H})$. Expanding the exponents in the integral

$$\mathcal{L}(g)[X] = \int_0^\infty (e^{ixg} e^{V^*x} X e^{Vx} - e^{V^*x} e^{Vx} \circ X) \frac{e^{-\beta x}}{x} dx,$$

we get the representation

$$\begin{aligned} \mathcal{L}(g)[X] = \sum_{j,k=1}^\infty (V^*)^j X V^k \int_0^\infty e^{ixg} \frac{x^{j+k-1}}{j!\,k!} e^{-\beta x} dx + \\ + \sum_{j=1}^\infty [(V^*)^j X + X V^j] \int_0^\infty (e^{ixg} - 1) \frac{x^{j-1}}{j!} e^{-\beta x} dx + \\ + X \int_0^\infty (e^{ixg} - 1) x^{-1} e^{-\beta x} dx + C^*X + XC, \end{aligned} \tag{5.8}$$

with

$$C = \frac{1}{2} \ln \left(1 - \frac{2 \operatorname{Re} V}{\beta}\right) - \ln \left(1 - \frac{V}{\beta}\right).$$

The "Gaussian" part in (5.8) is absent and the first three terms correspond to (5.7).

"Gaussian" functions of the form (5.6) for $G = \mathbb{R}^s$ appeared first in the work of Barchielli, Lanz, Prosperi [1],[2] in connection with the problem of continuous quantum measurement. More general expressions including "Poisson" terms where considered by Barchielli and Lupieri [18] and Parthasarathy [19] who introduced a class of functions close to (5.2). If $\mathcal{L}(g)$ is a scalar h.c.p.d. function, then the representation corresponding to the "factorizable" p.d. function $\Phi(g(\cdot)) = \exp \int \mathcal{L}(g(t))\,dt$ on the group of G-valued functions of t is equivalent to a special representation acting in the Fock space (the precise statement is the Araki-Woods embedding theorem [13],[8],[14]). From the present point of view the main mathematical result of [18],[19] is an extension of this embedding theorem to "factorizable" p.d. functions with values in $\mathcal{F}(\mathcal{L}(\mathcal{H}))$, given by the time-ordered exponentials

$$\Phi(g(\cdot)) = \mathcal{T}\exp \int \mathcal{L}(g(t))\,dt.$$

References

1. Barchielli, A., Lanz, L., Prosperi, G.M.: A model for macroscopic description and continuous observations in quantum mechanics. Nuovo Cimento, 72B, 79-121 (1982).
2. Barchielli, A., Lanz, L., Prosperi, G.M.: Statistics of continuous trajectories in quantum mechanics: operation-valued stochastic procrsses. Found.Phys., 13, 779-812 (1983).
3. Holevo, A.S.: Infinitely divisible measurements in quantum probability. Teor. veroyat. i ee primen. 31, 560-564 (1986) (In Russian)
4. Holevo, A.S.: Conditionally positive definite function in quantum probability. Proc. of the International Congress of Mathematicians, Berkeley, 1986.
5. Evans, E., Hanche-Olsen, H.: The generators of positive semigroups. Journ. Funkt. Anal. 32, 207-212 (1979).
6. Schoenberg, I.J.: Metric spaces and positive definite functions. Trans. Amer. Math. Soc., 4, 522-530 (1938).
7. Gelfand, I.M., Vilenkin, N.Ya.: Generalized functions, vol.4. NY-London: Academic Press 1964.
8. Parthasaraty, K.R., Schmidt, K.: Positive definite kernels, continuous tensor products, and central theorems of probability theory. Lecture Notes in Math., 272. Berlin-Heidelberg-NY:Springer 1972.
9. Lindblad, G.: On the generators of quantum dynamical semigroups. Commun. Math. Phys., 48, 119-130 (1976).

10. Evans, D.E., Lewis, J.T.: Dilations of irreversible evolutions in algebraic quantum theory. Commun. of the Dublin Institute of Advanced Studies, Ser.A, v.24, 1977.
11. Yuan, J.: On the construction of one-parameter semigroups in topological semigroups. Pacif. J.Math., 65, 285-292 (1976).
12. Heyer, H.: Probability measures on locally compact groups. Berlin-Heidelberg-NY: Springer 1977.
13. Araki, H.: Facrotizable representations of current algebra. Publ. RIMS Kyoto Univ. 5, 361-422 (1970).
14. Guichardet, A.: Symmetric Hilbert spaces and related topics. Lecture Notes in Math. 261. Berlin-Heidelberg-NY: Springer 1972.
15. Christensen, E., Evans, D.E.: Cohomology of operator algebras and quantum dynamical semigroups. J. London Math. Soc. 20, 358-368 (1979).
16. Nakai, M.: Some expectations in C^*-algebras. Proc. Japan Acad. 34, 411-416 (1958).
17. Holevo, A.S.: Levy-Khinchin-type representations in quantum probability. Teor. veroyat. i ee primen. 32, 142-146 (1987) (in Russian).
18. Barchielli, A., Lupieri, G.: Dilations of operation-valued stochastic processes. Lect. Notes Math., 1136, 57-66 (1985).
19. Parthasarathy, K.R.: On parameter semigroups of completely positive maps on groups arising from quantum stochastic differential equations. Bull. della Unione Matem. Italiana (6), 5-A, (1986).

CONTRACTION SEMIGROUPS IN L^2 OVER A VON NEUMANN ALGEBRA

by

Ryszard JAJTE

Institute of Mathematics, Łódź University, ul. Banacha 22, 90-238 Łódź, POLAND

1. We are going to discuss the asymptotic behaviour of some contraction semigroups in a Hilbert space which are generated by the quantum-dynamical systems in von Neumann algebras. The asymptotics will be described in the spirit of the classical individual ergodic theorems. The results presented here are related to [2], [3], [6]. We begin with some notation. Let M be a von Neumann algebra acting in a Hilbert space H with a cyclic and separating unit vector ξ. We consider a weak*-continuous semigroup $\alpha = (\alpha_t)_{t\geq 0}$ of linear maps of M with $\alpha_0 = I$ (identity map), $\alpha_t(1) = 1$, and such that α_t are Schwarz maps, i.e. $\alpha_t(x^*x) \geq \alpha_t(x)^*\alpha_t(x)$, for all $x \in M$ and $t \geq 0$. We assume that the state $\Phi = \omega_\xi$ is α-invariant. In particular, it follows that all α_t are positive normal maps of M.

2. With the dynamical system (M,α,Φ) we associate a semigroup $\beta = (\beta_t)_{t\geq 0}$ of contractions in H. Namely, we put $\beta_t(x\xi) = \alpha_t(x)\xi$, for $x \in M$, $t \geq 0$. Then after a standard unique extension to H, we obtain a strongly continuous contraction semigroup $\beta = (\beta_t)_{t\geq 0}$. In the sequel, $\|.\|$ denotes the norm in H and $\|.\|_\infty$ is the norm in M. We put, for $x \in M$, $|x|^2 = x^*x$. For $\eta \in H$ and a projection $p \in M$, we put $\|\eta\|_p = \inf\{\|\sum_{k=1}^{\infty} x_k p\|_\infty\}$, where the infimum is taken over all sequences (x_k) M such that $\sum_{k=1}^{\infty} x_k p$ converges (in norm) in M and $\sum_{k=1}^{\infty} x_k\xi = \eta$ in H. We adopt the following definition introduced in [2].

<u>Definition</u>. For $\eta, \eta_n \in H$, we say that $\eta_n \to \eta$ almost surely (a.s.) if for every strong neighbourhood U of the unity in M one can find a projection $p \in U$ such that $\|\eta_n - \eta\|_p \to 0$ as $n \to \infty$.

In the commutative case of $M = L_\infty(\Omega,\mu)$ the convergence just defined coincides (via Egorov's theorem) with the usual almost everywhere convergence.

3. For the semigroup $\beta = (\beta_t)_{t\geq 0}$ defined above we shall prove the following two theorems.

THEOREM 1. For every $h \in H$, there exists the limit

$$\lim_{T\to\infty} T^{-1} \int_0^T \beta_t(h) = Eh \quad \text{almost surely,}$$

where E is the (unique) extension of the conditional expectation of M onto the von Neumann algebra M^α of α-invariant elements of M.

THEOREM 2. For every $h \in H$, there exists the limit

$$\lim_{T\to 0} T^{-1} \int_0^T \beta_t(h)dt = h \quad \text{a.s.}$$

We start with some extension of Goldstein's maximal ergodic lemma [1] or [4], Th. 2.2.12

LEMMA. Let (α_t) be the semigroup defined in Section 1. Let $(x_n) \subset M_+$, $\varepsilon_n > 0$ $(n = 1,2,\ldots)$. Then there exists a projection $p \in M$ such that

$$\Phi(1 - p) \leq 2\Sigma\varepsilon_n^{-1}\Phi(x_n)$$

and

$$\|p \int_0^T \alpha_t(x_n)dtp\|_\infty \leq 4T\varepsilon_n \quad \text{for } T \geq 1, \quad n = 1,2,\ldots .$$

P r o o f. Put $A_n = \int_0^1 \alpha_t(x_n)dt$ and $s_n = n^{-1} \sum_{k=0}^{n-1} \alpha_1^k$. Then we have, for $N \leq T < (N + 1)$

$$T^{-1} \int_0^T \alpha_t(x_n)dt = T^{-1}[A_n + \alpha_1(A_n) + \ldots + \alpha_1^{N-1}(A_n) + \int_N^T \alpha_t(x_n)dt] \leq ((N + 1)/N)s_{N+1}(A_n).$$

By the Goldstein's maximal lemma [1], there is a projection $p \in M$ such that $\Phi(1-p) \leq \sum_{n=1}^{\infty} \varepsilon_n^{-1}\Phi(A_n) = \sum_{n=1}^{\infty} \varepsilon_n^{-1}\Phi(x_n)$ and $\|ps_{N+1}(A_n)p\|_\infty < 2\varepsilon_n$, $N = 1,2,\ldots$, $n = 1,2,\ldots$. Moreover, $\|pT^{-1}\int_0^T \alpha_t(x_n)dtp\|_\infty \leq 2\|ps_{N+1}(A_n)p\|_\infty \leq 4\varepsilon_n$, for $n = 1,2,\ldots$; $T \geq 1$, which ends the proof.

P r o o f of Theorem 1. Let (α_t) and (β_t) be the semigroups defined in Sections 1 and 2. Take $(\xi_k) \subset H$, $\varepsilon_k > 0$. Then there is a projection $p \in M$ such that $\Phi(1-p) \leq 4\sum_{k=1}^{\infty} \varepsilon_k^{-1}\|\xi_k\|^2$ and $\|\int_0^T \beta_t(\xi_k)dt\|_p < 5T\varepsilon_k^{\frac{1}{2}}$, for $k = 1,2,\ldots$; $T \geq 1$. Indeed, let us fix $(x_{kl}) \subset M$ in such a way that $\xi_k = \sum_{l=1}^{\infty} x_{kl}\xi$ and $\|x_{kl}\xi\| < 2^{-l+1}\|\xi_k\|$ for $k,l = 1,2,\ldots$ Put $\varepsilon_{kl} = \varepsilon_k 2^{-l+1}$. Then we have $\sum_{k=1}^{\infty}\sum_{l=1}^{\infty} \varepsilon_{kl}^{-1}\Phi(|x_{kl}|^2) \leq 2\sum_{k=1}^{\infty} \varepsilon_k^{-1}\|\xi_k\|^2$. By Lemma, there exists a projection $p \in M$ such that $\Phi(1-p) \leq 4\sum_k \varepsilon_k^{-1}\|\xi_k\|^2$ and

$$(*) \qquad \|p\int_0^T \alpha_t(|x_{kl}|^2)dtp\|_\infty < 2^{-l+2}T\varepsilon_k$$

holds for $k,l = 1,2,\ldots$; $T \geq 1$.

Using $(*)$ and [2], Lemma 1.3, we obtain

$$\|\int_0^T \beta_t(\xi_k)dt\|_p \leq \sum_{s=1}^{\infty} \|\int_0^T \alpha_t(x_{ks})dtp\|_\infty = \sum_{s=1}^{\infty} \|p|\int_0^T \alpha_t(x_{ks})dt|^2 p\|_\infty^{\frac{1}{2}} \leq$$

$$\leq T^{\frac{1}{2}} \sum_{s=1}^{\infty} \|p\int_0^T \alpha_t(|x_{ks}|^2)dtp\|_\infty^{\frac{1}{2}} < 5T\varepsilon_k^{\frac{1}{2}},$$

for $T \geq 1$ and $k = 1,2,\ldots$.

Let $\varepsilon > 0$. Fix $\varepsilon_k > 0$ with $\sum_{k=1}^{\infty} \varepsilon_k < \varepsilon/4$. Put $Eh = \tilde{h}$. We have $H \ominus E(H) = [(x - \alpha_t x)\xi : t \geq 0, x \in M]^-$. Thus we can find $(\eta_k) \subset H$, $(J_k) \subset M$, $t_k > 0$ such that $\|\eta_k\| < \varepsilon_k$ and

$$h = \tilde{h} + \eta_k + (J_k - \alpha_{t_k} J_k)\xi \quad \text{for } k = 1,2,\ldots .$$

We have $\sum_k \varepsilon_k^{-1}\|\eta_k\|^2 < \varepsilon/4$, thus there exists a projection $p \in M$ with $\Phi(1-p) < \varepsilon$ and such that

$$\|T^{-1}\int_0^T \beta_t(\eta_k)dt\|_p < 5\varepsilon_k^{\frac{1}{2}} \quad \text{for } T \geq 1 \text{ and } k = 1,2,\ldots .$$

Let $\delta > 0$. Fixing k large enough we obtain

$$\|T^{-1}\int_0^T (\beta_t(h) - \tilde{h})dt\|_p \leq \|T^{-1}\int_0^T \beta_t(\eta_k)dt\|_p +$$
$$+ \|T^{-1}\int_0^T \beta_t(J_k - \alpha_{t_k} J_k)\xi\|_p \leq \frac{\delta}{2} + \frac{1}{T}\|\int_0^T \alpha_t(J_k - \alpha_{t_k} J_k)dtp\|_\infty <$$
$$< \delta \qquad \text{for } T \text{ large enough.}$$

The application of [5], Prop. 5.3 ends the proof of the almost sure convergence. The projection E is given by the mean ergodic theorem. The fact that E is the extension of the conditional expectation onto M^α is well-known (see for ex. [6]).

P r o o f of Theorem 2. Let $h \in H$. Put $B_T(h) = T^{-1}\int_0^T \beta_t(h)dt$. Take $x_k \in M$ such that, for $J_k = x_k\xi$, we have $\sum_k \|h - J_k\|^2 < \infty$. Next, we fix the positive integers $n_s^{(k)}$ such that for $J_s^{(k)} = n_s^{(k)} \int_0^{(n_s^{(k)})^{-1}} \alpha_t(x_k)dt$ we have $\sum_s \|J_k - J_s^{(k)}\xi\|^2 < 2^{-k}$, $(k = 1,2,\dots)$. For every projection $p \in M$, we have the following estimation

$$(*) \qquad \|B_T(h) - h\|_p \leq \|B_T(h - J_k)\|_p + \|B_T(J_k - J_s^{(k)}\xi)\|_p +$$
$$+ \|B_T(J_s^{(k)}\xi) - J_s^{(k)}\xi\|_p + \|J_s^{(k)}\xi - J_k\|_p + \|J_k - h\|_p.$$

Arranging the elements $h - J_k$, $J_k - J_s^{(k)}\xi$, $B_T(h - J_k)$, $B_T(J_k - J_s^{(k)}\xi)$ $(k,s=1,2,\dots$ into one sequence $\{\eta_1,\eta_2,\dots\}$ and applying [2], Corollary 1.6 to $\{\eta_k\}$, for every $\varepsilon > 0$ we find a projection $p \in M$ with $\Phi(1 - p) < \varepsilon$ and such that $\|\eta_k\|_p \to 0$ as $k \to \infty$.

Moreover, by [6], Lemma 6.2, we have

$$\|B_T(J_s^{(k)}\xi) - J_s^{(k)}\xi\|_p \leq T^{-1}\|\int_0^T (\alpha_t(J_s^{(k)}) - J_s^{(k)})dt\|_\infty \to 0$$

as $T \to 0$, for all k and s.

All this implies that $B_T(h) \to h$ a.s. as $T \to 0$. Indeed, for $\varepsilon > 0$ we find p with $\Phi(1 - p) < \varepsilon$ and such that $\|\eta_k\|_p \to 0$ as $k \to \infty$. Now, for $\delta > 0$, to obtain $\overline{\lim_{T\to 0}} \|B_T(h) - h\|_p \leq \delta$, it suffices to fix in (*) k large enough and then take s large enough and at last to pass with T to zero. The proof is completed.

REFERENCES

[1] M.S. Goldstein, Theorems in almost everywhere convergence in von Neumann algebras (Russian), J. Oper. Theory 6 (1981), 233-311.

[2] E. Hensz and R. Jajte, Pointwise convergence theorems in L_2 over a von Neumann algebra, Math. Z. 193 (1986), 413-429.

[3] R. Jajte, Ergodic theorems in von Neumann algebras, Semesterbericht Funktionalanalysis Tübingen, Sommersemester 86, Band 10, 135-144.

[4] R. Jajte, Strong limit theorems in non-commutative probability, Lecture Notes in Math., vol. 1110, Springer-Verlag, Berlin-Heidelberg-New York-Tokyo 1985.

[5] M. Takesaki, Theory of operator algebras I, Springer - Verlag, Berlin-Heidelberg-New York 1979.

[6] S. Watanabe, Ergodic theorems for dynamical semigroups on operator algebras, Hokkaido Math. J 8 (1979), 176-190.

Survey on a Theory of Non-Commutative Stationary Markov Processes *)

by

Burkhard Kümmerer
Mathematisches Institut
Universität Tübingen
Auf der Morgenstelle 10
D-7400 Tübingen
West Germany

Abstract: Dilations are discussed from the point of view of a theory of non-commutative stationary Markov processes. We show that a theory of stationary Markov processes can be formulated in terms of certain dilations, and we put various results on dilations into the systematic context of such a theory. Finally, we discuss a class of non-commutative Poisson processes.

Introduction. In a series of papers ([Kü 1] to [Kü 7], [Kü 9] to [Kü 11]) we have considered various aspects of dilations for completely positive operators on W*-algebras. In the present survey we try to give a unified view of these results which shows how they fit into the context of a theory of non-commutative stationary Markov processes. The motivation for these investigations are manifold:
From a mathematical point of view it is natural to study the structure of a dilation. As a shining example of a dilation theory there is the theory of unitary dilations (see [Sz.-N]) which forms an important tool for the investigation of contractions on a Hilbert space. Therefore, dilations for completely positive operators seem to form a natural part of a theory of completely positive operators.
Our main impetus, however, came from two interpretations of the structure of a dilation.
In a statistical theory of irreversible processes, the structure of a dilation seems appropriate in order to describe the relation between an irreversible open thermodynamical system and the larger reversible system in which the open system is contained. This point of view was first

*) This paper is part of a research project which is supported by the Deutsche Forschungsgemeinschaft.

worked out by J.T. Lewis and raised the interest in a non-commutative dilation theory. For the early development of this theory we refer to [Lew] and [Da 1], [Da 2], [Em 2], [Ev 1] ,[Ev 3], [Ev 4].
Another interpretation of a dilation is a probabilistic one: The Markov process for a semigroup of transition operators can easily brought into the form of a dilation [Ker]. From this point of view, dilations on C*-algebras are part of a theory of non-commutative stochastic processes as it is needed for a probabilistic description of quantum systems. In this context our axioms may be considered as the axioms for a theory of non-commutative stationary (Markov) processes. It is this interpretation which we want to emphasize in the present survey.
Let us briefly discuss the contents of this paper.
After having introduced the notation we state the axioms in § 1. In § 2 we work out an interpretation of a dilation in terms of a theory of non-commutative stationary processes. Here we discuss the notions of non-commutative probability spaces, random variables, stationary processes, dilations, and Markov processes.
The paragraphs § 3, § 4, and § 5 give a survey on various results on dilations from the point of view of a theory for stationary processes: In § 3 we report on some general theory. In particular, an asymptotic analysis can be carried through such that it parallels the commutative theory. In §4 we collect some results on the existence of non-commutative stationary Markov processes while §5 discusses their general structure. Finally, in § 6 an account is given on a class of non-commutative Poisson processes which has recently been introduced.
We also mention that under the same title there exists as a preprint an extended version of the present paper.

Acknowledgements: Considerable parts of this paper originate from a year which the author spent at King's College, London. This stay was financed by a research grant of the Deutsche Forschungsgemeinschaft which is gratefully acknowledged. It is a pleasure to thank Prof. E.B. Davies and all his colleagues, particularly Dr. M. Lindsay and Dr. I. Wilde for their warm hospitality throughout that year.

§ 1 The Axioms

1.1 Notation. Let us first agree on some standard notational conventions. Given a W*-algebra A we denote its identity by 1, or 1_A if necessary, and its predual by A_* . The elements in A_* are called normal.

On a W*-algebra $\mathcal{A}$ beside its norm or uniform topology we will mainly consider the weak* topology $\sigma(\mathcal{A},\mathcal{A}_*)$ and the strong topology which is generated by the seminorms $\psi(x^*x)^{1/2}$ for $x \in \mathcal{A}$, where ψ runs through the normal states in $\mathcal{A}_*$.
If $\mathcal{B}$ is another W*-algebra, a linear operator $T: \mathcal{A} \to \mathcal{B}$ is said to be normal if it is continuous for the weak* topologies on $\mathcal{A}$ and $\mathcal{B}$. In this case its preadjoint, being an operator from $\mathcal{B}_*$ into $\mathcal{A}_*$, is denoted by T_*. Id or $\mathrm{Id}_{\mathcal{A}}$ will stand for the identity operator on $\mathcal{A}$.
A topology $\mathcal{T}$ on $\mathcal{A}$ induces canonically the topology of pointwise $\mathcal{T}$ convergence on the operators on $\mathcal{A}$, i.e., pointwise $\mathcal{T}\text{-}\lim_\alpha T_\alpha = T$ if $\mathcal{T}\text{-}\lim_\alpha T_\alpha(x) = T(x)$ for all $x \in \mathcal{A}$.
If $\mathcal{A}$ and $\mathcal{B}$ are W*-algebras then $\mathcal{A}\otimes\mathcal{B}$ will always denote their W*-tensor product.
For further standard notation we refer to the books of G.K. Pedersen ([Ped]), S. Sakai ([Sak]), and M. Takesaki ([Ta 2]). In particular, the definitions and basic properties of completely positive operators may be found in ([Ta 2], IV.3).

1.2 Morphisms and dynamical systems. As the objects of a category we consider pairs $(\mathcal{A},\phi)$ consisting of a W*-algebra $\mathcal{A}$ and a faithful normal state ϕ on $\mathcal{A}$. An object $(\mathcal{A},\phi)$ will sometimes be called a non-commutative probability space.
A morphism $T: (\mathcal{A}_1,\phi_1) \to (\mathcal{A}_2,\phi_2)$ is a completely positive operator $T: \mathcal{A}_1 \to \mathcal{A}_2$ satisfying $T(1) = 1$ and $\phi_2 \circ T = \phi_1$. In particular, a morphism is a normal operator since its adjoint T^* satisfies $T^*((\mathcal{A}_2)_*) \subset (\mathcal{A}_1)_*$.
If T is a morphism of $(\mathcal{A},\phi)$ into itself then we call T a morphism of $(\mathcal{A},\phi)$ and by $\mathcal{M}(\mathcal{A},\phi)$ we denote the set of all morphisms of $(\mathcal{A},\phi)$.
If $P \in \mathcal{M}(\mathcal{A},\phi)$ satisfies $P^2 = P$ then it follows that $\mathcal{B} := P(\mathcal{A})$ is a W*-subalgebra of $\mathcal{A}$ ([Kü 8], 2.4) and hence P is a conditional expectation (cf. [Ta 2], III.3.4). In particular, $P(x\cdot y\cdot z) = x\cdot P(y)\cdot z$ for $x, z \in \mathcal{B}$, $y \in \mathcal{A}$. In such a situation we will say that $P: (\mathcal{A},\phi) \to \mathcal{B}$ is a conditional expectation.
As we will sometimes consider discrete and continuous times simultaneously, we write $\mathbb{T}$ for $\mathbb{R}$ or $\mathbb{Z}$ while $\mathbb{T}_+ = \mathbb{R}_+$ or $\mathbb{Z}_+$.
If $(T_\tau)_{\tau\in\mathbb{T}_+} \subset \mathcal{M}(\mathcal{A},\phi)$ is a semigroup of morphisms which is continuous in the pointwise weak* topology then we call $(\mathcal{A},\phi,(T_\tau)_{\tau\in\mathbb{T}_+})$ a (continuous, resp. discrete) dynamical system which will usually be denoted simply by $(\mathcal{A},\phi,T_\tau)$, or by $(\mathcal{A},\phi,T)$ with $T := T_1$ if the time is discrete. In case that $(T_\tau)_{\tau\in\mathbb{T}_+}$ are automorphisms, we call the corresponding dynamical system reversible.

1.3 Definition. For a dynamical system $(A,\phi,(T_\tau)_{\tau\in\mathbb{T}_+})$ consider the following diagram

$$\begin{array}{ccc} (A,\phi) & \xrightarrow{\;T_\tau\;} & (A,\phi) \\ {\scriptstyle i}\downarrow & & \uparrow{\scriptstyle P} \\ (\hat{A},\hat{\phi}) & \xrightarrow[\;\hat{T}_\tau\;]{} & (\hat{A},\hat{\phi}) \end{array}$$

where $(\hat{A},\hat{\phi},(\hat{T}_\tau)_{\tau\in\mathbb{T}})$ is a reversible dynamical system and $i:(A,\phi)\to(\hat{A},\hat{\phi})$, $P:(\hat{A},\hat{\phi})\to(A,\phi)$ are morphisms.

Then $(\hat{A},\hat{\phi},\hat{T}_\tau;P)$ or $(\hat{A},\hat{\phi},\hat{T}_\tau;i,P)$ is called a

(i) (stationary) process over (A,ϕ) if the diagram commutes for $\tau = 0$,

(ii) dilation of first order of (A,ϕ,T) if $\mathbb{T} = \mathbb{Z}$, $T = T_1$, and the diagram commutes for $\tau = 0$ and $\tau = 1$,

(iii) dilation of (A,ϕ,T_τ) if the diagram commutes for all $\tau \in \mathbb{T}_+$.

We sometimes talk about a discrete or continuous process accordingly to whether $\mathbb{T} = \mathbb{Z}$ or $\mathbb{T} = \mathbb{R}$.

1.4 Remark. Let $(\hat{A},\hat{\phi},\hat{T}_\tau;P)$ be a process over (A,ϕ). Then we have $P\circ i = Id_A$ and hence $(i\circ P)^2 = i\circ P$. Therefore, $i\circ P$ is a conditional expectation from $(\hat{A},\hat{\phi})$ onto $i(A)$. It easily follows that i is an injective *-homomorphism.

Given P then i is uniquely determined and conversely. While constructing a process it is therefore enough to define either of them. We will call i (resp. P) the injection (resp. projection) corresponding to P (resp. i). In particular, it suffices to denote a process by $(\hat{A},\hat{\phi},\hat{T}_\tau;P)$ and we will use an expression like $(\hat{A},\hat{\phi},\hat{T}_\tau;i,P)$ only if we need to name the injection explicitly.

1.5 Markov Property. Let $(\hat{A},\hat{\phi},\hat{T}_\tau;i,P)$ be a process over (A,ϕ). For $I\subset\mathbb{T}$ we denote by A_I the W*-subalgebra of $\hat{A}$ generated by $\bigcup_{\tau\in I}\hat{T}_\tau\circ i(A)$. In ([Kü 1],2.1.3) it is shown that there exists a conditional expectation $P_I: (\hat{A},\hat{\phi}) \to A_I$.

Definition. A process $(\hat{A},\hat{\phi},\hat{T}_\tau;P)$ over (A,ϕ) is called minimal if $\hat{A} = A_{\mathbb{T}}$. It is called a Markov process if for all $x \in A_{[0,\infty)}$:
$P_{\{0\}}(x) = P_{(-\infty,0]}(x)$.

Note that from a process $(\hat{A},\hat{\phi},\hat{T}_\tau;P)$ one obtains a minimal process by restricting to $A_{\mathbb{T}}$.

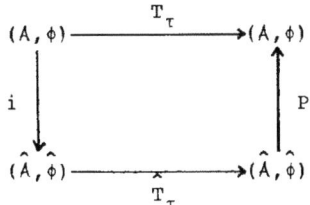

§ 2 Non-Commutative Stationary Processes

The purpose of this paragraph consists in showing that our axioms may indeed serve as the basis for a non-commutative theory of stationary processes. In a more general context a related discussion can be found in [Ac 1], [Ac 3] .

2.1 Probability spaces. In classical probability a system of interest is generally described by a probability space (Ω,Σ,μ). The set Ω represents the possible states of the system, Σ is the σ-algebra of all possible events and μ is a probability measure on (Ω,Σ), assigning to an element of Σ the probability of its occurence.
It is well known, that in a quantum mechanical description of a physical system the set of events can no longer be identified with the elements of a Boolean lattice. Instead, in traditional quantum mechanics the events are usually identified with the lattice of closed subspaces of some complex Hilbert space $\mathcal{H}$, or equivalently, the lattice of projections in the W*-algebra $\mathcal{B}(\mathcal{H})$ (cf., e.g., [v.N], [Var]).
More recently, the needs to deal with systems having an infinite number of degrees of freedom, as they occur in quantum statistical mechanics and quantum field theory, and the search for a unified description of classical mechanics and quantum mechanics led to a description of the events by the lattice of projections in a W*-algebra (von Neumann algebra) $\mathcal{A}$ (cf., e.g., [Bra], [Em 1], [Pri]).
The description by a probability space (Ω,Σ,μ) can be integrated into this approach by turning to the W*-algebra $\mathcal{A} = L^{\infty}(\Omega,\Sigma,\mu)$. A C*-algebraic description of some physical system can be subsumed by turning to an appropriate completion.
Therefore, in the following we adopt the point of view that the events of some system are described by the projections in a W*-algebra.
In such a description the probability measure μ is replaced by a function τ on the set of projections in $\mathcal{A}$ with values in $[0,1] \subset \mathbb{R}$ satisfying certain axioms. Gleason type theorems assure that there exists a normal state ϕ on $\mathcal{A}$ such that $\tau(p) = \phi(p)$ for all projections p in $\mathcal{A}$. For a discussion of Gleason's theorem on $\mathcal{B}(\mathcal{H})$ we refer to [Var], the general case is treated in [Chr], [Ye 1], [Ye 2].
In the following we assume, in addition, that with respect to a a priori reference probability measure ϕ each event occurs with a non-vanishing probability, i.e., $\phi(p) \neq 0$ for each projection $p \in \mathcal{A}$, which means that ϕ is faithful.
Thus our considerations equip the object $(\mathcal{A},\phi)$ with an interpretation as a non-commutative probability space.

2.2 Random variables. In classical probability theory a world, described by some probability space $(\hat{\Omega},\hat{\Sigma},\hat{\mu})$, influences a system of interest, described by a state space (Ω,Σ), via a measurable random variable $X: \hat{\Omega} \to \Omega$ which induces a probability measure μ on (Ω,Σ). The observation of the system (Ω,Σ,μ) is described by observables, i.e., measurable functions $f: (\Omega,\Sigma) \to \mathbb{R}$.

This situation may be sketched as follows:

$$(\hat{\Omega},\hat{\Sigma},\hat{\mu}) \xrightarrow{\quad X \quad} (\Omega,\Sigma,\mu) \xrightarrow{\qquad f \qquad} \mathbb{R}$$

a world influences a system which is observed on a scale

An algebraic reformulation of a random variable can be obtained as follows ([Ac 1], [Ac 3]): The random variable X induces an injective $*$-homomorphism $i_X: L^\infty(\Omega,\Sigma,\mu) \to L^\infty(\hat{\Omega},\hat{\Sigma},\hat{\mu}): f \mapsto f\circ X$.

It may be considered as an embedding of the uniformly bounded observables of our system into the uniformly bounded observables of $(\hat{\Omega},\hat{\Sigma},\hat{\mu})$.

Conversely, if (Ω,Σ,μ) is a standard Borel space then an injective $*$-homomorphism $i: L^\infty(\Omega,\Sigma,\mu) \to L^\infty(\hat{\Omega},\hat{\Sigma},\hat{\mu})$ with $\int_{\hat{\Omega}} i(f)\cdot d\hat{\mu} = \int_\Omega f\cdot d\mu$ for all $f \in L^\infty(\Omega,\Sigma,\mu)$ determines $\hat{\mu}$-almost everywhere a random variable X such that $i = i_X$ (cf. [Ac 1]). Therefore, we have obtained an equivalent formulation of the notion of a random variable which allows the following generalization.

Definition. Given a non-commutative probability space $(\hat{A},\hat{\phi})$ and a W*-algebra A then a random variable ("with values in A") is an injective $*$-homomorphism $i: A \to \hat{A}$, such that there exists a conditional expectation from $(\hat{A},\hat{\phi})$ onto $i(A)$.

Remarks. 1. If $\hat{A}$ is commutative then the conditional expectation onto $i(A)$ automatically exists, hence our definition generalizes the commutative notion of a random variable. For non-commutative $\hat{A}$, however, the existence of this conditional expectation is a non-trivial condition (cf. [Ta 1]), which is not required in [Ac 1], [Ac 3].

2. Since the conditional expectation is normal, it follows that $i(A)$ is a W*-algebra, hence i is normal (cf. [Ta 2], III.3.10).

Given a random variable $i: A \to \hat{A}$, then we define the state ϕ on A by $\phi(x) := \hat{\phi}(i(x))$ for $x \in A$. It follows that ϕ is a faithful normal state on A and i is a morphism from (A,ϕ) into $(\hat{A},\hat{\phi})$.

Therefore, in our definition of a stationary process in 1.3 the morphism i occuring there is a random variable. Indeed, the required commutativity of this diagram for $\tau = 0$ just reproduces our definition of a random variable as we discussed in 1.4.

2.3 Stationary processes. Given these notions of non-commutative probability spaces and random variables it is canonical to proceed to other basic notions of probability theory.
A process will be a family $(i_\tau)_{\tau\in\mathbb{T}_+}$ of random variables $i_\tau: A \to (\hat{A},\hat{\phi})$. It will be called a stationary process, if the multiple time correlations depend only on the time differences, i.e., $\hat{\phi}(i_{\tau_1+s}(x_1)\cdot \ldots \cdot i_{\tau_n+s}(x_n))$ does not depend on s for $x_j \in A$, τ_j, $s \in \mathbb{T}_+$, $1 \leq j \leq n$. In particular, the process is a family of injective *-homomorphisms $i_\tau: (A,\phi) \to (\hat{A},\hat{\phi})$ for some fixed ϕ.
Given a stationary process $(i_\tau)_{\tau\in\mathbb{T}_+}$ with $i_\tau: (A,\phi) \to (\hat{A}_+,\hat{\phi}_+)$ such that $\hat{A}_+$ is generated by the subalgebras $i_\tau(A)$, $\tau \in \mathbb{T}_+$, then one proceeds analogously to the commutative theory in order to prove the following:

1. There exists a time translation, i.e., there is a semigroup $(\hat{T}_{+,\tau})_{\tau\in\mathbb{T}_+}$ of injective *-homomorphisms of $(\hat{A}_+,\hat{\phi}_+)$ such that $i_\tau = (\hat{T}_{+,\tau})\circ i_0$ for $\tau \geq 0$ (cf. [Ac 3]).
2. The time translation can be extended to negative times, i.e., there exists a non-commutative probability space $(\hat{A},\hat{\phi})$ and a group $(\hat{T}_\tau)_{\tau\in\mathbb{T}}$ of *-automorphisms of $(\hat{A},\hat{\phi})$, such that $\hat{A}_+ \subset \hat{A}$, $\hat{\phi}_+ = \hat{\phi}|\hat{A}_+$, $\hat{T}_{+,\tau} = \hat{T}_\tau|\hat{A}_+$. Moreover, assuming that $\hat{A}$ is generated by the algebras $\hat{T}_\tau\circ i(A)$, $\tau \in \mathbb{T}$, (this can always be acchieved by restriction) then $\hat{A}$, $\hat{\phi}$, and $(\hat{T}_\tau)_{\tau\in\mathbb{T}}$ are uniquely determined by $\hat{A}_+$, $\hat{\phi}_+$, and $(\hat{T}_{+,\tau})_{\tau\in\mathbb{T}_+}$. The construction of this extension means to construct a dilation for $(A_+,\phi_+,T_{+,\tau})$ which has been done in ([Kü 6], 2.1.9).
3. If $\tau \to i_\tau(x)$ is weak* continuous for $x \in A$ then the automorphism group $(\hat{T}_\tau)_{\tau\in\mathbb{T}}$ is pointwise weak* continuous ([Kü 6], 1.3.5).

Therefore, using these three results we are led to our definition of a stationary process $(\hat{A},\hat{\phi},\hat{T}_\tau;i,P)$ over (A,ϕ) in 1.3. The injection i represents the random variable at time 0 and the automorphism group $(\hat{T}_\tau)_{\tau\in\mathbb{T}}$ implements the time translation of the process from which the random variable at time τ can be recovered by $i_\tau = \hat{T}_\tau\circ i$. In particular, for commutative $\hat{A}$ the definition in 1.3 is equivalent to the usual definition of a stationary process.
In our definition we made, for continuous time, the pointwise weak* continuity of the process already to a part of its definition since other stationary processes are of little interest. However, this condition could easily be omitted if necessary.

2.4 Transition operators. Consider now a stationary process $(\hat{A},\hat{\phi},\hat{T}_\tau;i,P)$ over (A,ϕ). A transition operator $T_{s,t}$ which describes the transitions between time s and time t $(0 \leq s \leq t)$ will be an operator $T_{s,t}: A \to A$

which should at least satisfy the relation

$$\phi(p\cdot T_{s,t}(q)\cdot p) = \hat{\phi}(\hat{i}_s(p)\cdot \hat{i}_t(q)\cdot \hat{i}_s(p)) \quad (p,q \text{ projections in } A) \qquad (*)$$

Since $i_\tau = \hat{T}_\tau \circ i$ and $\hat{\phi}$ is invariant under $\hat{T}_\tau$, our definition of a stationary process yields for the identity (*)

$$\begin{aligned}\phi(p\cdot T_{s,t}(q)\cdot p) &= \hat{\phi}(i_s(p)\cdot i_t(q)\cdot i_s(p)) = \hat{\phi}(\hat{T}_s\circ i(p)\cdot \hat{T}_t\circ i(q)\cdot \hat{T}_s\circ i(p))\\ &= \hat{\phi}(i(p)\cdot \hat{T}_{t-s}\circ i(q)\cdot i(p)) = \phi(P(i(p)\cdot \hat{T}_{t-s}\circ i(q)\cdot i(p)))\\ &= \phi(p\cdot P\circ \hat{T}_{t-s}\circ i(q)\cdot \ p) \quad .\end{aligned}$$

Since ϕ is faithful and this identity should hold for all projections p, q in A, it follows that $T_{s,t}$ is necessarily given by $T_{s,t} = P\circ \hat{T}_{t-s}\circ i$. In particular, $T_{s,t}$ is a morphism of (A,ϕ).

2.5 Dilations and Markov processes. If a stationary process $(\hat{A},\hat{\phi},\hat{T}_\tau;i,P)$ is a dilation of a dynamical system (A,ϕ,T_τ), then, by our foregoing considerations, the operators T_τ, $\tau \geq 0$, are the tansition operators $T_{0,\tau}$ of this process. Hence, by definition, a dilation is a stationary process, whose transition operators form a semigroup or, in a probabilistic language, it is a stationary process which satisfies the Chapman-Kolmogorov equations.

A dilation of first order fixes only the one step transition operator. In view of our interpretation so far the definition of the Markov property in 1.5 seems to be the canonical one.

It is well known that a stationary process may satisfy the Chapman-Kolmogorov equations without being a Markov process (cf. [Ros], III.1); a simple non-commutative example is provided by the dilation in [Var], see [Kü 9]. Conversely, however, as one would expect from the commutative theory, a Markov process automatically satisfies the Chapman-Kolmogorov equations, i.e., it is a dilation for some semigroup of transition operators (e.g. [Kü 1], 2.2.7).

Moreover, a repeated application of the Markov property shows that the time ordered multiple time correlations are reflected correctly:

$$\begin{aligned}&\hat{\phi}(\hat{T}_{t_1}\circ i(x_1)\cdot \hat{T}_{t_2}\circ i(x_2)\cdot \ \dots \ \cdot \hat{T}_{t_{n-1}}\circ i(x_{n-1})\cdot \hat{T}_{t_n}\circ i(x_n))\\ &\qquad = \phi(x_1\cdot T_{t_2-t_1}(x_2\cdot \ \dots \ \cdot T_{t_{n-1}-t_{n-2}}(x_{n-1}\cdot T_{t_n-t_{n-1}}(x_n))\dots)) \qquad (**)\end{aligned}$$

for $0 \leq t_1 \leq t_2 \leq \dots \leq t_{n-1} \leq t_n$ and $x_j \in A$, $1 \leq j \leq n$.

2.6 Discussion. Our considerations so far show that in the context of commutative W*-algebras our axioms reproduce the notions of classical probability theory. In particular, constructing Markov dilations means to construct a stationary Markov process for a given semigroup of transition operators.

There is one axiom which might not be considered as a "conditio sine qua non" for non-commutative probability theory. This is the existence of conditional expectations which is introduced by the requirements on the morphism P (cf. 1.4, 2.2). Let us finally discuss its advantages and disadvantages.

Its obvious disadvantage is the fact that it restricts the class of semigroups for which there does exist a corresponding Markov process (§ 4) . It is not clear so far, whether this class of semigroups allows a reasonable probabilistic or physical interpretation and there may be situations which demand for a more general theory. Indeed, there do exist approaches using generalized conditional expectations which are either unbounded ([Gud], [Vin]) or not idempotent ([Ac 2]).

On the other hand there are some reasons for imposing this restriction:

(i) It is difficult to formulate a Markov property which allows a reasonable interpretation in terms of predicting events in the future from events in the past or presence, if no conditional expectations are at hand.

(ii) The general theory developed so far for Markov dilations (cf. § 3) as well as their structure theory (cf. § 5) both heavily depend on the existence of conditional expectations.

(iii) Without the existence of conditional expectations there need not to exist any kind of transition operators satisfying (*). Conversely, for many situations it can be shown that the existence of transition operators satisfying (*) and (**) already implies the existence of conditional expectations (e.g. [Ac 3], 2.2.2).

(iv) As in classical probability theory one would like to investigate the behaviour of initial states different from ϕ since probability theory should be a theory for states rather than for observables. This requires an embedding of A_* into $\hat{A}_*$ which allows an extension of the process to the preduals, corresponding to the extension from L^∞ to L^1 in the classical theory (also, in 2.2 we have considered, somewhat arbitrarily, only uniformly bounded observables). Having conditional expectations this causes no problems as we can use their (pre-)adjoints. Conversely, however, it is easy to produce a number of results which show that any reasonable embedding of A_* into $\hat{A}_*$ already enforces the existence of conditional expectations (cf. e.g. [Kü 5]).

§ 3 Some General Theory

In this paragraph we mention some results from the general theory of stationary Markov processes.

3.1 Theorem. ([Kü 1], §3) Let (A,ϕ,T_τ) be a dynamical system and $(\hat{A},\hat{\phi},\hat{T}_\tau;P)$ a minimal Markov dilation of (A,ϕ,T_τ). Then the following holds.

(i) (A,ϕ,T_τ) is ergodic if and only if $(\hat{A},\hat{\phi},\hat{T}_\tau)$ is ergodic.

(ii) (A,ϕ,T_τ) is weakly mixing if and only if $(\hat{A},\hat{\phi},\hat{T}_\tau)$ is weakly mixing.

(iii) (A,ϕ,T_τ) is strongly mixing if and only if $(\hat{A},\hat{\phi},\hat{T}_\tau)$ is strongly mixing.

(iv) (A,ϕ,T_τ) is completely mixing ,i.e. $\lim_{\tau\to\infty} T_\tau(x) = \phi(x)\cdot 1$ strongly for all $x \in A$, if and only if $(\hat{A},\hat{\phi},\hat{T}_\tau)$ is purely non-deterministic, i.e. $\bigcap_{\tau\in\mathbb{T}_+} A_{(-\infty,-\tau]} = \mathbb{C}\cdot 1$.

The definitions of the above mixing properties may be found in any book on ergodic theory (e.g. [Wal], [Kre]) or in ([Kü 1], §3).
The above result, as an example, shows that as far as the asymptotic theory is concerned we obtain for the non-commutative situation the same results as one is used from the classical theory (cf. [Ros], IV.4, VI.2; [Ker], [Par]). Some further results on the asymptotic theory may be found in ([Kü 1], §3).
We remark that the "if part" of 3.4.iv is not explicitly stated in [Kü 1], but it follows immediately from the corresponding linear result (see [Sz.-N], II.1.2, II.2.1).

3.2 For a theory of stationary Markov processes it is natural to work in a W*-algebraic frame since this theory belongs to the measure theoretic part of probability theory. Nevertheless, in some situations it is also useful to consider C*-algebraic versions of dilations. In particular, some constructions naturally lead to a C*-dilation (e.g., the proofs of 3.3 and 4.3 involve such constructions). In ([Kü 6], 2.1.4) we provided a canonical procedure which allows to turn from certain C*-dilations to a W*-dilation.

3.3 Consider a sequence $(A,\phi,T_j)_{j\in\mathbb{N}}$ of dynamical systems. If A is commutative and the morphisms T_n converge in some sense to a morphism T, then one can use the Kolmogorov-Daniell construction in order to show that the corresponding minimal Markov processes converge in some sense to the minimal Markov process for (A,ϕ,T).
In our non-commutative context, however, a minimal Markov process is by no means determined by its semigroup of transition operators and therefore, we can not hope for an analogous theory. Using a convergence along free ultrafilters then, via an F-product technique, one can with a net of stationary processes still associate a limit process. Using this approximation theory the following result can be obtained.

Theorem. ([Kü 6], 2.3.6) Consider a fixed pair (A,ϕ).
(i) The set of morphisms of (A,ϕ) which have a dilation of first order is compact convex in the pointwise weak* topology.
(ii) Let J be some directed set and for $j \in J$ let $(A,\phi,(T_j)_\tau)$ be a discrete or continuous dynamical system which has a dilation.
If (A,ϕ,T_τ) is a dynamical system such that T_τ = pointwise weak* $\lim_j (T_j)_\tau$ for all $\tau \in \mathbb{T}_+$ then (A,ϕ,T_τ) has a dilation.
(iii) Given the situation of (ii) if, in addition, A is finite dimensional and $(A,\phi,(T_j)_\tau)$ has a Markov dilation, then (A,ϕ,T_τ) has a Markov dilation.

Remark. First applications of the above theorem are given in 4.6, 4.7. Combining this result with various composition techniques for dilations (e.g. [Kü 6], § 2.2), most problems on the existence of dilations for dynamical systems on hyperfinite factors can be reduced to existence problems on the $n \times n$-matrices.
Although the approximation along free ultrafilters seems to be rather abstract, in a number of concrete cases the limit dilation can be described explicitly.

3.4 We end this paragraph with a few remarks on the linear theory of stationary processes.
Consider a stationary process $(\hat{A},\hat{\phi},\hat{T}_\tau;i,P)$ over (A,ϕ). Denote by (H_ϕ,ξ_ϕ) and $(H_{\hat{\phi}},\xi_{\hat{\phi}})$ the Hilbert spaces with cyclic and separating vector arising from the GNS-construction for (A,ϕ) and $(\hat{A},\hat{\phi})$.
From the Schwarz inequality for completely positive operators it is immediate that a morphism T of (A,ϕ) induces a contraction T_ϕ on H_ϕ while $\hat{T}$ induces a unitary $\hat{T}_{\hat{\phi}}$ on $H_{\hat{\phi}}$. In the same way i induces an isometry from H_ϕ into $H_{\hat{\phi}}$, while P induces the adjoint of this isometry.
Therefore, a dilation $(\hat{A},\hat{\phi},\hat{T}_\tau;i,P)$ of a dynamical system (A,ϕ,T_τ) induces a unitary Hilbert space dilation of $(H_\phi,(T_\tau)_\phi)$. This unitary dilation, however, is not minimal in general, since usually the algebra generated by the translates $\hat{T}^k \circ i(A)$ will be larger than the linear span.
On the other hand, as long as we consider only elements in the linear span, we are essentially working in the frame of Hilbert space theory. In particular, the linear structure of this linear span is determined by the transition operators and all the classical linear theory of stationary processes (see, e.g., [Roz]), such as, e.g., linear prediction theory, applies as well to the non-commutative theory.
This was one motivation for our investigations on the structure of unitary dilations in [Kü 11] (cf. 5.5).

§ 4 On the Existence of Non-Commutative Markov Processes

In this paragraph we collect results on the problem for which semigroups of transition operators there exists a stationary Markov process. In particular, we present a result which reduces the existence problem for continuous times to the existence problem for discrete times.

4.1 The Kolmogorov-Daniell reconstruction of a Markov process from its transition probabilities provides us with the complete solution of the existence problem for commutative algebras: If (A,ϕ,T_τ) is a dynamical system where A is commutative, then it has a Markov dilation. Moreover, if this Markov dilation is minimal, it is uniquely determined up to equivalence.

4.2 In the non-commutative setting, we can not expect a positive solution of the general existence problem:

Proposition. ([Kü 1], 2.1.8). If (A,ϕ,T_τ) is a dynamical system which has a dilation, then T_τ, $\tau \geq 0$, commutes with the modular automorphism group σ_t^ϕ of the state ϕ.

For an account on modular automorphism groups see ([Ped], ch.8). This condition is trivial if the algebra A is commutative, and more generally, if ϕ is a trace. For a non-tracial state ϕ, however, the condition is non-trivial already on the algebra of 2×2-matrices ([Kü 1], 2.1.9).

In most cases this condition is still not sufficient for the existence of a dilation and we have yet no satisfactory description of the class of dynamical systems which admit a dilation. Some general properties of this class can be derived from the approximation theory (see 3.3).

4.3 Only in a few cases the converse of 4.2 holds (counter-examples on the 3×3-matrices are given in ([Kü 5], 3.3.3)):

Proposition. ([Kü 2], [Kü 3], [Kü 6], 2.1.8) Let (A,ϕ,T_τ) be a dynamical system. If either $(T_\tau)_{\tau\geq 0}$ is a semigroup of (injective) *-homomorphisms or $A = M_2$, then the following conditions are equivalent:

(a) The dynamical system (A,ϕ,T_τ) has a Markov dilation.

(b) The morphisms $(T_\tau)_{\tau\geq 0}$ commute with the modular automorphism group σ_t^ϕ.

4.4 Let us now summarize some further conditions which guarantee the existence of a dilation.

Theorems. 1. ([Em 2], [Ev 2]). Let (A,ϕ,T_τ) be a dynamical system and assume that A is a W*-algebra of canonical commutation relations or canonical anticommutation relations with a quasifree faithful normal state ϕ. If the semigroup $(T_\tau)_{\tau\geq 0}$ consists of quasifree operators then (A,ϕ,T_τ) has a Markov dilation.

2. ([Fr 3], [Fr 4]). Let (M_n,ϕ,T_t) be a continuous dynamical system and assume that T_t satisfies the detailed balance condition with respect to ϕ, i.e., its generator can be written as the sum of two generators L_1 and L_2 where L_1 generates a group of automorphisms of (M_n,ϕ) while L_2 generates a semigroup of morphisms of (M_n,ϕ) whose elements are self-adjoint with respect to ϕ. Then (M_n,ϕ,T_t) has a Markov dilation.

3. ([Kü 1], 4.3.3, [Kü 7], 1.1.1). Let (M_n,ϕ,T_τ) be a dynamical system. If T_τ is in the convex hull of the automorphisms of (M_n,ϕ) for all $\tau \geq 0$ then (M_n,ϕ,T_τ) has a Markov dilation.

Remarks: 1. The results in [Fr 3], [Fr 4] seemed to indicate that for (M_n,ϕ,T_t) the detailed balance condition might be also necessary for the existence of a dilation. The dynamical systems described in part 3, however, do not necessarily satisfy the detailed balance condition ([Kü 7], 2.6).

2. Part 3 holds also for more general W*-algebras (e.g. ([Kü 1], 4.3.3).

4.5 Finally, we report on a different approach to the existence problem for continuous dilations ([Kü 6]). If T is a morphism of (A,ϕ) then $e^{(T-Id)t}$ is a continuous semigroup of morphisms of (A,ϕ).
Using non-commutative Poisson processes we obtain the following result which assures the existence of a Markov dilation for many new semigroups. A complete proof is given in 6.2.

Theorem. ([Kü 6], 3.3.1) If the discrete dynamical system (A,ϕ,T) has a Markov dilation, then the continuous dynamical system $(A,\phi,e^{(T-Id)t})$ has a Markov dilation.

4.6 Combining this with the approximation result in 3.3 leads easily to the following result. Again we refer to 6.2 for the proof.

Theorem. ([Kü 6], 3.3.6) For a continuous dynamical system (A,ϕ,T_t) the following conditions are equivalent.

(a) The continuous dynamical system (A,ϕ,T_t) has a dilation.

(b) For each $t_0 \in \mathbb{R}_+$ the discrete dynamical system (A,ϕ,T_{t_0}) has a dilation.

(c) There exist discrete dynamical systems $(A,\phi,T_j)_{j\in J}$ each of which has a dilation, and there exist positive real numbers $(\alpha_j)_{j\in J}$ such that for all $t \geq 0$ T_t = pointwise weak* $\lim_j e^{\alpha_j(T_j - Id)t}$.

Moreover, if A is finite dimensional then "dilation" can be substituted by "Markov dilation" in the above conditions.

This result, which may be considered as a partial converse of 4.5 reduces the existence problem for continuous dilations to the existence problem for discrete dilations.

4.7 Another application of the approximation result in 3.3 gives the following general information on the set of dilatable semigroups:

Theorem ([Kü 6], 3.3.6). Given a non-commutative probability space (M_n,ϕ).

1. The set of morphisms of (M_n,ϕ) which have a discrete Markov dilation, forms a compact convex set which is also closed under multiplication.

2. The set of generators L for which (M_n,ϕ,e^{Lt}) has a Markov dilation forms a closed convex cone.

§ 5 On the Structure of Dilations

As already mentioned, in the non-commutative setting a minimal Markov dilation is no longer uniquely determined by its semigroup of transition operators. In this paragraph we summarize some results which nevertheless give information about the structure of any minimal Markov dilation for certain given semigroups. For simplicity, we mostly concentrate on the case of discrete time, although many results have their continuous counterparts.

5.1 In ([Kü 4], 2.2, 2.3) we gave some arguments why the commutative construction can not be generalized to the non-commutative case. Instead, in ([Kü 1], 4.2.2) we approached the problem of constructing non-commutative dilations as follows.

Let (A,ϕ,T) be a discrete dynamical system and assume that we have already found a "tensor dilation of first order", i.e., a dilation of first order of the form $(A\otimes C,\phi\otimes\psi,T_1;j,Q)$ where (C,ψ) is a non-commutative probability space, $j\colon (A,\phi)\to(A\otimes C,\phi\otimes\psi)\colon x\mapsto x\otimes 1$, $Q(x\otimes y)=\psi(y)\cdot x$ for $x\in A$, $y\in C$. The non-commutative probability space $\otimes_{\mathbb{Z}}(C,\psi)$ is obtained from the infinite algebraic tensor product $\odot_{\mathbb{Z}} C$ as the weak closure of its GNS-representation with respect to the infinite product state $\otimes_{\mathbb{Z}}\psi$.

Now put $(\hat{A},\hat{\phi}) := (A,\phi)\otimes(\otimes_{\mathbb{Z}}(C,\psi))$.

Defining σ as the tensor right shift on $\otimes_{\mathbb{Z}}(C,\psi)$ we put $\hat{\sigma} := Id_A\otimes\sigma$ on $\hat{A}$. Identifying $A\otimes C$ with the subalgebra of $\hat{A}$ which is the tensor product of A with the zero'th component of $\otimes_{\mathbb{Z}} C$ we can extend T_1 to an automorphism $\hat{T}_1$ of $(\hat{A},\hat{\phi})$ by taking the identity on all other factors of $\otimes_{\mathbb{Z}}(C,\psi)$.

Now put $\hat{T} := \hat{T}_1\circ\hat{\sigma}$

These definitions may be illustrated by the following sketch:

$$\left.\begin{matrix}(A,\phi)\\ \otimes\\ (C,\psi)\end{matrix}\right\}T_1$$

$$\ldots\otimes(C,\psi)\otimes(C,\psi)\otimes\underbrace{(C,\psi)}\otimes(C,\psi)\otimes(C,\psi)\otimes\ldots$$

$$\xrightarrow{\quad\sigma\quad}$$

Finally, define $i\colon(A,\phi)\to(\hat{A},\hat{\phi})\colon x\mapsto x\otimes 1$ and P as the corresponding projection.

Now it is easy to verify that $(\hat{A},\hat{\phi},\hat{T};P)$ is a Markov process and hence a Markov dilation of (A,ϕ,T).

5.2 This construction can be used in order to show that already simple dynamical systems on M_2 have uncountably many non-equivalent Markov dilations ([Kü 1], 5.8) (see also 5.9) and that even commutative dynamical systems have also non-commutative Markov dilations ([Kü 1], 5.12).

5.3 The next result shows that the W*-algebra which is generated by a Markov process, is not completely arbitrary.

<u>Theorem</u>. ([Kü 1], 2.3.3) Let (A,ϕ,T_τ) be a dynamical system with a minimal Markov dilation $(\hat{A},\hat{\phi},\hat{T}_\tau;P)$ and assume that the semigroup $(T_\tau)_{\tau\in\mathbb{T}_+}$ has no non-trivial fixed points in the center of A.

(i) $\hat{A}$ is finite if and only if there exists a T_τ-invariant faithful normal trace on A.

(ii) $\hat{A}$ is semifinite if and only if there exists a T_τ-invariant faithful normal semifinite trace on A.

(iii) $\hat{A}$ is of type III if and only if there exists no T_τ-invariant faithful normal semifinite trace on A.

The assertions (i), the "if" part of (ii) and the "only if" part of (iii) are true without the restriction on T_τ .
Therefore, if, e.g., we start with a dynamical system (M_2,ϕ,T_τ) where the trace is not fixed under T_τ then any Markov dilation of this dynamical system will generate a W*-algebra of type III.

5.4 The structure of the dilation in 5.1 may be described as "coupling a tensor dilation of first order to the zero'th component of a non-commutative Bernoulli-shift". It looks differently from the commutative dilation, but it is typical for the non-commutative theory (see 5.7).
If we consider for the moment only the linear theory of a Markov process then we obtain essentially a unitary dilation (3.4). This unitary dilation is uniquely determined by its semigroup and in [Kü 11] we have shown that the structure of a unitary dilation can always be understood as a coupling to a (linear) shift.

5.5 Returning to the algebraic setting, the non-commutative Bernoulli-shift in 5.1 can be replaced by any generalized Bernoulli-shift which is defined as follows.

Definition. A minimal process $(\hat{A},\hat{\phi},\hat{T};P)$ over (A,ϕ) is a generalized Bernoulli-shift if for all $I,J \subset \mathbb{Z}$ with $n < m$ for all $n \in I$, $m \in J$ we have $\hat{\phi}(x\cdot y) = \hat{\phi}(x)\cdot\hat{\phi}(y)$ for $x \in A_I$, $y \in A_J$.

Remarks. (i) By definition, generalized Bernoulli-shifts are stationary processes whose observables belonging to disjoint time intervals are stochastically independent under $\hat{\phi}$.
(ii) If $T \in M(A,\phi)$ is given by $T(x) = \phi(x)\cdot 1$ for $x \in A$ then any Markov dilation of (A,ϕ,T) is a generalized Bernoulli-shift over (A,ϕ) .
(iii) If $\hat{A}$ is commutative then the definition of a generalized Bernoulli-shift is equivalent to the definition of a classical Bernoulli-shift.
(iv) The non-commutative Bernoulli-shift σ in 5.1 is a generalized Bernoulli-shift, and the quasifree shift on the CAR algebra over $l^2(\mathbb{Z})$ leads to another example for a generalized Bernoulli-shift over M_2 .
The above Remark (ii) can be used in order to construct different examples of generalized Bernoulli-shifts (e.g. [Kü 5], 4.2.4).

5.6 Theorem. ([Kü 4], 3.6) Let (A,ϕ,T) be a discrete dynamical system which has an inner tensor dilation of first order $(A\otimes C,\phi\otimes\psi,\mathrm{Ad}\,U;j,Q)$ for some unitary $U \in A\otimes C$. Then the construction in 5.1 can be generalized by substituting the non-commutative Bernoulli-shift over (C,ψ) by any generalized Bernoulli-shift over (C,ϕ) .

For the precise form of this construction we refer to ([Kü 4],3.6).

5.7 In a number of situations there exists a converse of 5.6.

<u>Theorem</u>. ([Kü 5], 4.3.5) Let (M_n,ϕ,T) be a dynamical system and suppose that there exists a projection $p \in M_n$ which commutes with the trace class operator representing ϕ, such that ϕ is a trace on $p \cdot M_n \cdot p$ and $\phi(p) \gneq 1/2$.
Then for any minimal Markov dilation $(\hat{A},\hat{\phi},\hat{T};i,P)$ of (M_n,ϕ,T) there exists an inner tensor dilation of first order $(M_n \otimes C,\phi\otimes\psi,\mathrm{Ad}\,U;j,Q)$ and a generalized Bernoulli-shift over (C,ϕ) such that the Markov dilation is a coupling of the inner tensor dilation of first order to the zero′th component of the generalized Bernoulli-shift.

For a number of special cases this result has been obtained before: If ϕ is a trace then the assumption on ϕ is trivially fulfilled. For this case the result can be found in ([Kü 4], 3.9). On the 2×2-matrices M_2 the assumption of the theorem is always fulfilled since in this case ϕ is either the trace or its representing trace class operator has an eigenvalue which is larger than 1/2 . In this case special versions of the above result have been obtained in ([Kü 2], § 3) and in ([Kü 3], 3.6). On M_3 there exist states which do not fulfill the above assumptions and indeed there exists an example of a dynamical system on M_3 for which the above result does not hold.
We should also note that in the cases governed by the theorem there always exists an inner dilation of first order, while by ([Kü 1], 3.1.5) a non-trivial Markov dilation will never be inner.
This structure theorem shows that in many non-commutative situations every Markov dilation looks like the dilation in 5.1 where now any generalized Bernoulli-shift can replace the non-commutative Bernoulli-shift.
In [Kü 10] (see also [Kü 2], 2.2) we have worked out a detailed physical interpretation of a such a coupling dilation which shows that it typically describes a physical system which is coupled via the dilation of first order to a stochastic surroundings whose time evolution is represented by the generalized Bernoulli-shift. There can be worked out also a probabilistic interpretation which shows that with the generalized Bernoulli-shift we have isolated the innovation process from the Markov process.
In ([Kü 5], § 4.4) we have shown that also for commutative dynamical systems there can be constructed a "coupling dilation" which, in particular, gives an alternative procedure for constructing commutative Markov processes. There exist also analogues for continuous times of the above

structure theorem (see, e.g. [Kü 2] § 3, [Kü 7], 1.2.2). Conversely, the constructions of quantum stochastic processes in ,e.g., [Hud] and [Fr 1], [Fr 2] are precisely of the form which is derived in the (continuous analogues of the) structure theorem.

5.8 In the cases where theorem 5.7 applies, the dilation can be decomposed into its two constituents, an inner tensor dilation of first order and a generalized Bernoulli-shift. Correspondingly, the classification problem splits into the problems of classifying these two constituents. A classification of generalized Bernoulli-shifts seems, presently, out of reach. On the other hand many interesting features of the Markov process are already determined by the dilation of first order contained in it, and in some cases we can obtain information on the dilations of first order.

Denote by $\mathcal{M}(M_2,\phi;\sigma_t^\phi)$ the set of morphisms of (M_2,ϕ) which commute with the modular automorphism group σ_t^ϕ .

Theorem. ([Kü 3], 3.8) If T is an extreme point of $\mathcal{M}(M_2,\phi;\sigma_t^\phi)$ then its dilation of first order is uniquely determined. If, moreover, T is not an automorphism then this dilation of first order generates the algebra $M_2 \otimes \mathcal{B}(\mathcal{H})$ for some infinite dimensional Hilbert space $\mathcal{H}$.

Remarks. (i) For the precise form of the dilations of first order in the first theorem we refer to the cited paper. They have an interpretation as describing a spin-1/2-particle coupled to a single harmonic oscillator.
(ii) For the morphisms which are not extreme points but are on the boundary of $\mathcal{M}(M_2,\phi;\sigma_t^\phi)$ the dilation of first order is no longer unique but they still can be classified ([Kü 1], 5.10). For the morphisms which correspond to interior points no simple classification of dilations of first order is possible. In fact one can show that for any of these morphisms a dilation of first order can generate essentially every W*-algebra.
(iii) If T is an extreme point of $\mathcal{M}(M_2,\phi;\sigma_t^\phi)$ which is not an automorphism then by the above theorem and 5.7 every minimal Markov dilation of (M_2,ϕ,T) is a coupling of its dilation of first order to a generalized Bernoulli-shift over $(\mathcal{B}(\mathcal{H}),\psi)$ for a certain state ψ. Since $\mathcal{B}(\mathcal{H})$ is infinite dimensional it follows that the algebra generated by a Markov dilation can never be generated by the time ordered products (since the linear space generated by time ordered products from observables corresponding to finite time intervals is always finite dimensional, due to the finite dimensionality of M_2). In particular, this shows the insufficiency of generalized Stinespring constructions for producing non-commutative stationary Markov processes.

5.9 The following consideration may shed some more light upon the geometrical problems which are involved in a classification of dilations.

Theorem. ([Kü 5], 3.4.1, [Kü 7], 1.1.1, 1.5.1). For a dynamical system (M_n, tr, T_τ) the following conditions are equivalent.

(a) $T_\tau \in \text{co Aut}(M_n, tr)$ for all $\tau \geq 0$.

(b) There exists a weak*-continuous convolution semigroup $\{\rho_\tau\}_{\tau\geq 0}$ of probability measures on the group $\text{Aut}(M_n, tr)$ such that
$T_\tau(x) = \int_{\text{Aut}(M_n,tr)} \alpha(x)\, d\rho_\tau(\alpha) \quad (x \in M_n,\ \tau \geq 0)$.

(c) (M_n, tr, T_τ) admits a Markov dilation which generates an algebra $M_n \otimes C$ with C commutative.

Moreover, there is a canonical one-to-one correspondence between minimal Markov dilations generating $M_n \otimes C$ with C commutative and weak*-continuous convolution semigroups $\{\rho_\tau\}_{\tau\geq 0}$ of probability measures such that $T_\tau(x) = \int_{\text{Aut}(M_n,tr)} \alpha(x)\, d\rho_\tau(\alpha) \quad (x \in M_n,\ \tau \geq 0)$.

This result shows that a classification of these dilations is equivalent to a complete understanding of the geometrical structure of $\text{co Aut}(M_n)$. Presently, this convex set is not very well understood. We even do not have a handy criterion which allows to decide whether a given morphism fulfills condition (a). Some results on the geometry of $\text{co Aut}(M_n)$ are contained in ([Kü 5], § 3.4, 3.5).

§ 6 A Class of Non-Commutative Poisson Processes

In this final paragraph we review the construction of a class of Poisson processes which has been introduced in [Kü 6] .

6.1 We start with introducing a suitable description of the classical Poisson process.

A Markov process is frequently represented on the space of its possible paths, endowed with a certain measure. On the other hand, since the Poisson process, as Brownian motion, is a process with stationary independent increments it allows also a canonical realization on the derivatives of its paths (e.g. [Sha], [Maa], the corresponding realization of Brownian motion may be found ,e.g., in [Hid]):

We denote by $\mathcal{D}'(\mathbb{R})$ the space of distributions on $\mathbb{R}$ and let $\delta_t \in \mathcal{D}'(\mathbb{R})$ be the Dirac measure in $t \in \mathbb{R}$. As usual, χ denotes a characteristic function. Now define

$\Omega := \{\omega \in \mathcal{D}'(\mathbb{R}) : \omega = \sum_{n\in\mathbb{Z}} \delta_{t_n} \ , \quad \ldots < t_{-1} < t_0 < t_1 < \ldots\}$,

$\langle \chi_{(a,b]}, \omega\rangle := \sum_{n\in\mathbb{Z}} \chi_{(a,b]}(t_n)$ for $\omega \in \Omega$, $(a,b] \subset \mathbb{R}$,

$\Omega_n(a,b] := \{\omega \in \Omega: \langle\chi_{(a,b]},\omega\rangle = n\}$ for $n \geq 0$, $(a,b] \subset \mathbb{R}$.

Σ is the σ-algebra generated by $\{\Omega_n(a,b] : n \geq 0,\ (a,b] \subset \mathbb{R}\}$,

μ is the probability measure on (Ω,Σ) which is determined by

(i) $\mu(\Omega_n(a,b]) = e^{-(b-a)} \cdot \frac{(b-a)^n}{n!}$

(ii) $\mu(\Omega_n(a,b] \cap \Omega_m(c,d]) = \mu(\Omega_n(a,b]) \cdot \mu(\Omega_m(c,d])$ for $b \leq c$, $n, m \in \mathbb{Z}_+$,

$\tau_s: \Omega \to \Omega: \omega = \sum_n \delta_{t_n} \mapsto \sum_n \delta_{t_n - s}$, $s \in \mathbb{R}$, is a group of automorphisms of (Ω,Σ,μ) . Then

$X_t: \Omega \to \mathbb{Z} : \omega \mapsto \langle\chi_{(0,t]},\omega\rangle$, $t \geq 0$, is a Poisson process .

It satisfies the cocycle identity $X_{s+t} = X_s + X_t \circ \tau_s$.

Turning to an algebraic description, put

$C := L^\infty(\Omega,\Sigma,\mu)$, let

$\psi := \psi_\mu$ be the state on $L^\infty(\Omega,\Sigma,\mu)$ which is induced by μ , and let

$\sigma_t(f) := f \circ \tau_t$, $t \in \mathbb{R}$, $f \in C$. Moreover, define

$p_n(a,b]$ as the projection in C which is the characteristic function of $\Omega_n(a,b]$ (up to points of measure zero) for $n \geq 0$, $(a,b] \subset \mathbb{R}$ and let

C_I be the W*-subalgebra of C which is generated by $\{p_n(a,b] : n \geq 0 ,\ (a,b] \subset I\}$.

In this algebraic description the cocycle identity for X_t turns into

$p_n(0,s+t] = \sum_{k=0}^n p_k(0,s] \cdot \sigma_s(p_{n-k}(0,t]) = \sum_{k=0}^n p_k(0,s] \cdot p_{n-k}(s,s+t]$.

In the forthcoming computations we will use this identity without further mentioning.

<u>Proposition</u>. ([Kü 6], 3.2.4) (C,ψ,σ_t) is a reversible dynamical system.

Proof: The only property which is not immediate from the properties of the Poisson process is the pointwise weak* continuity of $t \mapsto \sigma_t$:

By the property (i) for μ we have

strong - $\lim_{s \downarrow 0} p_0(t,t+s] = 1$ for all $t \in \mathbb{R}$ while

strong - $\lim_{s \downarrow 0} p_n(t,t+s] = 0$ for $n \geq 1$.

Since $p_0(0,s_1] \cdot p_n(s_1,t] \leq p_0(0,s_2] \cdot p_n(s_2,t] \leq p_n(0,t]$ for $0 < s_1 \leq s_2 < t$, $n \geq 0$, and

$\lim_{s \downarrow 0} \psi(p_0(0,s] \cdot p_n(s,t]) = \psi(p_n(0,t])$ for $n \geq 0$ by property (ii), we obtain $\lim_{s \downarrow 0} p_0(0,s] \cdot p_n(s,t] = p_n(0,t]$ strongly and hence

$\lim_{s \downarrow 0} p_n(s,t] = \lim_{s \downarrow 0} (p_0(0,s] \cdot p_n(s,t] + (1 - p_0(0,s]) \cdot p_n(s,t] = p_n(0,t]$ strongly for $n \geq 0$. Since for $s < t$

$$p_n(s,s+t] = \sum_{k=0}^n p_k(s,t] \cdot p_{n-k}(t,t+s]$$
$$= p_n(s,t] \cdot p_0(t,t+s] + \sum_{k=0}^{n-1} p_k(s,t] \cdot p_{n-k}(t,t+s] ,$$

and since multiplication is jointly strongly continuous on the unit ball of C, we find

$\lim_{s\downarrow 0} \sigma_s(p_n(0,t]) = \lim_{s\downarrow 0} p_n(s,t+s] = p_n(0,t]$, and

$\lim_{s\downarrow 0} \sigma_s(p_n(t_1,t_1+t]) = \sigma_{t_1}(\lim_{s\downarrow 0} \sigma_s(p_n(0,t])) = p_n(t_1,t_1+t]$

for $t > 0$, $t_1 \in \mathbb{R}$.

Therefore, $\lim_{s\downarrow 0} \sigma_s(x) = x$ strongly for x in the stronly dense algebra which is finitely generated by the projections $\{p_n(s,t] : n \geq 0 , (s,t] \subset \mathbb{R}\}$ and the conclusion follows from the equicontinuity of $\{\sigma_t : t \in \mathbb{R}\}$ in the pointwise strong topology.

6.2 <u>Compound Poisson Processes</u>. If (A,ϕ,T) is a discrete dynamical system then $(A,\phi,e^{(T-Id)t})$ is a continuous dynamical system and

$$e^{(T-Id)t} = e^{-Id\cdot t}\cdot\sum_{n=0}^{\infty} \frac{T^n\cdot t^n}{n!} = \sum_{n=0}^{\infty} e^{-t}\cdot\frac{t^n}{n!}\cdot T^n \quad .$$

<u>Theorem</u>. ([Kü 6], 3.3.1) If for some λ, $0 \leq \lambda < 1$, the discrete dynamical system $(A,\phi,\lambda Id+(1-\lambda)T)$ has a (Markov) dilation then the continuous dynamical system $(A,\phi,e^{(T-Id)t})$ has a (Markov) dilation.

As a first step we prove this result for the case that T is an automorphisms and $\lambda = 0$. In this case (A,ϕ,T) has trivially a Markov dilation.

<u>Lemma</u>. Let (B,χ,S) be a discrete reversible dynamical system. Then $(B,\chi,e^{(S-Id)t})$ has a Markov dilation.

<u>Construction</u>. We adopt the notation from the previous section. In particular, (C,ψ,σ_t) denotes the reversible dynamical system which is associated with the Poisson process.

We define the ingredients of the desired dilation $(\hat{B},\hat{\chi},\hat{S}_t;j,Q)$ as follows:

$\hat{B} := B \otimes C$,

$\hat{\chi} := \chi \otimes \psi$,

$R_t(x) := \sum_{n=0}^{\infty} (1\otimes p_n(0,t])\cdot(S^n\otimes Id)(x)$ for $x \in \hat{B}$, $t > 0$,

$\hat{S}_t := R_t\circ(Id\otimes\sigma_t)$ for $t > 0$, $\hat{S}_0 := Id$, $\hat{S}_{-t} := (\hat{S}_t)^{-1}$ for $t > 0$,

$j(x) := x \otimes 1$ for $x \in B$,

$Q(x\otimes f) := \psi(f)\cdot x$ for $x \in B$, $f \in C$.

Proof: The projections $1 \otimes p_n(0,t]$ are in the center of $\hat{B}$, they are fixed under $S \otimes Id$ and satisfy $\sum_{n=0}^{\infty} 1\otimes p_n(0,t] = 1$ ($t > 0$, $n \geq 0$). Therefore, R_t and hence $\hat{S}_t$ are automorphisms of $(\hat{B},\hat{\chi})$ for $t > 0$.

For verifying the group property of $(\hat{S}_t)_{t\in\mathbb{R}}$ we first simplify the notation and write
$\bar{\sigma}_s$ for $1 \otimes \sigma_s$, $\bar{p}_n(0,t]$ for $1\otimes p_n(0,t]$, and $\bar{S}$ for $S \otimes \mathrm{Id}$ ($s \in \mathbb{R}$, $t > 0$, $n \geq 0$).
For $x \in \hat{B}$, $s, t > 0$ we have

$$\begin{aligned}
\bar{\sigma}_s \circ R_t \circ \bar{\sigma}_{-s}(x) &= \bar{\sigma}_s\left(\textstyle\sum_{l=0}^{\infty} \bar{p}_l(0,t]\cdot \bar{S}^l \circ \bar{\sigma}_{-s}(x)\right) \\
&= \textstyle\sum_{l=0}^{\infty} \bar{p}_l(s,s+t]\cdot \bar{\sigma}_s \circ \bar{S}^l \circ \bar{\sigma}_{-s}(x) \\
&= \textstyle\sum_{l=0}^{\infty} \bar{p}_l(s,s+t]\cdot \bar{S}^l(x), \qquad \text{and therefore, we obtain} \\
R_s \circ \bar{\sigma}_s \circ R_t \circ \bar{\sigma}_{-s}(x) &= \textstyle\sum_{k=0}^{\infty} \bar{p}_k(0,s]\cdot \bar{S}^k\left(\sum_{l=0}^{\infty} \bar{p}_l(s,s+t]\cdot \bar{S}^l(x)\right) \\
&= \textstyle\sum_{k,l=0}^{\infty} \bar{p}_k(0,s]\cdot \bar{p}_l(s,s+t]\cdot \bar{S}^{k+l}(x) \\
&= \textstyle\sum_{n=0}^{\infty} \left(\sum_{k=0}^{n} \bar{p}_k(0,s]\cdot \bar{p}_{n-k}(s,s+t]\right)\cdot \bar{S}^n(x) \\
&= \textstyle\sum_{n=0}^{\infty} \bar{p}_n(0,s+t]\cdot \bar{S}^n(x) \\
&= R_{s+t}(x).
\end{aligned}$$

From this cocycle identity follows immediately the group property
$\hat{S}_{s+t} = R_{s+t} \circ \bar{\sigma}_{s+t} = R_s \circ \bar{\sigma}_s \circ R_t \circ \bar{\sigma}_t = \hat{S}_s \circ \hat{S}_t$;
therefore, $(\hat{S}_t)_{t\geq 0}$ is a semigroup of automorphisms, hence $(\hat{S}_t)_{t\in\mathbb{R}}$ is a group.
For verifying its continuity we use
strong $\lim_{t\downarrow 0} p_0(0,t] = 1$ and strong $\lim_{t\downarrow 0} \sum_{n=1}^{\infty} p_n(0,t] = 0$.
From this we obtain

$$\lim_{t\downarrow 0} R_t(x) = \lim_{t\downarrow 0} \left(\bar{p}_0(0,t]\cdot x + \textstyle\sum_{n=1}^{\infty} \bar{p}_n(0,t]\cdot \bar{S}^n(x)\right) = x$$

strongly for $x \in \hat{B}$.
Since $\bar{\sigma}_t$ converges to the identity pointwise strongly as $t \to 0$, we have pointwise strong $\lim_{t\downarrow 0} \hat{S}_t = \mathrm{Id}$ which implies the pointwise strong continuity of the group.
For $x \in B$ the dilation property is obtained as

$$\begin{aligned}
Q \circ \hat{S}_t \circ j(x) &= Q \circ \hat{S}_t(x\otimes 1) = Q \circ R_t(x\otimes 1) = Q\left(\textstyle\sum_{n=0}^{\infty} S^n(x)\otimes p_n(0,t]\right) \\
&= \textstyle\sum_{n=0}^{\infty} \psi(p_n(0,t])\cdot S^n(x) = \sum_{n=0}^{\infty} e^{-t}\cdot \frac{t^n}{n!} S^n(x) \\
&= e^{(S-\mathrm{Id})t}(x).
\end{aligned}$$

It remains to prove the Markov property.
From the definitions of $\hat{S}_t$ and $C_{[s,t]}$ it immediately follows that $B_{[0,\infty)} \subset B \otimes C_{[0,\infty)}$ and $B_{(-\infty,0]} \subset B \otimes C_{(-\infty,0]}$.
Denote by $P_{(-\infty,0]}$ the conditional expectation from (C,ψ) onto $C_{(-\infty,0]}$; then the conditional expectation from $(\hat{B},\hat{\chi}) = (B\otimes C, \chi\otimes\psi)$ onto $B\otimes C_{(-\infty,0]}$ is given by $x \otimes f \mapsto x \otimes P_{(-\infty,0]}(f)$ for $x \in B$, $f \in C$.
Therefore, if $x \otimes f \in B \otimes C_{[0,\infty)}$ with $x \in B$, $f \in C_{[0,\infty)}$ then its conditional expectation onto $B \otimes C_{(-\infty,0]}$ is given by
$x \otimes P_{(-\infty,0]}(f) = x \otimes \psi(f)\cdot 1 \in B \otimes 1$. By linearity and continuity, the conditional expectation onto $B \otimes C_{(-\infty,0]}$ of any element in $B \otimes C_{[0,\infty)}$ is contained in $B \otimes 1$. Hence the same is true for the conditional expectation onto the subalgebra $B_{(-\infty,0]}$ since
$B \otimes 1 \subset B_{(-\infty,0]} \subset B \otimes C_{(-\infty,0]}$.

Remark. If the algebra $\mathcal{B}$ has separable predual then we may canonically identify $\mathcal{B} \otimes \mathcal{C}$ with $L^{\infty}(\Omega,\Sigma,\mu;\mathcal{B})$, the essentially bounded $\mathcal{B}$-valued functions on (Ω,Σ,μ) (cf. [Sak], 1.22.13). In this case the automorphism $Id \otimes \sigma_t$ is induced by the base space transformation τ_s on Ω while R_t acts "pointwise" on the fibres by $(R_t x)(\omega) = S^{X_t(\omega)}(x)$ for $\omega \in \Omega$, $x \in L^{\infty}(\Omega,\Sigma,\mu;\mathcal{B})$, and $t > 0$. The above proved cocycle identity for R_t is then only a reformulation of the cocycle identity $X_{s+t} = X_s + X_t \circ \tau_s$ for the Poisson process.

Proof of the Theorem. For $0 < \lambda \leq 1$, $t \geq 0$, the identity $(T - Id)\cdot t = ((\lambda\cdot T + (1-\lambda)\cdot Id) - Id)\cdot t/\lambda$ reduces, by scaling the time, the proof to the case where the dynamical system (A,ϕ,T) itself has a (Markov) dilation.

Therefore, assume that $(\hat{A},\hat{\phi},\hat{T};i,P)$ is a discrete (Markov) dilation for (A,ϕ,T). The dilation property $T^n = P\circ\hat{T}^n\circ i$ yields

$$e^{(T-Id)t} = \sum_{n=0}^{\infty} e^{-t}\cdot\frac{t^n}{n!}\cdot T^n = P\circ(\sum_{n=0}^{\infty} e^{-t}\cdot\frac{t^n}{n!}\cdot\hat{T}^n)\circ i = P\circ(e^{(\hat{T}-Id)t})\circ i$$

for $t \geq 0$.

We now use the Markov dilation $(\hat{A}\otimes C,\hat{\phi}\otimes\psi,\hat{\hat{T}}_t;j,Q)$ of $(\hat{A},\hat{\phi},e^{(\hat{T}-Id)t})$ as constructed in the previous Lemma. Using now the standard procedure for composing dilations (cf. [Kü 5], 2.2.1) we obtain a continuous dilation $(\hat{A}\otimes C,\hat{\phi}\otimes\psi,\hat{\hat{T}}_t;j\circ i,P\circ Q)$ for the continuous dynamical system $(\hat{A},\hat{\phi},e^{(\hat{T}-Id)t})$.

Denote by $A_{[0,\infty)}$ (resp. $A_{(-\infty,0]}$) the algebra of the future (resp. past) of the discrete dilation $(\hat{A},\hat{\phi},\hat{T};i,P)$. Given $x \in A$ one immediately checks from the definitions that $\hat{\hat{T}}_t\circ j\circ i(x) \in A_{[0,\infty)} \otimes C_{[0,\infty)}$ for $t \geq 0$ while $\hat{\hat{T}}_t\circ j\circ i(x) \in A_{(-\infty,0]} \otimes C_{(-\infty,0]}$ for $t \leq 0$.

Assuming now the Markov property for the dilation $(\hat{A},\hat{\phi},\hat{T};i,P)$ we conclude as in the previous Lemma that for $x \otimes f$ with $x \in A_{[0,\infty)}$, $f \in C_{[0,\infty)}$ and hence for all $x \in A_{[0,\infty)} \otimes C_{[0,\infty)}$ the conditional expectation onto $A_{(-\infty,0]} \otimes C_{(-\infty,0]}$ is already contained in $j\circ i(A)$. This implies the Markov property for the continuous dilation, assuming that the discrete dilation has had the Markov property.

In the following we will sometimes denote the continuous dilation $(\hat{A}\otimes C,\hat{\phi}\otimes\psi,\hat{\hat{T}}_t;j\circ i,P\circ Q)$ by $(\hat{\hat{A}},\hat{\hat{\phi}},\hat{\hat{T}}_t;j\circ i,P\circ Q)$.

Proof of Theorem 4.6. The implication (a) $\Rightarrow$ (b) is trivial.

(b) implies (c) since $T_t = \text{pointwise weak*} - \lim_{s\downarrow 0} e^{\frac{1}{s}(T_s - Id)\cdot t}$ for $t \geq 0$.

Assuming (c) then each of the continuous dynamical systems

$(A,\phi,e^{\alpha_j(T_j-Id)t})_{j\in J}$ have a dilation by the above theorem.
Now the implication (c) $\Rightarrow$ (a) is the content of part (ii) of Theorem 3.3. If A is finite dimensional then for the case of Markov dilations, part (iii) of Theorem 3.3 leads to the desired conclusion.

Remark. Let us assume, for simplicity, that (A,ϕ,T) is a discrete reversible dynamical system. Then a Markov dilation of $(A,\phi,\lambda Id + (1-\lambda)T)$ can be constructed on $(A\otimes C',\phi\otimes\psi')$ where now (C',ψ') is the algebra corresponding to the $(\lambda,1-\lambda)$-Bernoulli shift. Moreover, as λ approaches 1, the powers of $(\lambda Id + (1-\lambda)T)$ approximate the continuous semigroup $e^{(T-Id)t}$. Therefore, we could have used as well the approximation theory in 3.3 in order to obtain the corresponding continuous dilation. However, it can be shown that both procedures lead to the same continuous dilation as the corresponding Bernoulli-shifts approximate the Poisson process.

6.3 The process $(\hat{\hat{A}},\hat{\hat{\phi}},\hat{\hat{T}}_t;j\circ i,P\circ Q)$ obviously can be constructed from any stationary process $(\hat{A},\hat{\phi},\hat{T};i,P)$ over (A,ϕ) and we call $(\hat{\hat{A}},\hat{\hat{\phi}},\hat{\hat{T}}_t;j\circ i,P\circ Q)$ the generalized compound Poisson process over (A,ϕ) corresponding to the stationary process $(\hat{A},\hat{\phi},\hat{T};i,P)$.
If, in particular, $(\hat{A},\hat{\phi},\hat{T};i,P)$ is a generalized Bernoulli-shift over (A,ϕ) then we call $(\hat{\hat{A}},\hat{\hat{\phi}},\hat{\hat{T}}_t;j\circ i,P\circ Q)$ simply the generalized Poisson process over (A,ϕ) corresponding to $(\hat{A},\hat{\phi},\hat{T};i,P)$.
In ([Kü 6], 3.1.3) we have introduced the notion of a "generalized white noise" as the straightforward continuous analogue of our notion of a generalized Bernoulli-shift (5.6) and to the construction of a "coupling to a generalized Bernoulli-shift" there corresponds the "coupling to a generalized white noise" in continuous time ([Kü 6], 3.1.7).

Theorem. ([Kü 6], 3.4.2, 3.4.4) (i) If $(\hat{A},\hat{\phi},\hat{T};i,P)$ is a generalized Bernoulli-shift over (A,ϕ) then the corresponding generalized Poisson process $(\hat{\hat{A}},\hat{\hat{\phi}},\hat{\hat{T}}_t;j\circ i,P\circ Q)$ is a generalized white noise.
(ii) If $(\hat{A},\hat{\phi},\hat{T};i,P)$ is a coupling to a generalized Bernoulli-shift then the corresponding generalized compound Poisson process $(\hat{\hat{A}},\hat{\hat{\phi}},\hat{\hat{T}}_t;j\circ i,P\circ Q)$ is a coupling to the generalized Poisson process which corresponds to this generalized Bernoulli-shift.

Of particular interest is the first part of this theorem: Starting from our various examples of generalized Bernoulli-shifts we obtain the first examples of generalized white noises which are not built from the classical Brownian motions, Poisson processes or from the examples arising from the canonical commutation/anticommutation relations.

6.4 Interpretation. The construction in 6.2 is closely related to the compound Poisson process in classical probability theory. A treatment of compound Poisson processes may be found in ([Kar], ch. 16). The Gaussian jump processes described in [Sha] also fit into this frame.
We remark that for $A = \mathbb{C}$ we retain from the construction in 6.2 the classical Poisson process in its algebraic reformulation described in 6.1.
Now consider the following simple dynamical system (A,ϕ,T) where A is given by $l^\infty(\{A,B\})$, ϕ is induced by the probability measure $(\lambda,1-\lambda)$ on $\{A,B\}$ and T is given by $T(x) = \phi(x)\cdot 1$ for $x \in A$.
The Markov process for (A,ϕ,T) is the $(\lambda,1-\lambda)$-Bernoulli-shift $(\hat{A},\hat{\phi},\hat{T};i,P)$ with $(\hat{A},\hat{\phi}) = \otimes_{\mathbb{Z}} (A,\phi)$, $\hat{T}$ is the tensor right shift and i is the injection of A into the zero'th factor of $\hat{A}$ while P projects onto this factor.
The interpretation of the compound Poisson process $(\hat{\hat{A}},\hat{\hat{\phi}},\hat{\hat{T}}_t;j\circ i,P\circ Q)$ is based on the following simple observation: Consider, e.g., two decaying materials A and B, which decay independently from each other and whose decay laws are described by the Poisson processes with decay rates λ and $1-\lambda$ for some λ , $0 < \lambda < 1$ (we can always choose the time unit accordingly). Then there are two ways of describing the compound system:
(i) A path of the compound system is a pair of paths (ω_A,ω_B) where ω_A , resp. ω_B , is a path of the Poisson process describing the decay of A , resp. B . The probability space of the compound system is $(\Omega_A\times\Omega_B,\mu_\lambda\otimes\mu_{(1-\lambda)})$, the product of the single systems, and the time evolution is $\tau_t\otimes\tau_t$, the tensor product of the time evolutions of the single systems.
(ii) A path of the compound system is a pair of paths (ω_1,ω_2). But now $\omega_1 \in \Omega$ is a path of the Poisson process describing the decay of the compound system , i.e., there is a jump in ω_1 whenever either of the materials A or B decays. The path ω_2 then gives the information about which of the materials is decaying, i.e., ω_2 is a path in $\{A,B\}^{\mathbb{Z}}$ and the value of ω_2 at position $n \in \mathbb{N}$ tells whether the n-th jump in ω_1 is caused by A or by B .
The corresponding probability space will now be $(\Omega\times\{A,B\}^{\mathbb{Z}},\mu\otimes(\lambda,1-\lambda)^{\mathbb{Z}})$.
The time evolution on a path (ω_1,ω_2) will be the time shift τ_t on ω_1 and whenever a jump of ω_1 arrives at zero under τ_t then ω_2 is shifted one unit to the left, i.e., $(\omega_1,\omega_2) \mapsto (\tau_t(\omega_1),s^{X_t(\omega_1)}(\omega_2))$ where s denotes the left shift on $\{A,B\}^{\mathbb{Z}}$.
The equivalence between these descriptions can be made rigorously by establishing a canonical isomorphism between the two dynamical systems. The second description corresponds to our construction of the compound Poisson process. Therefore, the interpretation of such a compound process

is now straightforward. The compound Poisson process corresponding to a Bernoulli-shift over any n-dimensional commutative algebra is isomorphic to the tensor product of n Poisson processes and hence describes n systems decaying independently according to Poisson laws. Equivalently, we may say that a classical system has a phase space consisting of n points and each of its "degrees of freedom" decays independently.

If we have, instead, a quantum system, say, a spin particle, whose algebra of observables is given by the 2×2-matrices M_2 and the equilibrium distribution is given by a state ϕ on M_2, then the compound Poisson process constructed from the non-commutative Bernoulli-shift or, more generally, any generalized Bernoulli-shift over (M_2, ϕ), describes a spin particle whose spin polarization is decaying. The decay is induced by events which are distributed according to a Poisson law and which change the spin polarization in an unpredictable way, subject only to the preservation of the mean expectations described by ϕ (cf. [Kü 7], 2.6).

Therefore, a generalized Poisson process over (A, ϕ) has a natural interpretation as a Poisson process with non-commutative state space A_*. In some sense it may be considered as a non-commutative tensor product of classical Poisson processes.

The observables of a generalized Poisson process over (A, ϕ), which correspond to disjoint time intervals, commute, if and only if the corresponding generalized Bernoulli-shift is a non-commutative Bernoulli-shift, i.e., the right shift on the infinite tensor product of copies of (A, ϕ). In this case we can take any two dimensional subalgebra of A generated by a non-trivial projection in A and by restricting the generalized Poisson process over (A, ϕ) to the generated commutative subalgebra we retain the situation which we have discussed in the beginning of this section. In particular, we recover the classical Poisson distributions. This will be no longer true, if the algebras corresponding to disjoint time intervals do not commute. Nevertheless, every generalized Poisson process does have a Poisson type behaviour. In particular, in its center there is always contained a classical Poisson process.

Finally, if we return to the more general construction of a generalized compound Poisson process corresponding to any stationary process, then the interpretation still remains essentially the same: A Poisson process describes the times when a non-commutative discrete ("jump") process takes place. The only difference is that now the outcomes depend on the previous outcomes in the way as it is implemented by the discrete stationary process.

We finally remark that the Poisson processes in [Hud] and the Poisson processes described in the contribution of A. Frigerio in these proceedings are special cases of the processes described in this paragraph.

References.

[Ac 1] L. Accardi: Nonrelativistic quantum mechanics as a noncommutative Markov process. Advances in Math. 20 (1976), 329-366.

[Ac 2] L. Accardi, C. Cecchini: Conditional expectations in von Neumann algebras and a theorem of Takesaki. J. Funct. Anal. 45 (1982), 215-274.

[Ac 3] L. Accardi, A. Frigerio, J.T. Lewis: Quantum stochastic processes. Publ. RIMS, Kyoto Univ. 18 (1982), 97-133.

[Bra] O. Bratteli, D.W. Robinson: "Operator Algebras and Quantum Statistical Mechanics I, II". Springer-Verlag, New York 1979.

[Chr] E. Christensen: Measures on projections and physical states. Commun. Math. Phys. 86 (1982), 529-538.

[Da 1] E.B. Davies: "Quantum Theory of Open Systems". Academic Press, London 1976.

[Da 2] E.B. Davies: Dilations of completely positive maps. J. London Math. Soc. (2), 17 (1978), 330-338.

[Em 1] G.G. Emch: "Algebraic Methods in Statistical Mechanics and Quantum Field Theory". John Wiley & Sons Inc. ,New York 1972.

[Em 2] G.G. Emch, S. Albeverio, J.P. Eckmann: Quasi-free generalized K-flows. Reports Math. Phys. 13 (1978), 73-85.

[Ev 1] D.E. Evans: Positive linear maps on operator algebras. Commun. Math. Phys. 48 (1976), 15-22.

[Ev 2] D.E. Evans: Completely positive quasi-free maps on the CAR algebra. Comm. Math. Phys. 70 (1979), 53-68.

[Ev 3] D.E. Evans, J.T. Lewis: Dilations of dynamical semigroups. Commun. Math. Phys. 50 (1976), 219-227.

[Ev 4] D.E. Evans, J.T. Lewis: "Dilations of Irreversible Evolutions in Algebraic Quantum Theory." Comm. Dublin Inst. Adv. Stud. Ser. A 24 (1977).

[Fr 1] A. Frigerio: Construction of stationary quantum Markov processes through quantum stochastic calculus. In L. Accardi, W. v. Waldenfels (Eds.): "Quantum Probability and Applications II". Proceedings, Heidelberg 1984, Lecture Notes in Mathematics 1136, pp 207-222. Springer-Verlag, Heidelberg 1985.

[Fr 2] A. Frigerio: Covariant Markov dilations of quantum dynamical semigroups. Publ. RIMS, Kyoto University, 21 (1985), 657-675.

[Fr 3] A. Frigerio; V. Gorini: On stationary Markov dilations of quantum dynamical semigroups. In L. Accardi, A. Frigerio, V. Gorini (Eds): "Quantum Probability and Applications to the Quantum Theory of Irreversible Processes". Proceedings, Villa Mondragone 1982, Lecture Notes in Mathematics 1055, pp 119-125. Springer-Verlag, Heidelberg 1984.

[Fr 4] A. Frigerio, V. Gorini: Markov dilations and quantum detailed balance. Commun. Math. Phys. 93 (1984), 517-532.

[Gud] S.Gudder, J.-P. Marchand: Conditional expectations in von Neumann algebras: A new approach. Rep. Math. Phys. 12 (1977), 317-329.

[Hid] T. Hida: "Brownian Motion". Applications of Mathematics 11, Springer-Verlag, New York 1980.

[Hud] R. Hudson, K.R. Parthasarathy: Quantum Ito's formula and stochastic evolutions. Commun. Math. Phys. 93 (1984), 301-323.

[Kar] S. Karlin, H.M. Taylor:"A Second Course in Stochastic Processes". Academic Press, New York 1981.

[Ker] M. Kern, R. Nagel, G. Palm: Dilations of positive operators: Construction and ergodic theory. Math. Z. 156 (1977), 265-277.

[Kre] U. Krengel: "Ergodic Theorems", Walter de Gruyter, Berlin 1985.

[Kü 1] B. Kümmerer: Markov dilations on W*-algebras. Journ. Funct. Anal. 63 (1985), 139-177.

[Kü 2] B. Kümmerer: Examples of Markov dilations over the 2×2-matrices. In L. Accardi, A. Frigerio, V. Gorini (Eds.): "Quantum Probability and Applications to the Quantum Theory of Irreversible Processes", Villa Mondragone 1982, Lecture Notes in Mathematics 1055, Springer-Verlag, Heidelberg 1984, 228-244.

[Kü 3] B. Kümmerer: Markov dilations on the 2×2-matrices. In H. Araki, C.C. Moore, S. Stratila, D. Voiculescu (Eds.): "Operator Algebras and their Connections with Topology and Ergodic Theory". Proceedings, Busteni 1983, Lecture Notes in Mathematics 1132, pp 312-323. Springer-Verlag, Heidelberg 1985.

[Kü 4] B. Kümmerer: On the structure of Markov dilations on W*-algebras. In L. Accardi, W. v. Waldenfels (Eds.): "Quantum Probability and Applications II". Proceedings, Heidelberg 1984, Lecture Notes in Mathematics 1136, pp 318-331. Springer-Verlag, Heidelberg 1985.

[Kü 5] B. Kümmerer: "Construction and Structure of Markov Dilations on W*-Algebras". Habilitationsschrift, Tübingen 1986.

[Kü 6] B. Kümmerer: Markov dilations and non-commutative Poisson processes. Preprint.

[Kü 7] B. Kümmerer, H. Maassen: The essentially commutative dilations of dynamical semigroups on M_n. Commun. Math. Phys. 109 (1987), 1-22.

[Kü 8] B. Kümmerer, R. Nagel: Mean ergodic semigroups on W*-algebras. Acta Sci. Math. 41 (1979), 151-159.

[Kü 9] B. Kümmerer, W. Schröder: A Markov dilation of a non-quasifree Bloch evolution. Comm. Math. Phys. 90 (1983), 251-262.

[Kü 10] B. Kümmerer, W. Schröder: A survey of Markov dilations for the spin-$\frac{1}{2}$-relaxation and physical interpretation. Semesterbericht Funktionalanalysis, Tübingen, Wintersemester 1981/82, 187-213.

[Kü 11] B. Kümmerer, W. Schröder: a) On the structure of unitary dilations. Semesterbericht Funktionalanalysis. Tübingen, Wintersemester 1983/84, 177-225.
b) A new construction of unitary dilations: singular coupling to white noise. In L. Accardi, W. v. Waldenfels (Eds.):"Quantum Probability and Applications II". Proceedings, Heidelberg 1984, Lecture Notes in Mathematics 1136, pp 332-347. Springer-Verlag, 1985.

[Lew] J.T. Lewis, L.C. Thomas: How to make a heat bath. In A.M. Arthurs (Ed): "Functional Integration". Proceedings, Cumberland Lodge 1974, pp 97-123. Oxford University Press (Clarendon), London 1975.

[Maa] H. Maassen: Quantum Markov processes on Fock space described by integral kernels. In L. Accardi, W.v. Waldenfels (Eds.): "Quantum Probability and Applications II". Proceedings, Heidelberg 1984, Lecture Notes in Mathematics 1136, pp 361-374. Springer-Verlag, Heidelberg 1985.

[Par] W. Parry: "Topics in Ergodic Theory". Cambridge University Press, Cambridge 1981.

[Ped] G.K. Pedersen:"C*-Algebras and their Automorphism Groups". Academic Press, London 1979.

[Pri] H. Primas: "Chemistry, Quantum Mechanics, and Reductionism", Lecture Notes in Chemistry 24, Springer-Verlag, Berlin-Heidelberg 1981.

[Ros] M Rosenblatt: "Markov Processes. Structure and Asymptotic Behaviour". Grundlehren Vol. 184, Springer-Verlag, Heidelberg 1971.

[Roz] Yu. A. Rozanov. "Stationary Random Processes". Holden Day, San Francisco 1967.

[Sak] S. Sakai: "C*-Algebras and W*-Algebras", Springer-Verlag, Heidelberg 1971.

[Sha] D. Shale: Analysis over discrete spaces. Journ. Funct. Anal. 16 (1974), 258-288.

[Sz-N] B. Sz.-Nagy, C. Foias: "Harmonic Analysis of Operators on Hilbert Spaces", North Holland, Amsterdam 1970.

[Ta 1] M. Takesaki: Conditional expectations in von Neumann algebras. J. Funct. Anal. 9 (1971), 306-321.

[Ta 2] M. Takesaki: "Theory of Operator Algebras I", Springer-Verlag, New York 1979.

[Var] V.S. Varadarajan: "Geometry of Quantum Theory I". D. van Nostrand Comp., Inc., Princeton 1968.

[Vin] G.F. Vincent-Smith: Dilations of a dissipative quantum dynamical system to a quantum Markov process, Proc. London Math. Soc. (3), 49 (1984), 58-72.

[v.N] J. von Neumann: "Mathematische Grundlagen der Quantenmechanik". Springer-Verlag, Berlin-Heidelberg 1932 and 1971.

[Wal] P. Walters: "An Introduction to Ergodic Theory". Graduate Texts in Mathematics, Springer-Verlag, New York 1982.

[Ye 1] F.J. Yeadon: Measures on projections in W*-algebras of type II_1. Bull. London Math. Soc. 15 (1983), 139-145.

[Ye 2] F.J. Yeadon: Finitely additive measures on projections in finite W*-algebras. Bull. London Math. Soc. 16 (1984), 145-150.

DYNAMICAL ENTROPY FOR QUANTUM SYSTEMS

Göran Lindblad
Department of Theoretical Physics
Royal Institute of Technology
S-100 44 Stockholm, Sweden

INTRODUCTION

This talk will give an outline of an attempt to define a non-commutative counterpart to the Kolmogorov-Sinai entropy, an approach which is different from the Connes-Størmer one [1,2] as well as that of Emch [3]. The important points are:

a. It contains the commutative KS entropy as a simple special case.
b. It is based on the time-ordered quantum correlation kernels. This time order aspect is essential for non-commutative systems but seems to be completely lacking in the Connes-Størmer work.
c. It is closely related to an earlier work of mine in this direction [4] , but it seems to be more general and promising.
d. It is still restricted essentially to tracial states.
e. Some of the desired properties of the new definition are still uncertain conjectures.

For a non-commutative dynamical system we take a W*-algebra A, a *-automorphism T (and hence the group T^n, $n \in Z$) and a stationary normal state ρ. We can identify the system with its GNS representation

$$(A,T,\rho) = (\pi(A),W^+ . W, \Omega)$$

i.e. the lowest order correlation kernel is of the form

$$\rho(X^+T(Y)) = (\Omega,\pi(X)^+W^+\pi(Y)W \Omega).$$

In addition we need to generalize the following two notions of commutative ergodic theory: A finite partition and the join V of two such partitions.

The most evident thing to try is to use finite-dimensional subalgebras of A as finite partitions, and the algebra generated by two such subalgebras as the join

$$A_1 \vee A_2 = \left\{ A_1 \cup A_2 \right\}'' .$$

This gives the right thing for commutative A, but in the non-commutative case $A_1 \vee A_2$

will be ∞ - dimensional in general, and it may be the whole of A even if A_1 and A_2 are very 'small' subalgebras of A.

Another rather natural approach is to define a partition to be a partition of unity (PU), i.e. a decomposition of the unit 1 in A into a finite set of orthogonal projections in A

$$1 = \sum_k P_k , \qquad P_k^+ = P_k$$

$$P_i P_j = \delta_{ij} P_j .$$

However, the 'V' is not so easy to define for non-commuting PUs, and there is no maximal PU in a given finite-dimensional algebra A_1. The quantum measurements associated with the PUs:

$$\rho \rightarrow p_k^{-1} P_k \rho P_k \qquad \text{with probability } p_k = \rho(P_k),$$

do not form a closed set under composition. The fact that they need not commute means that there is in general a randomness in the outcomes of sequences of such measurements which ultimately comes from the measurement apparatus rather than from the dynamics of the system itself. Furthermore, in general the stationary state will not be invariant under the measurements, i.e. the equality

$$\sum_k \rho(P_k X P_k) = \rho(X), \quad \text{all } X \in A$$

will not hold, and this will lead to a lack of stationarity for the process. If the P_k are fixed under the modular automorphism group of ρ , then the equality holds, but there are examples where there is no subalgebra A_1 of finite dimension left pointwise invariant [2].

The definition of a generalized KS entropy in [2] avoids all the problems indicated above in a highly ingenuous way. This subtle approach, however, seems difficult to interpret in a physically meaningful way. The entropy associated with a finite-dimensional subalgebra A_1 and its successive time translates $A_n = T^{n-1}(A_1)$ turns out to be a symmetric function of the A_ns. Thus there is no intrinsic time order in this scheme, contrary to what one naturally expects to hold in the quantum case. The time translations are treated in precisely the same way as space translations. In the approach advocated here there is a considerable difference between space and time translations due to the causal time order in the latter case.

After this talk was given, a continuation of the work of Connes and Størmer was published by Connes, Narnhofer and Thirring in [5], where a refinement of [2] is given with full proofs. It is shown that in some cases the so defined KS entropy coincides with the entropy density of of quantum statistical mechanics (which deals with space translations, of course).

NON-COMMUTATIVE PARTITIONS AND ENTROPY

By a partition (operational partition of unity = OPU) associated with a finite-dimensional subalgebra $A_1 \subset A$ we will understand the same thing as an instrument [4]

$$\mathcal{E} = \left\{ V_i \in A_1 ; \ \sum_{i \in I} V_i^+ V_i = 1, \ I \text{ finite} \right\}.$$

In the development below we will have to assume in addition that the state ρ is left invariant by the OPU

$$\rho\left(\sum_i V_i^+ X V_i \right) = \rho(X), \quad \text{all } X \in A.$$

Provided that $\sum_i V_i V_i^+ = 1$ this holds when ρ is a tracial state, and more generally when each V_i is fixed under the modular automorphism group of ρ .

If $\mathcal{E}_1$, $\mathcal{E}_2$ are OPUs, then the composition $\mathcal{E} = \mathcal{E}_1 \cdot \mathcal{E}_2$ is defined in a straightforward way through operator multiplication of the corresponding V_is. This construction replaces the join V. Here there is a natural time order reflected in the operator multiplication. However, if $\mathcal{E}_1$ and $\mathcal{E}_2$ belong to different, non-commuting subalgebras, then $\mathcal{E}$ need not be associated with a finite-dimensional subalgebra.

The entropy of an OPU relative to the state ρ is defined to be

$$S(\mathcal{E};\rho) := S(\sigma) = - \operatorname{Tr}(\sigma \ln \sigma)$$

$$\sigma \equiv \sigma(\mathcal{E};\rho) \sim \sigma_{ij} := \rho(V_i^+ V_j)$$

and the entropy associated with the subalgebra A_1 is

$$S(A_1;\rho) = \sup_{\mathcal{E}} S(\mathcal{E};\rho)$$

where the supremum is over all OPUs associated with A_1. We now look at the meaning of this definition in the commutative and quantum cases.

For a commutative system the definition reduces to the standard one. Let A_1 be the algebra of measurable functions defined by a finite partition $\alpha = \{X_\beta\}$ of a probability space (X,μ), and let χ_β be the characteristic function of the subset X_β of X. The functions χ_β generate A_1, and any $f \in A_1$ is of the form

$$f(x) = \sum_\beta f^\beta \chi_\beta(x), \qquad x \in X ,$$

$$p_\beta f^\beta = \int_X f(x)\chi_\beta(x)\, d\mu(x), \qquad p_\beta = \int_X \chi_\beta(x)\, d\mu(x).$$

The OPU $\{V_i \in A_1\}$ defines the density matrix

$$\sigma_{ij} = \int_X V_i^+ V_j \, d\mu = \sum_\beta V_j^\beta \int_X V_i^+ \chi_\beta \, d\mu = \sum_\beta p_\beta \bar{V}_i^\beta V_j^\beta$$

where $\sum_i |V_i^\beta|^2 = 1$, all β. Thus σ is a convex combination of pure density matrices with weights p_β. From standard properties of the quantum entropy [6] it follows that

$$S(\sigma) \leq H(\alpha) = - \sum_\beta p_\beta \ln p_\beta$$

with equality iff the pure components are orthogonal. It is easy to see that the maximum is obtained (for a given α) when we choose

$$\mathcal{E}_{max} = \{V_\beta = X_\beta\}$$

$$S(\mathcal{E}_{max};\mu) = S(A_1;\mu) = H(\alpha).$$

$H(\alpha)$ is the standard definition of the entropy of a finite partition, thus we obtain the claimed identity.

The case $A_1 = A = M_n(C)$ turns out to have a simple form as well. Let ρ_1 be a faithful state on A_1, e.g. the tracial state. There is a GNS representation

$$\rho_1(X) = (\Omega,\pi(X)\Omega)$$

in a Hilbert space K such that

$$\pi(A_1) = A_1 \otimes 1$$

$$\pi(A_1)' = 1 \otimes A_1 .$$

With a proper choice of Ω the expression

$$\omega(X \otimes Y) = (\Omega, X \otimes Y\, \Omega)$$

defines a pure state ω on B(K) such that

$$\rho_1(X) = \omega(X \otimes 1) = \omega(1 \otimes X).$$

Then the following expression defines a pure state:

$$\mu_{ij} = (V_i \otimes 1)\ \omega\ (V_j \otimes 1)^+ .$$

The two complementary traces, over K and over the index set I, respectively, give density operators

$$\sigma_{ij} = Tr_K\, \mu_{ji} = \rho_1(V_i^+ V_j),$$

$$\tau = \sum_i \mu_{ii} .$$

Due to the the triangle inequality for the quantum entropy [6] we know that

$$S(\sigma) = S(\tau).$$

In [4] this common value was called 'the entropy of the CP map E' where

$$E(X) = \sum_i V_i^+ X V_i .$$

Among the CP maps on A_1 which conserve ρ_1 we claim that the one which gives the maximal entropy is

$$E_{max}(X) = \rho_1(X)\cdot 1 .$$

It follows that

$$\tau = (E^*_{max} \otimes id)(\omega) = \rho_1 \otimes \rho_1 .$$

It is easy to write down a corresponding OPU $\mathcal{E}_{max}$. It is not uniquely defined by E_{max} but the entropy is

$$S(\mathcal{E}_{max};\rho_1) = 2S(\rho_1).$$

Any other OPU conserving ρ_1 will satisfy

$$E \cdot E_{max} = E_{max} \cdot E = E_{max} .$$

From the non-negativity of the relative entropy [6] (ρ,σ any two states)

$$S(\rho|\sigma) = Tr(\rho \ln \rho - \rho \ln \sigma) \geq 0$$

and the projective and trace-preserving property of E_{max} (it is a conditional expectation) it follows that

$$S(\mathcal{E}_{max};\rho_1) - S(\mathcal{E};\rho_1) = S((E^* \otimes id)(\omega)|\rho_1 \otimes \rho_1) \geq 0 ,$$

which verifies the maximality, i.e.

$$S(A_1;\rho_1) = 2S(\rho_1).$$

Note that $S(\rho_1) < \infty$ in this finite-dimensional case. This means that the entropy of any OPU representing an infinite sequence of non-commuting measurements on this finite system will be finite, in spite of the fact that the sequence of outcomes may be perfectly random. Thus the entropy of the OPU does not measure the randomness which has its origin in the measuring apparatus. Of course, this is related to the fact that in a certain sense we have a maximal element in the set of OPUs.

For a general finite-dimensional algebra we can make a central decomposition and obtain the entropy of an arbitrary OPU associated with this algebra as a contribution from the center Z (of the commutative form) plus a weighted contribution from the finite type I factors :

$$A = \sum_k^{\oplus} A_k , \qquad \rho = \sum_k p_k \rho_k ,$$

$$S(A;\rho) = S(Z;\rho) + \sum_k p_k S(A_k;\rho_k) .$$

For tensor products we have additivity

$$S(A_1 \otimes A_2;\rho_1 \otimes \rho_2) = S(A_1;\rho_1) + S(A_2;\rho_2).$$

For composition of state-preserving OPUs it holds that

$$\max_i S(\mathcal{E}_i;\rho_1) \leq S(\mathcal{E}_1 \cdot \mathcal{E}_2;\rho_1) \leq S(\mathcal{E}_1;\rho_1) + S(\mathcal{E}_2;\rho_1) ,$$

where the first inequality follows from the monotonic property of the relative entropy under CP maps, the second from the triangle inequality [6].

DYNAMICAL ENTROPY

Consider a W*-dynamical system (A,T,ρ), a finite-dimensional subalgebra $A_1 \subseteq A$ and OPUs $\mathcal{E}$ associated with A_1 which leave ρ invariant. The last restriction is serious and may lead to trivial results unless ρ is tracial. A given $\mathcal{E}$ is time translated in the natural way

$$\mathcal{E} = \{V_i\} \rightarrow T^n(\mathcal{E}) \equiv \mathcal{E}_n = \{T^n(V_i)\}.$$

Introduce the time-ordered composition of these OPUs

$$\mathcal{E}(0,1,..,n) = \mathcal{E}_n \cdot \mathcal{E}_{n-1} \cdot \,\ldots\, \cdot \mathcal{E}_1 \cdot \mathcal{E}_0 ,$$

which is an OPU on A. It immediately follows that the density matrix

$$\sigma_n \equiv \sigma(\mathcal{E}(0,..,n-1);\rho)$$

defined in the previous section involves the n-th order time-ordered correlation kernel [7-9] when A_1 represents the observed subsystem of the whole system described by the algebra A.

We note that there is time translation invariance due to the stationarity of the state under T:

$$\sigma(\mathcal{E}(0,..,n-1);\rho) = \sigma(\mathcal{E}(1,..,n);\rho)$$

hence we are allowed to use the notation σ_n for both. Furthermore, due to the invariance of ρ under the OPUs, σ_{n-1} is obtained from σ_n by taking the partial trace over the first or the last OPU. It follows that $\{\sigma_n\}_1^\infty$ form a translation invariant state of a lattice system (with the algebra $A_1 \otimes A_1$ in each lattice point [4] for A_1 = B(H) [4]). Standard arguments prove the existence of the mean entropy

$$h(\mathcal{E},T;\rho) = \lim_{n\to\infty} n^{-1} S(\sigma_n),$$

where lim can be replaced by inf (it is a decreasing sequence). Define the dynamical entropy associated with A_1 as the sup over all OPUs associated with A_1

$$h(A_1,T;\rho) = \sup_{\mathcal{E}} h(\mathcal{E},T;\rho).$$

From the results of the previous section follows that the RHS $\leq 2S(\rho_1)$, $\rho_1 = \rho \mid A_1$. Clearly the function h satisfies

$$A_1 \subseteq A_2 \Rightarrow h(A_1,T;\rho) \leq h(A_2,T;\rho)$$

Finally we can remove the dependence on the subalgebra A_1 and introduce a non-commutative counterpart of the KS entropy:

$$h(T;\rho) = \sup_{A_1 \text{ finite} \subseteq A} h(A_1,T;\rho).$$

It is clear from the previous section that when A is commutative, A_1 the subalgebra defined by a finite partition α, then

$$h(A_1,T;\rho) = h(\alpha,T)_{KS} .$$

Consequently we find, taking the supremum over all finite partitions, that

$$h(T;\rho) = h(T)_{KS}$$

in this case, confirming that we have a genuine generalization of the commutative Kolmogorov-Sinai scheme.

The definition above is closely related to the one given in [4], where a stationary quantum stochastic process was defined, essentially through the time-ordered correlation kernels over a fixed algebra A_1. The stationarity was defined in such a way that an OPU over A_1 has this necessary property if it leaves the partial state ρ_1 of A_1 invariant, a condition which is easily satisfied. Then the entropy defined for the n-th order correlation kernel, called S_n in [4], is, for $A_1 = B(H)$,

$$S_n = S(\sigma(\mathcal{E}_{max}(0,..,n);\rho) - 2S(\rho_1)$$

in the present notation. The mean entropy defined in [4] is the same as

$$h(\mathcal{E}_{max},T;\rho)$$

but the dynamics T was not explicitly introduced in [4] , just the kernels. It is not obvious that the supremum in the definition of $h(A_1,T;\rho)$ must be achieved by $\mathcal{E}_{max}$, and there is therefore not an immediate identification with the present definition for fixed A_1. For particular examples it is indeed possible to see that for fixed n, $\sigma(\mathcal{E}(0,..,n))$ need not have maximal entropy for the choice $\mathcal{E} = \mathcal{E}_{max}$ (this seems to be one aspect of the 'quantum Zeno paradox'). It is not clear what happens for the mean entropy $h(\mathcal{E},T;\rho)$.

OPEN PROBLEMS

The definition of a generalized KS entropy given above seems to have some conceptual advantages over the Connes-Størmer one, namely that it is based on quantities with an obvious physical interpretation, and that it has a rather simple structure.

The main fault of the scheme is the restriction to OPUs leaving the state of the system invariant. This condition may be too strong to allow non-trivial OPUs when the state is not tracial. It may be worthwhile to recall that there is a very basic physical fact behind this, namely that observations of a quantum system perturbs the state in general. This can not be changed by clever mathematical definitions. The problem may look a bit different to a matematician who wants to use the KS entropy in the classification of W*-automorphisms and to a physicist interested in things like quantum chaos [8].

A possible line of development is to use OPUs not belonging to finite-dimensional subalgebras. The index set I in the definition of the OPU was assumed finite, which means that the mean entropy for a given OPU is finite

$$h(\mathcal{E},T;\rho) \leq \ln(\text{card } I) < \infty \quad .$$

It may be easier to satisfy the invariance property with this relaxation of the restrictions on the OPUs, but instead several other problems turn up. The OPU may now contain a non-trivial inner dynamics of A which could perhaps give

$$h(\mathcal{E},T;\rho) > 0$$

even for a trivial dynamics T = id. Such a contribution would have to be subtracted in some way.

In order to have the possibility of calculating the entropy for simple models it seems necessary to have what Connes calls a Kolmogorov-Sinai theorem. This is a result of the form: If $\{A_k\}$ is an increasing sequence of finite-dimensional subalgebras such that $\{\cup A_k\}'' = A$, then

$$h(T;\rho) = \lim_{k\to\infty} h(A_k,T;\rho).$$

Such a property is satisfied by the Connes-Størmer entropy [2]. It is not yet clear if the same is true of the present scheme.

In [9] is described a special class of systems with a deterministic property. For them the GNS dynamics, given by the group of unitaries $\{W^n\}$ in the Hilbert space K, has no shift component. By analogy with classical results [10] we expect such a system to have zero entropy. It is a natural conjecture that a system with positive entropy has a part with countable Lebesgue spectrum. It is still an open problem if the present scheme has this highly desirable property.

REFERENCES

1. A. Connes, E. Størmer: Acta Math. 134, 289 (1975)
2. A. Connes: C.R. Acad. Sci. Paris 301, ser 1, no 1 (1985)
3. G.G. Emch: Commun. Math. Phys. 49, 191 (1976)
4. G. Lindblad: Commun. Math. Phys. 65, 281 (1979)
5. A. Connes, H. Narnhofer, W. Thirring: Commun. Math. Phys. 112, 691 (1987)
6. A. Wehrl: Rev. Mod. Phys. 50, 221 (1978)
7. G. Lindblad, in:'Quantum Probability and Applications II', L. Accardi, W. von Waldenfels, eds., Springer Lecture Notes in Mathematics 1136, 348 (1985)
8. G. Lindblad, in:'Fundamental Aspects of Quantum Theory', V. Gorini, A. Frigerio, eds. NATO ASI B144, 199, Plenum Press 1986
9. G. Lindblad:'A reconstruction theorem for quantum dynamical systems'. Preprint Stockholm 1985
10. V.A. Rohlin: Russian Math. Surveys 22(5),(1967)

AN INTEGRAL KERNEL APPROACH TO NOISE

Martin Lindsay
Department of mathematics, King's College
Strand, London WC2R 2LS

Hans Maassen
Instituut voor Theoretische Fysica, KU Nijmegen
Toernooiveld, 6525 ED Nijmegen, the Netherlands

Abstract. A stochastic calculus based on integral kernels is developed for the Wiener process. The application of integral kernels to other types of noise is indicated.

Introduction.

Given a process of stationary independent increments of zero mean (a "noise") in classical or in quantum probability theory, its representation Hilbert space can be given the structure of a Fock space in a natural way: the "n-particle" subspace is the space of all n-fold stochastic integrals. It is for this reason that Fock spaces occur quite often in probability theory. From the above consideration it is clear that the Fock space structure does not betray which noise is being considered. However, by transferring the *-algebraic structure to Fock space (i.e. by defining an appropriate involution and product on it), one settles for a definite type of noise. The product is always associative, but may be commutative (in the case of classical noise) or noncommutative (for a quantum noise). There are essentially two types of classical noise: the Wiener and the (compensated) Poisson process, whereas the following quantum noises are known: Clifford noise [BSW], the quantum Wiener process [CoH] or Bose noise, Fermi noise [ApH] and quantumPoisson processes [Küm], [Fri], [FrM].

The Fock space or "integral kernel" approach to noise has the advantage that all non-algebraic, linear aspects, such as stochastic integration and differentiation bacome simple and independent of the type of noise under consideration. A disadvantage is that the product usually takes a complicated form, so that algebraic aspects tend to get obscured.
In this paper we first introduce a convenient kernel notation for Fock spaces (§1), and then heuristically sketch the above ideas (§2). In §3 we treat in detail the integral kernel calculus for one particular case: the Wiener process in classical probability. (§3 may be read independently of §2). We briefly comment on quantum noises in §4. The case of Bose noise will be treated in detail elsewhere [LiM].

§1 Fock space in set notation.

Let I denote the interval [0,1]. The symmetric and antisymmetric Fock spaces over $L^2(I)$ are usually defined as the Hilbert direct sums

$$F_s = C \oplus L^2(I) \oplus L^2(I^2)_{sym} \oplus \cdots$$

and

$$F_a = C \oplus L^2(I) \oplus L^2(I^2)_{antisym} \oplus \cdots$$

respectively. In view of the total ordering of the real line and the non-atomicity of Lebesgue measure both symmetric and antisymmetric functions on I^n are determined (up to null sets) by their values on the strictly ordered simplex $I^n_< := \{ (t_1,\cdots,t_n) \in I^n \mid t_1<t_2<\cdots<t_n \}$. Therefore restriction to $I^n_<$ followed by (anti-)symmetric extension defines a linear bijection between $L^2_{sym}(I^n)$ and $L^2_{antisym}(I^n)$. In view of the obvious isometry of these bijections, F_s and F_a are *naturally isomorphic*.
Since the points of $I^n_<$ are in a one to one correspondence with the subsets of I consisting of n elements, let us agree to *identify* $I^n_<$ and $\Gamma_n(I) := \{ \sigma \subset I \mid \#\sigma = n \}$. The intermediate object

$$C \oplus L^2(I) \oplus L^2(I^2_<) \oplus \cdots$$

in the isomorphism between F_s and F_a is thereby identified with

$$F(I) := L^2(\Gamma(I),\mu),$$

where $\Gamma(I) = \{ \sigma \subset I \mid \#\sigma < \infty \}$ and the measure μ is given by

$$\mu(\{\emptyset\}) = 1 \ , \quad (\mu \mid I^n_<) = \text{n-dimensional Lebesgue measure.}$$

We write $d\sigma$ for $d\mu(\sigma)$ and from now on refer to $F(I)$ as Fock space.

An immediate advantage of this set notation is its combinatorial convenience. For instance, if $h \in L^1(\Gamma\times\Gamma)$ then

$$\int_\Gamma \int_\Gamma h(\alpha,\beta)\, d\alpha\, d\beta = \int_\Gamma \{ \sum_{\alpha\in\sigma} h(\alpha,\sigma\backslash\alpha) \}\, d\sigma \ . \tag{1.1}$$

The *product* (or *coherent*) *vectors*

$$\pi_\phi : \ \tau \rightarrow \prod_{t\in\tau} \phi(t) \ , \qquad (\phi \in L^2(I))$$

satisfy

$$\langle \pi_\phi, \pi_\psi \rangle = \int \pi_{\bar\phi\psi}(\sigma)\, d\sigma = e^{\langle \phi,\psi\rangle}, \tag{1.2}$$

and the creation, number and annihilation processes of [HuP] are given respectively by:

$$A_t^\star g(\sigma) = \sum_{s\in\sigma\cap[0,t]} g(\sigma\backslash\{s\}) \ ; \ A_t g(\sigma) = \int_0^t g(\sigma\cup\{s\})\, ds \ .$$

$$\Lambda_t g(\sigma) = \#(\sigma\cap[0,t])\cdot g(t) \ ;$$

for $g \in F(I)$. Factorials usually present in such formulae have been subsumed by the isomorphism $f\in F \rightarrow \{ (n!)^{-\frac{1}{2}} f_n^{a/s} \}$, (a/s denoting (anti-)symmetric extension).

Remark: While the natural isomorphism between $F(X)$ and $F_s(L^2(X))$ extends from $X=I$ to $X=I^n$, for anti-symmetric Fock space there is no such natural extension: the isomorphism between $F_s(I)$ and $F_a(I)$ is an accident of one dimension. (We note however that the Fock operator stochastic calculus [HuP] does yield isomorphisms between the symmetric and antisymmetric Fock spaces intertwining irreducible CAR representations for multidimensional X [PaS].)

§2 Fock space and noise (*heuristic*).

Starting from postulated "Ito rules" and commutation relations for infinitesimal increments of a "noise" (a zero mean independent increment stochastic process) we shall derive *in an entirely formal way*, bilinear products on Fock space which, in the classical cases, will reflect the algebraic structure of pointwise multiplication between random variables in the corresponding probability spaces. Brownian motion, the (compensated) Poisson processes and the Clifford process [BSW] are treated here, quantum Brownian motion will be mentioned in §5 and treated in [LiM].

Let $(N_t)_{0\le t\le 1}$ be a noise. Writing

$$N(f) = \int_\Gamma f(\sigma)\, dN_\sigma$$

for the sum of repeated stochastic integrals

$$\sum_{n=0} \quad \int_{0\le t_1\le\cdots\le t_n\le 1} \cdots\int f(\{t_1,\cdots,t_n\})\, dN_{t_1}\cdots dN_{t_n}\ , \tag{2.1}$$

we call f the *kernel* of the "random variable" N(f). The map N is linear and respects complex conjugation, $E(N(f)) = f(\emptyset)$ and, for suitible f and g, $N(f)\cdot N(g) = N(f\star g)$, where the product $\star$ depends on the noise:

Brownian motion: $N_t = B_t$, where

$$dB_t^{\,2} = dt\ ;\ dB_t dB_s = dB_s dB_t\ ,$$

so that $dB_\sigma dB_\tau = d\delta\, dB_\gamma$, where $\delta = \sigma\cap\tau$ and $\gamma = \sigma\Delta\tau$. Thus, putting $\alpha = \sigma\backslash\tau$ and $\beta = \tau\backslash\sigma$, and applying (1,1):

$$\int_\Gamma f(\sigma)\, dB_\sigma \cdot \int_\Gamma g(\tau)\, dB_\tau = \int_\Gamma \iint_{\alpha\cap\beta=\emptyset} f(\alpha\cup\delta)\, g(\beta\cup\delta)\, dB_\alpha dB_\beta\, d\delta$$

$$= \int_\Gamma \Big\{ \sum_{\gamma=\alpha\cup\beta} \int_\Gamma f(\alpha\cup\delta)\, g(\beta\cup\delta)\, d\delta \Big\}\, dB_\gamma\ .$$

In other words, the *Wiener product* $\star^W$ is given by

$$f\star^W g\ :\ \sigma \to \sum_{\alpha\cup\beta=\sigma} \int_\Gamma f(\alpha\cup\delta)\, g(\beta\cup\delta)\, d\delta\ .$$

Isometry of the map B is now easily verified:

$$E(\,|B(f)|^2\,) = E(\,B(\bar{f}\star f)\,) = (\bar{f}\star f)(\emptyset) = \int_\Gamma \bar{f}(\delta)f(\delta)\,d\delta = \int_\Gamma |f|^2 d\mu.$$

The Clifford process: $N_t = C_t$ with

$$dC_t^2 = dt \;\; ; \;\; dC_t dC_s = -dC_s dC_t.$$

Let $\epsilon(\sigma,\tau)$ denote the sign of the permutation $\left\{\begin{matrix}\sigma,\tau\\ \sigma\cup\tau\end{matrix}\right\}$. The same computation as above yields the *Clifford product* [Mey]:

$$f \star_c g \;:\; \sigma \to \sum_{\alpha\cup\beta=\sigma} \int_\Gamma \epsilon(\alpha,\beta\cup\delta)\, f(\alpha\cup\delta)\, g(\beta\cup\delta)\, d\delta\,.$$

The Poisson process: $N_t = P_t - \lambda t$, where P_t is a Poisson process with density $\lambda>0$:

$$dP_t^2 = dP_t \;\; ; \;\; dP_t dP_s = dP_s dP_t\,,$$

so that $dN_\upsilon dN_\tau = \sum_{\rho\cup\eta=\delta} \lambda^{\#\eta} d\eta\; dN_\rho dN_\gamma$ with γ,δ as before. This yields the *Poisson product*

$$f \star^P g \;:\; \sigma \to \sum_{\alpha\cup\beta\cup\rho=\sigma} \int_\Gamma f(\alpha\cup\rho\cup\eta)\, g(\beta\cup\rho\cup\eta)\, \lambda^{\#\eta}\, d\eta\,.$$

The map N is thus isometric from $L^2(\,\Gamma(I)\,,\,\lambda^{\#\eta}d\eta\,) =: F_\lambda(I)$.

If $(f_s)_{0\le s\le 1}$ is a family of kernels, then the corresponding stochastic process $(N(f_s))_{0\le s\le 1}$ will be adapted to the filtration of the noise if each f_s vanishes outside $\Gamma[0,s]$. The family is then called an *(adapted) kernel process.* We next derive the transformation on kernel processes which corresponds to *stochastic integration* with respect to the noise. Thus we seek the kernel of $\int_0^1 N(f_s)dN_s$, where f is an adapted kernel process. The stochastic integral we have in mind is one of Ito type: the increments are "in the future": $dN_s = N_{s+ds}-N_s$ $(ds>0)$. Using the change of variable $\sigma = \tau\cup\{s\}$:

$$\int_0^1 N(f_s)dN_s = \int_0^1 \int_{\Gamma[0,s]} f_s(\tau)\, dN_\tau dN_s = \int_{\Gamma[0,1]} f_{\max\sigma}(\sigma\setminus\{\max\sigma\})\, dN_\sigma$$

Thus the kernel of $\int_0^1 N(f_s)dN_s$ is $\sigma \mapsto f_{\max\sigma}(\sigma\setminus\{\max\sigma\})$. Note that this transformation is quite independent of the type of the noise.

The operation on kernels corresponding to *conditional expectation* with respect to the σ-algebra generated by the noise up to time t is the multiplication operator $M_{\chi_{\Gamma[0,t]}}$, as a moment's reflection on (2.1) reveals.

Finally we calculate the effect on kernels corresponding to multiplication of "random variables" by the noise itself. Since $N_t = N(\chi_{\Gamma_1[0,t]})$, and

$$\chi_{\Gamma_1[0,t]} \star^W g\,(\sigma) = \sum_{\alpha\cup\beta=\sigma} \int_\Gamma \chi_{\Gamma_1[0,t]}(\alpha\cup\gamma)\, g(\beta\cup\gamma)\, d\gamma$$
$$= \sum_{s\in\sigma\cap[0,t]} g(\sigma\setminus\{s\}) + \int_0^t g(\sigma\cup\{s\})\, ds\,,$$

multiplication by Brownian motian is effected by the sum $A_t^\star + A_t$ on F(I) . On the other hand,

$$\chi_{\Gamma_1[0,t]} \star^P g\,(\sigma) = \sum_{\alpha\cup\beta\cup\gamma=\sigma} \int_\Gamma \chi_{\Gamma_1[0,t]}(\alpha\cup\gamma\cup\delta)\, g(\beta\cup\gamma\cup\delta)\, \lambda^{\#\delta}\, d\delta$$
$$= \sum_{s\in\sigma\cap[0,t]} g(\sigma\setminus\{s\}) + \sum_{s\in\sigma\cap[0,t]} g(\sigma) + \int_0^t g(\sigma\cup\{s\})\, ds,$$

so that multiplication by the Poisson process $P_t = N_t + \lambda t$ is effected by the operator $A_t^\star + \Lambda_t + A_t + \lambda t 1$ on F(I) [HuP]. (A difference of factors $\lambda^{\frac{1}{2}}$ is due to our choice of the norm in $F_\lambda(I)$.)

§3 Kernel calculus of the Wiener process.

We now forget the heuristic introduction in §2. Starting from the Fock space F[0,1] in §1 and the product $\star = \star^W$, we build up the stochastic calculus for the Wiener process $(B_t)_{0\le t\le 1}$.

3.1. *An algebra of integral kernels.*

Let Λ denote the operator on F, given on its domain $\mathrm{Dom}(\Lambda) = \{ f\in F \mid \int (\#\tau)^2 |f(\tau)|^2 d\tau < \infty \}$ by multiplication by the particle number:

$$(\Lambda f)(\tau) = (\#\tau)\cdot f(\tau)\,.$$

Lemma 3.1. *Let* $f,g \in \mathrm{Dom}(3^{\frac{1}{2}\Lambda})$. *Then the map*

$$h : \tau \to \sum_{\alpha \subset \tau} \int_\Gamma f(\alpha \cup \beta)\, g((\tau \backslash \alpha) \cup \beta)\, d\beta$$

is well-defined for almost all $\tau \in \Gamma$ *and yields a function* $h \in F$ *satisfying*

$$\|h\| \le \|3^{\frac{1}{2}\Lambda} f\| \cdot \|3^{\frac{1}{2}\Lambda} g\| . \tag{3.1}$$

We shall denote h by $f \star g$. Note that $(\bar{f} \star f)(\emptyset) = \|f\|^2$.

proof. By (1.1) we have for all $f \in \mathrm{Dom}(3^{\frac{1}{2}\Lambda})$:

$$\|3^{\frac{1}{2}\Lambda} f\|^2 = \int_\Gamma 3^{\#\tau} |f(\tau)|^2 \, d\tau = \int_\Gamma (|f(\tau)|^2 \sum_{\alpha \subset \tau} 2^{\#\alpha})\, d\tau$$

$$= \int_{\Gamma \times \Gamma} 2^{\#\alpha} |f(\alpha \cup \beta)|^2 \, d\alpha \, d\beta .$$

It follows that $f^\alpha : \beta \to f(\alpha \cup \beta)$ is square-integrable for almost all $\alpha \in \Gamma$ and that

$$\int_\Gamma 2^{\#\alpha} \|f^\alpha\|^2 \, d\alpha = \|3^{\frac{1}{2}\Lambda} f\|^2 . \tag{3.2}$$

Now, applying the inequality $|\sum_{j=1}^J c_j|^2 \le J \cdot \sum_{j=1}^J |c_j|^2$, (valid for all $J \in \mathbb{N}$ and $c_1, \cdots, c_J \in \mathbb{C}$), we find for all $f,g \in \mathrm{Dom}(3^{\frac{1}{2}\Lambda})$ and almost all $\tau \in \Gamma$:

$$|\sum_{\alpha \subset \tau} \langle f^\alpha, g^{\tau \backslash \alpha} \rangle|^2 \le 2^{\#\tau} \cdot \sum_{\alpha \subset \tau} |\langle f^\alpha, g^{\tau \backslash \alpha} \rangle|^2$$

$$\le 2^{\#\tau} \cdot \sum_{\alpha \subset \tau} \|f^\alpha\|^2 \cdot \|g^{\tau \backslash \alpha}\|^2 .$$

It follows that $h(\tau) = \sum_{\alpha \subset \tau} \langle \overline{f^\alpha}, g^{\tau \backslash \alpha} \rangle$ is well-defined for almost all $\tau \in \Gamma$. Moreover, applying (1.1) and (3.2) we conclude that

$$\int_\Gamma |h(\tau)|^2 d\tau \le \int_\Gamma 2^{\#\tau} \sum_{\alpha \subset \tau} \|f^\alpha\|^2 \cdot \|g^{\tau \backslash \alpha}\|^2 d\tau$$

$$= \int_{\Gamma \times \Gamma} (2^{\#\alpha} \|f^\alpha\|^2) \cdot (2^{\#\beta} \|g^\beta\|^2)\, d\alpha\, d\beta = \|3^{\frac{1}{2}\Lambda} f\|^2 \cdot \|3^{\frac{1}{2}\Lambda} g\|^2 .$$

□

Now let $K := \bigcap_{t>0} \mathrm{Dom}(e^{t\Lambda})$ $\;(= \bigcap_{t>0} e^{-t\Lambda} F)$.

Corollary. $(K,\star)$ *is a commutative algebra.*

proof. By lemma 3.1 for $t \geq \frac{1}{2}\log 3$ the operation $\star$ maps $\mathrm{Dom}(e^{t\Lambda}) \times \mathrm{Dom}(e^{t\Lambda})$ to $\mathrm{Dom}(e^{t'\Lambda})$ with $t' = t - \frac{1}{2}\log 3$. Hence K is closed under $\star$. Commutativity is obvious and associativity is shown by repeated use of (1.1). □

3.2. *The duality transform.*

Let $\Omega = C[0,1]$ and let P be the Wiener measure on Ω . For $0 \leq t \leq 1$ let $B_t : \Omega \to \Omega : \omega \mapsto \omega(t)$ be the Wiener process itself. We denote the *Wiener space* $L^2(\Omega,P)$ by $W[0,1]$ or W . For $\phi \in L^2[0,1]$ let the *exponential vector* $\epsilon_\phi \in W$ be given by the exponentiated stochastic integral

$$\epsilon_\phi(\omega) = \exp\left\{ \int_0^1 \phi(t)\, d\omega(t) - \tfrac{1}{2} \int_0^1 \phi(t)^2 dt \right\} .$$

Let Ψ denote the linear span of the vectors $\epsilon_{i\phi}$ with real-valued $\phi \in L^2[0,1]$. Note that Ψ consists of bounded functions.
On the other hand, let Π be the linear span of the product vectors $\pi_{i\phi} \in F$ with real- valued ϕ (cf. §1).

Proposition *There exists a unique unitary equivalence* $B: F \to W$ *mapping* $\pi_{i\phi}$ *to* $\epsilon_{i\phi}$.
For $f \in F$ we shall often denote $B(f)$ by $\hat{f}$.

proof. Π and Ψ are dense in F and W respectively. The statement follows from the fact that

$$\langle \epsilon_{i\phi}, \epsilon_{i\psi} \rangle = e^{-\langle \phi,\psi \rangle} = \langle \pi_{i\phi}, \pi_{i\psi} \rangle . \qquad \square$$

Following Segal, this unitary equivalence is called the *duality transform*, since it connects a Fock space, associated with "particles" to the Wiener space, associated with a random "field".
The following theorem motivates the introduction of $(K,\star)$:

Theorem 3.2. *For all* $f,g \in K$ *one has*

$$(f \star g)^{\wedge} = \hat{f} \cdot \hat{g} . \tag{3.3}$$

proof. One calculates that, for $\phi,\psi \in L^2[0,1]$:

$$(\pi_\phi \star \pi_\psi)(\tau) = \sum_{\alpha\subset\tau} \int_\Gamma \pi_\phi(\alpha\cup\beta)\, \pi_\psi((\tau\backslash\alpha)\cup\beta)\, d\beta$$

$$= \left\{ \sum_{\alpha\subset\tau} \pi_\phi(\alpha)\, \pi_\psi(\tau\backslash\alpha) \right\} \cdot \left\{ \int_\Gamma \pi_{\phi\psi}(\beta)\, d\beta \right\}$$

$$= \pi_{\phi+\psi}(\tau) \exp(\int_0^1 \phi\psi\, dt), \tag{3.4}$$

and

$$\epsilon_\phi \cdot \epsilon_\psi = \epsilon_{\phi+\psi} \exp(\int_0^1 \phi\psi\, dt).$$

It follows that (3.3) holds for $f=\pi_\phi$ and $g=\pi_\psi$, and hence for all $f,g \in \Pi$ by linearity. We shall extend this result to all of K in two steps.
Define the linear operators

$$S_f : K \to K : g \to f\star g , \qquad (f \in K) ;$$

$$M_h : \mathrm{Dom}(M_h) \to W : k \to h\cdot k , \qquad (h \in W) ;$$

where $\mathrm{Dom}(M_h)$ is the natural domain $\{ k \mid h\cdot k \in W \}$ of the multiplication operator M_h . By $\hat{S}_f$ we denote the operator BS_fB^{-1} on BK.

<u>Lemma</u> $\hat{S}_f \subset M_{\hat{f}}$ *for all* $f \in \Pi$.

proof. For $f \in \Pi$, $\hat{f} \in \Psi$ is bounded, hence $M_{\hat{f}}$ is a bounded operator. On the other hand, for $f = \pi_{i\phi}$ ($\phi \in L^2[0,1]$ real-valued), one sees that S_f is bounded from the following calculation based on (3.4) and the facts that $(\bar{g}\star g)(\emptyset) = \|g\|^2$ and $\pi_0 \star g = g$. For $g \in K$:

$$\|S_f g\|^2 = \|f\star g\|^2 = (\bar{g}\star\pi_{-i\phi}\star\pi_{i\phi}\star g)(\emptyset) = e^{\|\phi\|^2}(\pi_0\star\bar{g}\star g)(\emptyset) =$$

$$= e^{\|\phi\|^2}\|g\|^2.$$

It follows that S_f , and hence $\hat{S}_f$, is bounded for $f \in \Pi$. But we know $\hat{S}_f$ and $M_{\hat{f}}$ to agree on Ψ. Therefore $\hat{S}_f \subset M_{\hat{f}}$, as $M_{\hat{f}}$ is everywhere defined.

□

Lemma $\hat{S}_f \subset M_{\hat{f}}$ *for all* $f \in \mathbf{K}$.

proof. The previous lemma implies that (3.3) holds for all $f \in \Pi$ and $g \in \mathbf{K}$, hence for all $f \in \mathbf{K}$ and $g \in \Pi$ by commutativity. This means that for $f \in \mathbf{K}$: $(M_{\hat{f}} \mid \Psi) \subset \hat{S}_f$, and it follows that

$$\hat{S}_f^{\star} \subset (M_{\hat{f}} \mid \Psi)^{\star}, \qquad (f \in \mathbf{K}).$$

Now, $S_f \subset S_{\bar{f}}^{\star}$ since $\langle g, f{\star}h\rangle = (\bar{g}{\star}f{\star}h)(\emptyset) = \langle \bar{f}{\star}g, h\rangle$ for all $f,g,h \in \mathbf{K}$ by associativity. On the other hand, since Ψ is a dense set, we have $(M_h \mid \Psi)^{\star} \subset M_{\bar{h}}$ for all $h \in \mathbf{W}$. We conclude that

$$\hat{S}_f \subset (\hat{S}_{\bar{f}})^{\star} \subset (M_{\bar{\hat{f}}} \mid \Psi)^{\star} \subset M_{\hat{f}}.$$

□

Thus theorem (3.2) is proved.

3.3. *Stochastic integration.*

In the theory of stochastic processes on Wiener space a curve $t \mapsto h_t \in \mathbf{W}$ is called an "adapted process" if for all $t \in [0,1]$ the random variable $\omega \mapsto h_t(\omega)$ does not depend on $\omega|(t,1]$; i.e.:

$$h_t \in \mathbf{W}_{t]} := \text{closed span of } \{\, \epsilon_\phi \mid \phi{=}0 \text{ on } (t,1] \,\}.$$

Now, $\mathbf{W}_{t]} = \hat{\mathbf{F}}_{t]}$, where

$$\mathbf{F}_{t]} := \text{closed span of } \{\, \pi_\phi \mid \phi{=}0 \text{ on } (t,1] \,\}$$
$$= \{\, f{\in}\mathbf{F} \mid f(\tau){=}0 \text{ unless } \tau \subset [0,t] \,\}.$$

For this reason we call a curve $t \mapsto f_t \in \mathbf{F}$ an *adapted kernel process* if $f_t(\tau)=0$ unless $\tau \subset [0,t]$. It is convenient to introduce the measure space (Γ_{ad}, μ_{ad}) as follows:

$$\Gamma_{ad} = \{\, (\tau,t) \in \Gamma{\times}[0,1] \mid \tau \subset [0,t) \,\}; \quad \mu_{ad} = j(\mu) = \mu \circ j^{-1},$$

where $j: \Gamma\backslash\{\emptyset\} \to \Gamma_{ad} : \tau \mapsto (\tau\backslash\{\max\tau\}, \max\tau)$. We shall denote the space $L^2(\Gamma_{ad}, \mu_{ad})$ of square integrable kernel processes by $\mathcal{I}_2$. A process $f \in \mathcal{I}_2$ may be considered as a function in $L^2(\Gamma{\times}[0,1], \mu{\times}\lambda)$, vanishing outside Γ_{ad} . The prime example of an adapted kernel process is the process $b_t = \chi_{\Gamma_1[0,t]}$ of Brownian motion. (One sees that $\hat{b}_t = B_t$ for instance by considering the limits of $\lambda^{-1}\pi_{\lambda\chi_{[0,t]}}$ and $\lambda^{-1}\epsilon_{\lambda\chi_{[0,t]}}$ as $\lambda\to 0$).

A total set in $\mathcal{I}_2$ is formed by the functions

$$g\cdot\chi_J \ : \ (\tau,t) \to g(\tau)\chi_J(t) \ , \qquad (g \in F_{\inf J]}) \ ,$$

where J is some subinterval of $[0,1]$. We define the *kernel integral* of such elementary adapted processes by

$$I_0(g\cdot\chi_J) = g\star b_J \ ,$$

where $b_{[s,t]} := b_t - b_s$.

Proposition 3.3. *The operator* I_0 *extends by continuity to the isometry* $I_1 : \mathcal{I}_2 \to F$ *given by*

$$(I_1 f)(\tau) = \begin{cases} 0 & \textit{if } \tau = \emptyset \ , \\ f_{\max\tau}(\tau\setminus\{\max\tau\}) & \textit{otherwise} \ , \end{cases}$$

which under the duality transform corresponds to the Ito integral $h\to \int_0^1 h_t \, dB_t$.

proof. As j , given above, is measure-preserving $\Gamma\setminus\{\emptyset\} \to \Gamma_{ad}$, I_1: $f\to f\circ j$ is isometric. I_1 extends I_0 since, for $f=g\cdot\chi_J$ and $\tau\in\Gamma$ nonempty:

$$(I_0 f)(\tau) = (g\star b_J)(\tau) = \sum_{\alpha\cup\beta=\tau} \int_\Gamma g(\alpha\cup\gamma)\, b_J(\beta\cup\gamma)\, d\gamma$$

$$= \sum_{t\in\tau} g(\tau\setminus\{t\})\,\chi_J(t) = g(\tau\setminus\{\max\tau\})\,\chi_J(\max\tau) = f_{\max\tau}(\tau\setminus\{\max\tau\}) \ .$$

It follows that $\hat{I}_1 := BI_1B^{-1}$ is the linear and isometric extension of the map $\hat{I}_0 = BI_0B^{-1}$, acting on the elementary adapted processes as

$$\hat{I}_0(k\cdot\chi_{[s,t]}) \ : \ \omega \to k(\omega)\cdot(\,\omega(t) - \omega(s)\,) \ , \qquad (k\in W_{s]}) \ .$$

I.e., $\hat{I}_1$ is the Ito integral $h\to\int_0^1 h_t \, dB_t$.

□

3.4. *Smooth kernel processes.*

In order to consider differentiation of adapted kernel processes we introduce smooth processes as follows. On Γ_{ad} we define a metric ρ_0 :

$$\rho_0((\tau,t),(\sigma,s)) = \begin{cases} \min(1\ ,\ \max(|t_1-s_1|,\cdot\cdot,|t_n-s_n|,|t-s|)\ ) & \text{if } n=m\ ; \\ 1 & \text{otherwise,} \end{cases}$$

where $\tau = (t_1,\cdot\cdot,t_n)$ and $\sigma = (s_1,\cdot\cdot,s_m)$. Let (∇_{ad},ρ) denote the completion of the metric space (Γ_{ad},ρ_0) . An adapted kernel process $f\colon\ \Gamma_{ad}\to C$ will be called *smooth* if it satisfies the conditions

(P*i*) $\quad |f_t(\tau)| \le K^{\#\tau+1}$;

(P*ii*) $\quad \rho_0((\tau,t),(\sigma,s)) < 1 \implies |f_t(\tau)-f_s(\sigma)| \le \rho_0((\tau,t),(\sigma,s))\cdot K^{\#\tau+1}$.

We denote the class of smooth adapted processes by P_0 . Note that (P*i*) implies that $f_s \in \mathrm{Dom}(e^{t\Lambda})$ for all $s\in[0,1]$ and $t>0$, hence $f_s \in K$ for all s. It follows that the product

$$(f\star g)_t := f_t \star g_t$$

is well-defined on P_0 . A straightforward estimate shows that P_0 is closed under this product.

By (P*ii*) a smooth process is uniformly continuous on each $\Gamma_{ad,n} := \{(\tau,t)\in\Gamma_n\times[0,1]\ |\ \tau\subset[0,t)\}$, and therefore admits a unique continuous extension to $\nabla_{ad} = \cup_{n=0}^{\infty} \nabla_{ad,n}$.

A process $f\in P_0$ has "sample paths" $t\mapsto f_t(\tau)$ of the form

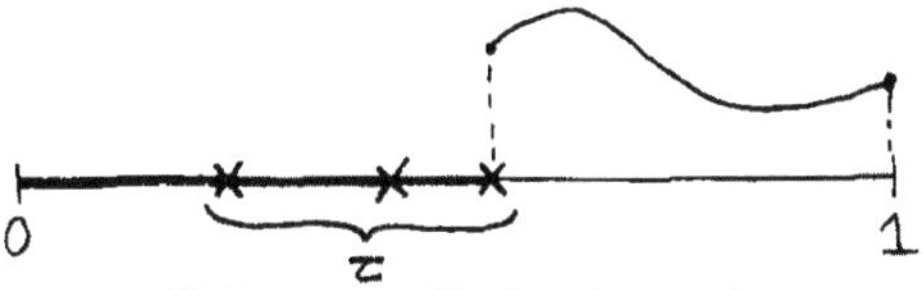

smooth kernel process

A process $f\in\mathcal{I}_2$ is called a *kernel martingale* if $\chi_{\Gamma[0,s]}\cdot f_t = f_s$ for $s\le t$. The sample paths of a martigale look as follows:

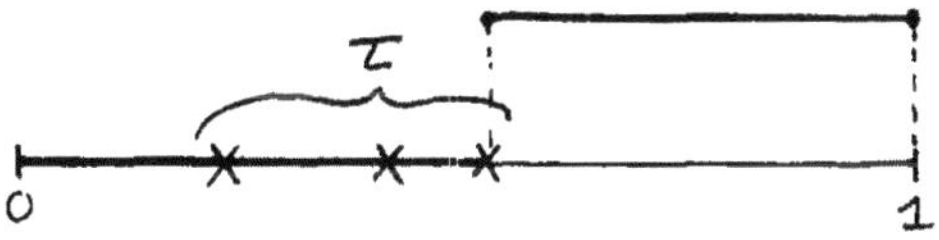

kernel martingale

3.5 *Kernel calculus*

The change in time of a P_0-process consists of jumps at the points of the argument τ (in fact only at $\max\tau$ by adaptedness) and smooth variation between the points (in fact only between $\max\tau$ and 1). One is thus naturally led to two different differentiation operators Δ and D defined below.

$$\Delta : P_0 \to P_0 : (\Delta f)_t(\tau) = \text{jump at } t \text{ of } s \mapsto f_s(\tau \cup \{t\}) , \qquad (t \notin \tau);$$

$$D : P_1 \to P_0 : (Df)_t(\tau) = d/ds\, f_s(\tau) \mid_{s=t} , \qquad (t \notin \tau),$$

where P_1 is the natural domain $\{ f \in P_0 \mid df_s/ds \in P_0 \}$ of D.

Proposition *For* $f \in P_0$ *one has*

$$\Delta I f = D \int_0^{\cdot} f_s ds = f \; ; \qquad (3.5)$$

and for $f \in P_1$ *we have the "fundamental theorem":*

$$f_t - f_0 = (I\Delta f)_t + \int_0^t (Df)_s ds . \qquad (3.6)$$

Here $I : \mathcal{I}_2 \to \mathcal{I}_2$ is defined by $(If)_t = I_1(\chi_{[0,t]} \cdot f)$.

proof. Obviously, $d/ds \int_0^s f_u(\tau) du \mid_{s=t} = f_t(\tau)$, whereas the jump at t in $s \to (If)_s(\tau\cup\{t\}) = \sum_{u \in \tau\cup\{t\}} \chi_{[0,t]}(s)\, f_u(\tau\cup\{t\}\setminus\{u\})$ is precisely $f_t(\tau)$. This proves (3.5). The "fundamental theorem" (3.6) expresses the difference $f_t(\tau)-f_0(\tau)$ as an integral over the derivative plus a sum over the jumps of the path $s \mapsto f_s(\tau)$ for s running from 0 to t . (Because of adaptedness only one jump occurs, and the derivative only becomes nonzero after it).

□

Remarks. Note that adaptedness does not play an essential role in the above proof. The relations (3.5) and (3.6) immediately generalise to non-adapted kernel processes. We do not consider these here, however.

Under the duality transform, (3.6) goes to the Doob-Meyer decomposition of a process $f \in P_0$. From (3.6) one may conclude that $\hat{\Delta}$ is the stochastic derivative, whereas $\hat{D}$ is Nelson's forward derivative.

Theorem 3.5 (Ito's lemma). *For* $f,g \in P_0$:

$$\Delta(f{\star}g) = \Delta f{\star}g + f{\star}\Delta g ,$$

and for $f,g \in P_1$:

$$D(f{\star}g) = Df{\star}g + f{\star}Dg + \Delta f{\star}\Delta g .$$

proof. For $f,g \in P_0$ we have by adaptedness:

$$(\Delta(f{\star}g))_t(\tau) = \lim_{s\downarrow t}(f{\star}g)_s(\tau\cup\{t\})$$

$$=\lim_{s\downarrow t}\int_\Gamma \sum_{\alpha\cup\beta=\tau}\Big\{f_s(\alpha\cup\{t\}\cup\gamma)\, g_s(\beta\cup\gamma) + f_s(\alpha\cup\gamma)\, g_s(\beta\cup\{t\}\cup\gamma)\Big\}\, d\gamma$$

$$= (\Delta f{\star}g)_t(\tau) + (f{\star}\Delta g)_t(\tau) .$$

Moreover, for $f,g \in P_1$:

$$(D(f{\star}g)_t)(\tau) = \sum_{\alpha\cup\beta=\tau}\frac{d}{ds}\Big\{\int_{\Gamma[0,s]} f_s(\alpha\cup\gamma)\, g_s(\beta\cup\gamma)\, d\gamma\Big\}\Big|_{s=t}$$

$$=\sum_{\alpha\cup\beta=\tau}\int_{\Gamma[0,t]}\Big\{\frac{d}{ds}\left(f_s(\alpha\cup\gamma)\, g_s(\beta\cup\gamma)\right)\Big|_{s=t}$$
$$+ f_{t+}(\alpha\cup\gamma\cup\{t\})\, g_{t+}(\beta\cup\gamma\cup\{t\})\Big\}\, d\gamma$$

$$= (Df{\star}g)_t(\tau) + (f{\star}Dg)_t(\tau) + (\Delta f{\star}\Delta g)_t(\tau) .$$

The first term in the differentiation of $s \rightarrow \int_{\Gamma[0,s]} f_s^{\ \alpha}\, g_s^{\ \beta}\, d\mu$ above, is due to the smooth change of the integrand; the second comes from the growth of the region of integration. □

We have now moved full circle. Formal Ito relations led in §2 to an algebraic structure on Fock space; stochastic calculus on Fock space when applied to this structure in turn yielded the Ito relations. What has been gained ? The treatment in this section is "classical", however the approach here highlighted the structure of noise in a fashion which makes it ripe for generalisation to quantum noises. Moreover its "set point of view" and notational compactness allows considerable abbreviation of previous arguments and also contributes an explicitness which is particularly effective when solving stochastic differential equations.

§4. Bose and Fermi noise.

In this section we heuristically derive, in the spirit of §2, the algebraic structures on the representation spaces of the quantum Wiener process (Bose noise) and Fermi noise.

4.1. *The Bose product*

Consider the following Ito relations for a noise $N = (A^+, A^-)$:

$$(dA_t^+)^2 = (dA_t^-)^2 = 0 \ ; \ dA_t^+ dA_t^- = c_- dt \ ; \ dA_t^- dA_t^+ = c_+ dt \ ;$$
$$dA_s^+ dA_t^+ = dA_t^+ dA_s^+ \ , \ (s \neq t) \ ; \ dA_t^- = (dA_t^+)^\star \ ,$$

where $c_\pm$ are a pair of strictly positive parameters $(c_+ \geq c_-$, say).
Let $\Gamma^\star(I)$ denote the set of finite "charged" subsets of $I = [0,1]$:

$$\{ \tau = ((t_1,\epsilon_1), \cdots, (t_n,\epsilon_n)) \mid n \in \mathbb{N}, \ \epsilon_i \in \{+,-\}, \ \underline{t} \in I^n_< \} \ ,$$

and μ_{c_+,c_-} the measure on $\Gamma^\star$ given by $d\mu = m d\tau$, where

$$m(\tau) = \Pi_i \, c_{\epsilon_i} \ .$$

For $f \in L^2(\Gamma^\star, \mu_{c_+,c_-})$ let $N(f)$ be given by

$$\int_{\Gamma^\star} f(\tau) \, dN_\tau = \sum_{n \in \mathbb{N}} \sum_{\epsilon \in \{+,-\}^n} \int\cdots\int_{0 \leq t_1 < \cdots < t_n \leq 1} f((\underline{t},\underline{\epsilon})) dA_{t_1}^{\epsilon_1} \cdots dA_{t_n}^{\epsilon_n}$$

Note that $N(f)^\star = N(\tilde{f})$, where $\tilde{f}(\tau) = \overline{f(\tilde{\tau})}$, $(\underline{t},\underline{\epsilon})^\sim = (\underline{t},-\underline{\epsilon})$.
We formally compute the kernel of $N(f)N(g)$: since $dN_\sigma dN_\tau = 0$ unless $\tau = \beta \cup \gamma$, $\sigma = \alpha \cup \tilde{\gamma}$ for some disjoint α, β, γ,

$$\int_{\Gamma^\star} f(\sigma) \, dN_\sigma \int_{\Gamma^\star} g(\tau) \, dN_\tau = \int\int\int f(\alpha \cup \tilde{\gamma}) \, g(\beta \cup \gamma) \, d\mu(\gamma) \, dN_\alpha dN_\beta$$
$$= \int \sum_{\alpha \cup \beta = \delta} \int f(\alpha \cup \tilde{\gamma}) \, g(\beta \cup \gamma) \, d\mu(\gamma) \, dN_\delta \ .$$

Thus the *Bose product* is given by :

$$f\star g : \tau \to \sum_{\alpha\cup\beta=\tau} \int f(\alpha\cup\tilde{\gamma})\, g(\beta\cup\gamma)\, d\mu(\gamma) .$$

Again "isometry" is immediate:

$$E(\, N(f)^{\star} N(f)\,) = \tilde{f}\star f(\emptyset) = \int_{\Gamma^\star} \overline{f(\gamma)} f(\gamma)\, d\mu(\gamma) = \|f\|^2_{L^2(\Gamma^\star)} .$$

4.2 *The Fermi product.*

The Fermi commutation relations have $dA_t^+ dA_s^+ = -dA_s^+ dA_t^+$ instead of $dA_t^+ dA_s^+ = dA_s^+ dA_t^+$, everything else being identical to the Bose case. For $\sigma,\tau \in \Gamma^\star$ we now have : $dN_\sigma dN_\tau = 0$ unless $\sigma=\alpha\cup\tilde{\gamma}$, $\tau=\beta\cup\gamma$ for disjoint α,β,γ , in which case

$$\begin{aligned} dN_\sigma dN_\tau &= \epsilon(\alpha,\tilde{\gamma})\, dN_\alpha\, dN_{\tilde{\gamma}}\, dN_\gamma\, dN_\beta\, \epsilon(\gamma,\beta) = \\ &= (-1)^{\frac{1}{2}\#\gamma(\#\gamma-1)}\, \epsilon(\alpha,\tilde{\gamma})\, \epsilon(\gamma,\beta)\, d\mu(\gamma)\, dN_\alpha\, dN_\beta , \end{aligned}$$

where $\epsilon(\alpha,\beta)$ is as defined in §2. Letting $\nu=\alpha\cup\beta$:

$$\begin{aligned} &\int\int f(\sigma)\, g(\tau)\, dN_\sigma\, dN_\tau = \\ &= \int\Big\{ \sum_{\alpha\subset\nu} \epsilon(\alpha,\nu\setminus\alpha) \int (-1)^{\frac{1}{2}\#\gamma(\#\gamma-1)}\, \epsilon(\alpha,\tilde{\gamma})\, \epsilon(\gamma,\nu\setminus\alpha) \\ &\qquad \times f(\alpha\cup\tilde{\gamma})\, g(\nu\setminus\alpha\cup\gamma)\, d\mu(\gamma) \Big\}\, dN_\nu . \end{aligned}$$

so the object between curly brackets is the Fermi product $f\star g(\nu)$.

The *involution* is given by

$$\tilde{f}(\sigma) = (-1)^{\frac{1}{2}\#\sigma(\#\sigma-1)}\, \overline{f(\tilde{\sigma})} .$$

in particular

$$\tilde{f}\star f(\emptyset) = \int \overline{f(\gamma)} f(\gamma)\, d\mu(\gamma) = \|f\|^2_{L^2(\mu)} .$$

ACKNOWLEDGEMENTS

H.M. is grateful for support from the Netherlands Organisation for the Advancement of Pure Research (ZWO). J.M.L. thanks the Science and

Engineering Research Council, U.K. for support under Research fellowship B/RF/7349 and also ZWO and King's College, London for travel grants.

REFERENCES

[ApH] D.B. Applebaum, R.L. Hudson: "Fermion Ito's formula and stochastic evolutions", Comm. Math. Phys. 96 (1984) 473-496.

[BSW] C. Barnett, R.F. Streater, I.F. Wilde: "The Ito-Clifford integral", Journ. Func. Anal. 48(1982)172-212.

[CoH] Cockroft, Hudson: "Quantum mechanical Wiener process", Journ. Multiv. Anal. 7(1977)107-124.

[Fri] A. Frigerio: "Quantum Poisson processes: Physical motivations and applications", this volume.

[FrM] A. Frigerio, H. Maassen: "Quantum Poisson processes and dilations of dynamical semigroups", preprint, Nijmegen.

[HuP] R.L. Hudson, K.R. Parthasarathy: "Quantum Ito's formula and stochastic evolutions", Comm. Math. Phys. 93(1984)301-323.

[Küm] B. Kümmerer: "Markov dilations and non-commutative Poisson processes", preprint, Tubingen.

[LiM] J.M. Lindsay, H. Maassen: "The stochastic calculus of Bose noise" preprint, Nijmegen.

[Mey] P.A. Meyer: "Eléments de probabilités quantiques", Exposés I à V, Inst. de Math., Université Louis Pasteur, F-67084 Strasbourg Cedex.

[PaS] K.R. Parthasarathy, K.B. Sinha: "Boson-fermion relations in several dimensions" Pramana Journ. Phys. 27(1986) 105-116.

A NOTE ON SHIFTS AND COCYCLES

P.A. Meyer

This note consists of very simple mathematics. It is above all a warning against an error which I committed myself, about analogies between classical Markov processes and quantum stochastic processes on Fock space. Consider for instance two recent papers, Parthasarathy-Sinha [3] and Journé [2] : their reader may be tempted to identify the "local clock" in [3] with a classical change of time using an additive functional, and the "unitary cocycles" in [2] wich classical multiplicative functionals (unitary operator valued). These identifications turn out to be misleading, because the shift operator of "quantum probabilists" doesn't correspond to the classical shift. I will make this explicit on a concrete example, and suggest some generalizations. This topic is closely related to Accardi's general definition of a "quantum Feynman-Kac formula" in [1].

Conversations with J.L. Journé on these topics are gratefully acknowledged : some of the main remarks below are due to him. I also thank L. Accardi for useful comments on the first version of this note.

1. THE SHIFT ON FOCK SPACE

1. Let Φ be the usual Fock space over time used by quantum probabilists.

To be concrete, we describe Φ as $L^2(\Omega,\mathcal{F},P)$ where P is Wiener measure on the canonical sample space $(\Omega,\mathcal{F})$ for a real valued brownian motion (B_t) starting from 0. Let $\mathcal{F}_{t]}$ and $\mathcal{F}_{[t}$ respectively be the σ-fields generated by $(B_s)_{s\leq t}$ and $(B_s-B_t)_{s\geq t}$, and let $\Phi_{t]}$, $\Phi_{[t}$ be the corresponding L^2 spaces, considered as subspaces of Φ. The vacuum vector (= the function 1) is denoted by $\mathbb{1}$. The shift is defined by

(1) $$\Gamma_t f = f\circ\gamma_t \quad \text{where} \quad B_s(\gamma_t\omega) = B_{s+t}(\omega)-B_t(\omega) .$$

We couple an initial Hilbert space J to Fock space, defining

(2) $$\Psi = J\otimes\Phi \ , \quad \Psi_{t]}=J\otimes\Phi_{t]}$$

We also extend the shift isometries by $I_J\otimes\Gamma_t$, which we again denote by Γ_t (this ambiguity isn't really dangerous). The image $\Gamma_t\Psi$ is denoted by $\Psi_{[t}$. It is convenient to identify J with $\Psi_{0]}$ ($j\in J$ corresponding to $j\otimes\mathbb{1}$). Elements of Ψ are denoted by roman letters, elements of Φ by greek ones, and a lot of $\otimes$ signs will be omitted. We denote by E_t the projection on $\Psi_{t]}$.

We don't consider unbounded operators in this note.

2. We have tensor product decompositions to the right and to the left

$$\Psi \approx \Psi_{t]}\otimes\Phi_{[t} \approx \Phi_{t]}\otimes\Psi_{[t} .$$

The first decomposition is used in the definition of a $\Psi_{t]}$-adapted operator, namely an operator A on Ψ which can be written as $B\otimes I_{\Phi_{[t}}$ for some operator B on $\Psi_{t]}$. The second decomposition is necessary to shift operators : let A be an operator on Ψ ; considering Γ_t as an isomorphism of Ψ on $\Psi_{[t}$, Γ_t^* as the inverse isomorphism, $\Gamma_t A\Gamma_t^*$ appears as an operator on $\Psi_{[t}$, which we tensorize with $I_{\Phi_{t]}}$ to get an operator on Ψ , denoted by $\overline{\Gamma_t A\Gamma_t^*}$ in Journé [2]. This operator is unitary if A is unitary, in contrast with the usual $\Gamma_t A\Gamma_t^*$.

Typical elements of Ψ are $u\alpha$, $u\beta\Gamma_t\alpha$ ($u\in\Psi_{0]}$, $\alpha\in\Phi$, $\beta\in\Phi_{t]}$), the second class being included in the first one and linear combinations of each being dense in Ψ . We have omitted $\otimes$ signs. With these notations

(3) $$\Gamma_t(u\alpha) = u\Gamma_t\alpha \quad , \quad E_0(u\alpha) = u\langle 1\!\!1,\alpha\rangle \ , \ E_t(u\beta\Gamma_t\alpha) = u\beta\langle 1\!\!1,\alpha\rangle$$
$$\Gamma_t^*(u\beta\Gamma_t\alpha) = u\langle 1\!\!1,\beta\rangle\alpha$$

Then the reader will be able to see explicitly the difference between the two operators $\Gamma_t A\Gamma_t^*$ and $\overline{\Gamma_t A\Gamma_t^*}$.

3. The coupling of J to Fock space is used to construct semigroups on $J=\Psi_{0]}$ via cocycles on Ψ . For clarity we distinguish between possible definitions of cocycles by a reference to papers which have used these definitions, of course without any pretence to making history ! In all three definitions, a cocycle is a family $(M_t)_{t\geq 0}$ of $\Psi_{t]}$-adapted operators. $*$-cocycles (star-cocycles, from Accardi [1]) satisfy

(4) $$M_{t+s} = M_t\Gamma_t M_s\Gamma_t^*$$

$^-$-cocycles (bar-cocycles, from Journé [2]), correspond to

(5) $$M_{t+s} = M_t\overline{\Gamma_t M_s\Gamma_t^*} \ ,$$

and finally weak cocycles satisfy

(6) $$M_{t+s}\Gamma_t = M_t\Gamma_t M_s \ .$$

Note that since $\Gamma_t^*\Gamma_t=I$ $*$-cocycles are weak cocycles, and since $\Gamma_t M_s\Gamma_s^*$ and $\overline{\Gamma_t M_s\Gamma_t^*}$ agree on the image of Γ_t, $^-$-cocycles are weak cocycles too. In [2], Journé describes unitary $^-$-cocycles as solutions of quantum s.d.e's in Ψ with coefficients which are linear (possibly unbounded) operators on J.

The main use of cocycles is given by the following lemma, essentially due to Accardi (it is a « quantum Feynman-Kac formula »).

LEMMA. Let (M_t) be a weak cocycle, and define P_t on $\Psi_{0]}$ by

(7) $$P_t = E_0 M_t\Gamma_t$$

(this could be written E_0M_t since $\Gamma_t=I$ on $\Psi_{0]}$, but Γ_t is necessary for the general situations to come). Then (P_t) is a semigroup.

Proof. $P_{s+t}=E_0M_{s+t}\Gamma_{s+t} = E_0M_{s+t}\Gamma_t\Gamma_s = E_0M_t\Gamma_tM_s\Gamma_s$ (weak cocycle property) while $P_tP_s = E_0M_t\Gamma_tE_0M_s\Gamma_s$. So it suffices to prove $E_0M_t\Gamma_t=E_0M_tE_0$. We prove a slightly more precise result, namely that $E_tM_t\Gamma_t=E_tM_tE_0$. To see this, we apply both sides to $u\alpha$ ($u\in\Psi_{0]}$, $\alpha\in\Phi$). Then $\Gamma_t(u\alpha)=u\Gamma_t\alpha$, then by adaptation to $\Psi_{t]}$ $M_t\Gamma_t(u\alpha)=M_tu.\Gamma_t\alpha$, and finally $E_tM_t\Gamma_t(u\alpha) = M_tu<\mathbb{1},\Gamma_t\alpha>$. On the other hand $E_0(u\alpha)=u<1,\alpha>$, $E_tM_tE_0(u\alpha)=M_tu<\mathbb{1},\alpha>$. $\square$

2. USING A DIFFERENT SHIFT

Let λ be Lebesgue measure on $\mathbb{R}$, and J be $L^2(\lambda)$. Let $(\Omega,\mathcal{G})$ be the canonical sample space for brownian motion $(X_t)_{t\geq 0}$ with arbitrary starting point, and let B_t be X_t-X_0 ; then for the measure P^λ corresponding to the initial measure λ, X_0 and (B_t) are independent, so $L^2(P^\lambda)\approx J\otimes L^2(P)$ can be considered as a $\Psi=J\otimes\Phi$ as in the first part, and $L^2(\mathcal{G}_{t]},P^\lambda)$ with $\mathcal{G}_{t]}$ generated by $(X_s)_{s\leq t}$ is identified with $\Psi_{t]}$. However, we are going to introduce on this space the usual shift for a Markov process

(8) $\qquad \Theta_tf=f\circ\theta_t \qquad$ where $X_s(\theta_t\omega) = X_{s+t}(\omega)$

Because of our choice of Lebesgue measure, this new shift is isometric. It is compatible with the old shift in the sense that we still have a right tensor product decomposition $\Psi=\Psi_{t]}\otimes\Phi_{[t}$, and that

(9) $\qquad \Theta_t(u\alpha) = \Theta_tu.\Gamma_t\alpha \qquad$ for $u\in\Psi_{0]}$, $\alpha\in\Phi$.

It is compatible with the filtration in the sense that Θ_t carries $\Psi_{s]}$ into $\Psi_{s+t]}$ for all s . There is no difficulty in defining $*$-cocycles (as in (4), Θ_t replacing Γ_t) and weak cocycles (as in (6)). The same proof as above shows that $P_t=E_0M_t\Gamma_t$ is a semi-group whenever (M_t) is a weak cocycle. The isometric property of Θ_t isn't used in the proof of the lemma itself.

What about $^-$-cocycles ? For this, we need to define $\overline{\Theta_tA\Theta_t^*}$, hence a left tensor product decomposition $\Psi\approx\Phi_{t]}\otimes\Psi_{[t}$, where $\Psi_{[t}=\Theta_t\Psi$ is different from the space denoted $\Psi_{[t}$ in the first part. Such a tensor product decomposition exists, taking for $\Phi_{t]}$ the L^2 space generated by the brownian motion $(X_{t-s}-X_t)_{s\leq t}$ (the fact that it is also generated by $(B_s)_{s\leq t}$ is misleadingly simple : if were working with the flow of a diffusion, generating the past $\mathcal{G}_{t]}$ by X_t and an independent brownian motion would be a non trivial problem). So me may also consider $^-$-cocycles. Have they any interest ?

To show this, we use for the first time the fact that our Hilbert spaces consist of random variables, and introduce the highly non-linear Feynman-Kac cocycles (classical multiplicative functionals)

(10) $\qquad M_t = \exp(\int_0^t v(X_s)ds)$

considered as multiplication operators, and check that they are indeed

$^{-}$-cocycles. Let $\beta f(X_t)\Gamma_t\alpha$ be the product of three r.v.'s, β being measurable w.r. to the reversed brownian motion $(X_{s-t}-X_t)_{s\le t}$ and $\Gamma_t\alpha$ measurable w.r. to the future brownian motion $(X_s-X_t)_{s\ge t}$; because of independence, tensor products and ordinary products are identified. Then $\overline{\Theta_t M_s \Theta_t^*}$ applied to this vector is the product of β and $\Theta_t M_s \Theta_t^*(f(X_t)\Gamma_t\alpha)$ $= \Theta_t(M_s(f(X_0)\alpha) = \Theta_t((\int_0^s v(X_r)dr)f(X_0)\alpha) = \int_s^{s+t} v(X_r)dr.f\circ X_t.\Gamma_t\alpha$ (ordinary products), and finally the complicated operator $\overline{\Theta_t M_t \Theta_t^*}$ is just the ordinary multiplication operator by $\int_s^{s+t} v(X_r)dr$! Then the $^{-}$-cocycle property is trivial.

From an abstract point of view, the possibility of using different shifts might lead to new solutions of the dilation problem : we may play on two parameters, the shift and the cocycle. However, the following very simple remark shows that the new shift itself is described by a $*$-cocycle of the old shift :

LEMMA. <u>Let</u> V_t <u>be</u> $\Theta_t\Gamma_t^*$; <u>then</u> (V_t) <u>is a</u> $*$-<u>cocycle relative to</u> (Γ_t). <u>Conversely, if</u> (W_t) <u>is a</u> $*$-<u>cocycle</u>, $W_t\Gamma_t = H_t$ <u>maps</u> $\Psi_{s]}$ <u>into</u> $\Psi_{t+s]}$ <u>and satisfies</u> $H_tH_s=H_{t+s}$ <u>and</u> (9).

<u>Proof</u>. Using the isometric property $\Gamma_t^*\Gamma_t=I$ we have

$$V_{t+s} = \Theta_t\Theta_s\Gamma_s^*\Gamma_t^* = \Theta_t(\Gamma_t^*\Gamma_t)\Theta_s\Gamma_s^*\Gamma_t^* = V_t\Gamma_tV_s\Gamma_t^*$$

We check that V_t is $\Psi_{t]}$-adapted. Set $f=u\beta\Gamma_t\alpha$, $u\in\Psi_{0]}$, $\alpha\in\Phi$, $\beta\in\Phi_{t]}$. Then $\Gamma_t^*f = u\langle 1,\beta\rangle\alpha$ and $V_tf=\Theta_tu\langle 1,\beta\rangle\Gamma_t\alpha = V_t(u\beta)\Gamma_t\alpha$, while $V_t(u\beta)\in\Psi_{t]}$.

Converse : $H_{t+s}=W_{t+s}\Gamma_{t+s}=W_t\Gamma_tW_s(\Gamma_t^*\Gamma_t)\Gamma_s = H_tH_s$. We have $H_t(u\alpha) = W_t(u\Gamma_t\alpha)$ for $u\in\Psi_{0]}$, $\alpha\in\Phi$. Since W_t is $\Psi_{t]}$-adapted this is $W_t(u)\Gamma_t\alpha$, and since $\Gamma_tu=u$ this is also $H_tu.\Gamma_t\alpha$, so (H_t) satisfies (9). Finally if $\alpha\in\Phi_{s]}$, $H_t(u\alpha)=W_tu.\Gamma_t\alpha$ belongs to $\Psi_{s+t]}$.

This lemma shows that each one of the classes of cocycles we have defined plays its own role : weak cocycles produce semi-groups, $*$-cocycles produce shifts, and $^{-}$-cocycles seem to represent the true analog of a multiplicative integral. Of course, a better terminology should be found.

REFERENCES

[1]. ACCARDI (L.). A quantum formulation of the Feynman-Kac formula. Coll. Math. Soc. J. Bolyai, Random Fields, Esztergom 1979, p. 25-35.

[2]. JOURNÉ (J.L.). Structure des cocycles markoviens sur l'espace de Fock. A paraître, ZW 1987.

[3]. PARTHASARATHY (K.R.) et SINHA (K.B.). Stop times in Fock space stochastic calculus. A paraître, ZW 1987.

IRMA, 7 rue René Descartes
67084 Strasbourg Cedex, France

Local Measures in Fock Space Stochastic Calculus and a Generalized Ito-Tanaka Formula

by

K.R. Parthasarathy

Indian Statistical Institute, Delhi Centre, 7, S.J.S. Sansanwal Marg, New Delhi - 110 016, India

1. Introduction

Suppose $\{w(t),\ t \geq 0\}$ denotes the sample path of a standard Brownian motion process on the real line. Consider the random variable $L(t,a,\varepsilon,w)$ which denotes the Lebesque measure of the set $\{s : 0 \leq s \leq t,\ a - \varepsilon < w(s) < a + \varepsilon\}$ and can be interpreted as the amount of time spent by the Brownian motion during the interval $[0,t]$ in the neighbourhood $(a - \varepsilon,\ a + \varepsilon)$ of the point a. P. Levy showed that $(2\varepsilon)^{-1} L(t,a,\varepsilon,w)$ converges to a limit $L(t,a,w)$ for all t with probability one as $\varepsilon \to 0$. The random variables $L(t,a) = L(t,a,w)$ define an adapted stochastic process with many interesting properties. It is possible to choose a version of $L(t, a)$ so that it is jointly continuous in (t, a). For fixed a, $L(\cdot,a)$ is continuous and increasing and can therefore be looked upon as $L(t, a) = L([0,t],a)$ where $L(\cdot,a)$ is a nonatomic Radon measure on $\mathbb{R}_+$. The random variable $L(t, a)$ is called the Levy local time of the Brownian motion on the set $\{s : w(s) = a\}$.

H. Tanaka identified the local time $L(t,a)$ as follows. Since $w(t) - a$ is a continuous martingale for any fixed a it follows that $|w(t) - a|$ is a continuous semimartingale and hence admits a unique Doob-Meyer decomposition $M(t) + A(t)$ where M is a continuous martingale and A is a continuous natural increasing process. Tanaka showed that this decomposition is given by

$$|w(t) - a| = \{|a| + \int_0^t \operatorname{sgn}(w(s)-a)\,dw(s)\} + L(t,a). \qquad (1.1)$$

If we write $f(x) = |x - a|$ then $f'(x) = \mathrm{sgn}\,(x - a)$ and $f''(x) = 2\delta(x - a)$ in the generalized sense of tempered distributions and therefore (1.1) can be interpreted as a generalized Ito's formula

$$d(f(w(t))) = f'(w(t))dw(t) + \frac{1}{2} f''(w(t))dt$$

so that

$$\frac{1}{2} f''(w(t))dt = \delta(w(t) - a)dt = L(dt, a)$$

in the generalized sense. For a comprehensive and pleasantly readable account of this circle of ideas we refer to the book [1] by K.L. Chung and R.J. Williams. For a historically interesting and analytically formidable account of local times we refer to the book [4] by Ito and Mckean.

With the classical setting briefly described above we now address ourselves to the quantum description. Suppose we replace the Brownian motion $\{w(t), t \geq 0\}$ in the above discussion by a family of selfadjoint operators $\{X(t), t \in T\}$ on a Hilbert space $\mathcal{H}$ where the time space T is a locally compact second countable metric space with Borel σ-algebra $\mathcal{F}$ and the Lebesque measure is replaced by a Radon measure ν on $(T, \mathcal{F})$. We denote by $\mathcal{F}_0 = \{C \in \mathcal{F}, \bar{C} \text{ is compact}\}$. The family $\{X(t), t \in T\}$ is interpreted as a quantum stochastic process of observables. Each operator $X(t)$ has a spectral decomposition $\int_{\mathbb{R}} x\, E_t(dx)$ and we may introduce the observables

$$L_\nu(C, a,\varepsilon) = (2\varepsilon)^{-1} \int_C E_t(a-\varepsilon, a+\varepsilon)\nu(dt),$$

provided $\{X(t), t \in T\}$ satisfy the obvious measurability conditions and $C \in \mathcal{F}_0$. Then $L_\nu(C, a, \varepsilon)$ is a bounded selfadjoint operator on $\mathcal{H}$. We may now ask the following question: under what conditions on the (scaling) measure ν the operator $L_\nu(C, a, \varepsilon)$ converges to a limiting operator $L_\nu(C, a)$ as $\varepsilon \to 0$? What are the properties of the family of operators $\{L_\nu(C, a), a \in \mathbb{R}, C \in \mathcal{F}_0\}$? We may call $L_\nu(C, a)$ the local (time) measure of the quantum process $\{X(t), t \in T\}$ during C with respect to the scale ν in the set of time points when X(t) visits a.

In the present exposition we present a very preliminary account of such local measures when $X(t) = i\{a^+(\phi(t)) - a(\phi(t))\}$ where a, a^+ denote the annihilation and creation fields over h and $\phi : T \to h$ is a strongly continuous map. As a corollary of such an investigation one obtains the existence of local measures in the case of a multiparameter Brownian motion and some stationary Gaussian processes as well as an operator version of the Ito-Tanaka formula. Our computations are too elementary and straightforward to throw light on the continuity properties of individual sample paths in the classical case. The main tool is an operator version of the classical Fourier inversion formula combined with the method of quantum stochastic integration.

The author wishes to thank K.B. Sinha for several useful conversations.

2. Notations and Preliminaries

Let h be a complex separable Hilbert space and let H be the boson Fock space over h defined by

$$H = \mathbb{C} \oplus h \oplus h^{\text{ⓢ}2} \oplus \dots \oplus h^{\text{ⓢ}n} \oplus \dots$$

where ⓢ^{n} denotes n-fold symmetric tensor product. For any $u \in h$ the coherent vector $\psi(u)$ associated with u is defined by

$$\psi(u) = 1 \oplus u \oplus (2!)^{-1/2} u^{\otimes^2} \oplus \dots \oplus (n!)^{-1/2} u^{\otimes^n} \oplus \dots \quad .$$

We denote by $<.,.>$ the inner product in any Hilbert space with the convention that it is conjugate linear in the first variable. The relation

$$<\psi(u), \psi(v)> = \exp <u,v> \text{ for all } u,v \in h \tag{2.1}$$

implies that $\{\psi(u), u \in h\}$ is a total set of linearly independent vectors in H. We denote by E the dense linear manifold generated by the set of all coherent vectors. For each $u \in h$ the associated unitary Weyl operator $W(u)$ in H is defined by the relations:

$$W(u)\psi(v) = e^{-1/2||u||^2 - <u,v>}\psi(v+u) \text{ for all } v \in h. \tag{2.2}$$

Then one has the Weyl commutation relations:

$$W(u)W(v) = e^{-i\,\mathrm{Im}\langle u,v\rangle}\, W(u+v). \tag{2.3}$$

In particular, for each $u \in h$, the map $x \to W(xu)$ is a strongly continuous group of unitary operators in H and hence by Stone's theorem there exists a selfadjoint operator $P(u)$ such that

$$W(xu) = e^{ix\,P(u)}, \quad x \in \mathbb{R},\ u \in h. \tag{2.4}$$

Suppose

$$P(u) = \int_{\mathbb{R}} x\, E_u(dx) \tag{2.5}$$

is the spectral resolution of $P(u)$. The Weyl commutation relations (2.3) imply that E_u is an absolutely continuous spectral measure on $\mathbb{R}$ for every fixed $u \neq 0$ in h. Furthermore,

$$P(u)\xi = i\{a(u) - a^+(u)\}\xi, \quad \xi \in E \tag{2.6}$$

where $a(u)$ and $a^+(u)$ are the annihilation and creation operators satisfying the relations:

$$a(u)\psi(v) = \langle u,v\rangle\psi(v),$$

$$a^+(u)\psi(v) = \frac{d}{dx}\,\psi(v+xu)\Big|_{x=0} \quad \text{for all } u,v \in h.$$

We write s.lim to denote strong limit and adopt the convention that integrals of operator valued functions with respect to measures will be in the strong Bochner sense. Our first proposition is a Fourier inversion formula.

Proposition 2.1: For any $a \in \mathbb{R}$, $\varepsilon > 0$, $u \in h$

$$E_u((a-\varepsilon, a+\varepsilon)) = \underset{n\to\infty}{\mathrm{s.lim}}\ \pi^{-1}\int_{-n}^{n} e^{-iax}\,\frac{\sin \varepsilon x}{x}\, W(xu)\,dx. \tag{2.7}$$

Proof: For any $\xi \in H$ let

$$\xi_n = \pi^{-1}\int_{-n}^{n} e^{-iax}\,\frac{\sin \varepsilon x}{x}\, W(xu)\xi\, dx. \tag{2.8}$$

Suppose μ is the totally finite measure on $\mathbb{R}$ whose Fourier transform $\hat{\mu}$ is given by

$$\hat{\mu}(x) = \int e^{ixy}\, \mu(dx) = \langle \xi, W(xu)\xi \rangle .$$

By elementary computation we obtain

$$\|\xi_m - \xi_n\|^2 = \pi^{-2} \int_{\mathbb{R}} \Big|\int_m^n 2\, \frac{\sin \varepsilon x}{x} \cos(a-z)x\, dx\Big|^2 \mu(dz) \tag{2.9}$$

for $m < n$. Since

$$\sup_{x,\alpha \in \mathbb{R}} \Big| \int_0^x \frac{\sin \alpha y}{y}\, dy\Big| = M < \infty, \tag{2.10}$$

$$\lim_{m,n \to \infty} \int_m^n \frac{\sin \alpha y}{y}\, dy = 0 \quad \text{for every } \alpha \in \mathbb{R},$$

(2.9) implies that the sequence $\{\xi_n\}$ defined by (2.8) is a Cauchy sequence and hence $\lim_{n \to \infty} \xi_n = \xi_\infty$ exists. For any $\eta \in H$ we have

$$\langle \eta, \xi_\infty \rangle = \lim_{n \to \infty} \pi^{-1} \int_{-n}^{n} e^{-iax}\, \frac{\sin \varepsilon x}{x} \langle \eta , W(xu)\xi \rangle dx.$$

Since $\langle \eta, W(xu)\xi \rangle$ is the Fourier transform of the complex valued, totally finite and absolutely continuous measure $\langle \eta, E_u(\cdot)\xi \rangle$ an application of the classical Fourier inversion theorem (cf. page 93, [2]) shows that

$$\xi_\infty = E_u((a-\varepsilon,\ a+\varepsilon))\xi.$$

This implies (2.7).

□

Corollary: For any $a \in \mathbb{R}$, $\varepsilon > 0$ *and positive integer* n

$$\Big\|\pi^{-1} \int_{-n}^{n} e^{-iax}\, \frac{\sin \varepsilon x}{x} W(xu)\, dx\Big\| \leq 2M\pi^{-1} \tag{2.11}$$

where M *is defined by* (2.10).

Proof: This is immediate from (2.9) if we put $m = 0$ and use (2.10).

□

3. Local Measures of Some Quantum Processes in Fock Space

Let T be a locally compact second countable metric space and let F denote its Borel σ-algebra. Consider a fixed strongly continuous map $\phi : T \to h$. Define the selfadjoint operators

$$P_\phi(t) = P(\phi(t)), \quad t \in T \tag{3.1}$$

where $P(u)$, $u \in h$ is determined by (2.4). If ρ is a nonnegative selfadjoint operator of unit trace in H we can look upon $\{P_\phi(t),\ t \in T\}$ as a "quantum stochastic process" in the state ρ obeying the commutation relations:

$$[P_\phi(s),\ P_\phi(t)] = 2i \ \mathrm{Im} \ \langle\phi(s),\phi(t)\rangle, \quad s,t \in T. \tag{3.2}$$

The central aim of this section is to construct a local operator valued measure for the process $\{P_\phi(t),\ t \in T\}$ in analogy with the classical notion of local time for the standard Brownian motion in the sense of P. Levy (cf. Section 2.2, [4]). It may be recalled that when $h = L_2(\mathbb{R}_+)$, $T = \mathbb{R}_+$ and $\phi(t) = \chi_{[0,t]}$, the indicator of the interval $[0,t]$ then $\{P_\phi(t),\ t \geq 0\}$ is, indeed, standard Brownian motion in the Fock vacuum state.

We introduce the following notations:

$$F_0 = \{C : C \in F, \text{ closure of } C \text{ is compact}\} \tag{3.3}$$

$$K_\phi(s,t) = \{||\phi(s)||^2 ||\phi(t)||^2 - (\mathrm{Re}\langle\phi(s),\phi(t)\rangle)^2\}^{-1/2} \tag{3.4}$$

$$R_\phi = \{\nu : \nu \text{ is a Radon measure on } (T,F),$$

$$\int_C K_\phi(s,t)\,\nu(ds)\,\nu(dt) < \infty \text{ for every compact set } C \subset T \times T\} \tag{3.5}$$

$$L_\nu(C,a,\varepsilon) = (2\varepsilon)^{-1} \int_C E_{\phi(t)}((a-\varepsilon,a+\varepsilon))\nu(dt) \tag{3.6}$$

$$\text{for } C \in F_o,\ a \in \mathbb{R},\ \varepsilon > 0,\ \nu \in R_\phi$$

where E_u, $u \in h$ is determined by (2.4) and (2.5). It is to be noted that $2\varepsilon\, L_\nu(C,a,\varepsilon)$ is a bounded nonnegative selfadjoint operator which can be interpreted as an observable measuring the amount of "time" from C spent by the process $\{P_\phi(t),\ t \in T\}$ in the neighbourhood $(a-\varepsilon,\ a+\varepsilon)$ when time is measured according to the scale ν in (T,F). To arrive at a notion of "local time" we shall analyse the asymptotic behaviour of $L_\nu(C,a,\varepsilon)$ as $\varepsilon \to 0$.

Proposition 3.1: For any $C \in F_o$, $a \in \mathbb{R}$ *and* $\varepsilon > 0$ *the following holds:*

$$L_\nu(C,a,\varepsilon) = \underset{n\to\infty}{\text{s.lim}} \int_C \{(2\pi)^{-1} \int_{-n}^{n} e^{-iax} \frac{\sin \varepsilon x}{\varepsilon x} W(x\phi(t))dx\}\nu(dt). \tag{3.7}$$

Proof: This is immediate from Proposition 2.1, (2.11) and dominated convergence theorem.

Proposition 3.2: For any $u \in h$, $C_1, C_2 \in F_o$, $a_1, a_2 \in \mathbb{R}$, $\varepsilon_1, \varepsilon_2 > 0$ *the following holds:*

$$\langle L_\nu(C_1,a_1,\varepsilon_1)\psi(u),\ L_\nu(C_2,a_2,\varepsilon_2)\psi(u)\rangle$$

$$= (2\pi)^{-2} e^{||u||^2} \int_{C_1\times C_2} \{\int e^{i(a_1x-a_2y)} \frac{\sin \varepsilon_1 x}{\varepsilon_1 x}$$

$$\cdot \frac{\sin \varepsilon_2 y}{\varepsilon_2 y} \eta_u(s,t,x,y)dx\, dy\}\nu(ds)\nu(dt) \tag{3.8}$$

where

$$\eta_u(s,t,x,y) = \exp\{ -\tfrac{1}{2}(x^2||\phi(s)||^2+y^2||\phi(t)||^2-2xy\langle\phi(s),\phi(t)\rangle)$$

$$+ 2\,ix\, \text{Im}\, \langle\phi(s),u\rangle + 2iy\, \langle u,\phi(t)\rangle\} . \tag{3.9}$$

Proof: By Proposition 3.1, (2.1) and (2.2) we obtain

$$\langle L_\nu(C_1,a_1,\varepsilon_1)\psi(u),\ L_\nu(C_2,a_2,\varepsilon_2)\psi(u)\rangle$$

$$= \lim_{n\to\infty} \int_{C_1\times C_2} \{(2\pi)^{-2} e^{||u||^2} \int_{-n}^{n}\int_{-n}^{n} e^{i(a_1x-a_2y)} \frac{\sin \varepsilon_1 x}{\varepsilon_1 x}$$

$$\cdot \frac{\sin \varepsilon_2 y}{\varepsilon_2 y} \eta_u(s,t,x,y)dx\, dy\}\nu(ds)\nu(dt)$$

where η_u is defined by (3.9). Since $|\frac{\sin\theta}{\theta}| \leq 1$ and

$$\int_{\mathbb{R}^2} |\eta_u(s,t,x,y)|dx\, dy = 2\pi\ K_\phi(s,t) \tag{3.10}$$

where K_ϕ is defined by (3.4) we obtain (3.8) from dominated convergence theorem and the integrability condition in (3.5).

□

Corollary: For any $u \in h$, $C_1, C_2 \in F_o$, $a_1, a_2 \in \mathbb{R}$

$$\lim_{\varepsilon_1,\varepsilon_2\to 0} \langle L_\nu(C_1,a_1,\varepsilon_1)\psi(u),\ L_\nu(C_2,a_2,\varepsilon_2)\psi(u)\rangle$$

$$= (2\pi)^{-2} e^{||u||^2} \int_{\mathbb{R}^2\times C_1\times C_2} e^{i(a_1x-a_2y)} \eta_u(s,t,x,y)dx\, dy\ \nu(ds)\nu(dt)\ . \tag{3.11}$$

Proof: This is immediate from (3.8), (3.10), the integrability condition in (3.5) and the dominated convergence theorem.

Proposition 3.3: Let $\nu \in R_\phi$, $C \in F_o$ *and* $a \in \mathbb{R}$. *Then there exists an operator* $L_\nu(C,a)$ *defined on the domain* E *such that*

$$\lim_{\varepsilon\to 0} L_\nu(C,a,\varepsilon)\xi = L_\nu(C,a)\xi \quad \text{for all } \xi \in E \tag{3.12}$$

Proof: From (3.11) we have

$$\lim_{\varepsilon_1,\varepsilon_2 \to 0} ||L_\nu(C,a,\varepsilon_1)\psi(u) - L_\nu(C,a,\varepsilon_2)\psi(u)||^2 = 0$$

for all $C \in F_o$, $a \in \mathbb{R}$. Thus (3.12) holds when $\xi = \psi(u)$, $u \in h$. Since E is the linear manifold generated by $\psi(u)$, $u \in h$ the proof is complete.

□

Proposition 3.4: Let the operators $L_\nu(C,a)$, $C \in F_o$, $a \in \mathbb{R}$ *be defined by (3.12) on the domain* E. *Then*

$$<\psi(u), L_\nu(C,a)\psi(v)>$$

$$= (2\pi)^{-1/2} e^{<u,v>} \int_C ||\phi(t)||^{-1}\exp(\tfrac{1}{2}||\phi(t)||^{-2}$$

$$\{<u,\phi(t)>- <\phi(t),v> - ia\})\nu(dt) \quad \text{for all } u,v \in h. \qquad (3.13)$$

Proof: By (3.12), Proposition 3.1, (2.1) amd (2.2) we have

$$<\psi(u), L_\nu(C,a)\psi(v)>$$

$$= \lim_{\varepsilon \to 0} \lim_{n \to \infty} (2\pi)^{-1} \int_C \{\int_{-n}^{n} e^{-iax} \frac{\sin \varepsilon x}{\varepsilon x} <\psi(u), W(x\phi(t))\psi(v)>dx\}\nu(dt)$$

$$= \lim_{\varepsilon \to 0} \lim_{n \to \infty} (2\pi)^{-1} \int_C \int_{-n}^{n} \frac{\sin \varepsilon x}{\varepsilon x} \exp\{- \frac{1}{2} x^2||\phi(t)||^2$$

$$- x(ia+<\phi(t),v> - <u,\phi(t)>) + <u,v>\}dx\ \nu(dt). \qquad (3.14)$$

Since $\nu \in R_\phi$, (3.4) amd (3.5) imply

$$\{\int_C||\phi(s)||^{-1} \nu(ds)\}^2 = \int_{C\times C} ||\phi(s)||^{-1} ||\phi(t)||^{-1} \nu(ds)\nu(dt)$$

$$\leq \int_{C\times C} ||\phi(s)||^{-1} ||\phi(t)||^{-1} \{1 - (\mathrm{Re} \frac{<\phi(s),\phi(t)>}{||\phi(s)||\ ||\phi(t)||})^2\}^{-\frac{1}{2}} \nu(ds)\nu(dt)$$

$$= \int_{C\times C} K_\phi(s,t)\nu(ds)\nu(dt) < \infty \quad \text{for } C \in F_o . \qquad (3.15)$$

The absolute value of the integrand in (3.14) does not exceed

$\exp\{-\frac{1}{2}x^2||\phi(t)||^2 + x\,\mathrm{Re}(\langle u,\phi(t)\rangle - \langle\phi(t),v\rangle) + \mathrm{Re}\,\langle u,v\rangle\} = g(x,t)$, say. Since for any $\sigma > 0$ and complex scalar α

$$\int_{-\infty}^{\infty} e^{-1/2\,\sigma^2 x^2 + \alpha x}\,dx = (2\pi)^{1/2}\sigma^{-1}\exp\frac{1}{2}\alpha^2\sigma^{-2} \tag{3.16}$$

we have

$$\int_{-\infty}^{\infty} g(x,t)dx = (2\pi)^{1/2}||\phi(t)||^{-1}\exp\frac{1}{2}\left(\mathrm{Re}\,\frac{\langle u,\phi(t)\rangle - \langle\phi(t),v\rangle}{||\phi(t)||}\right)^2.$$

Now using (3.15) we conclude

$$\int_C\int_{-\infty}^{\infty} g(x,t)dx\,\nu(dt) < \infty.$$

By dominated convergence theorem, (3.16) and (3.14) we obtain (3.13). □

Theorem 3.5: The operators $L_\nu(C,a)$, $C \in F_o$, $a \in \mathbb{R}$ *defined by (3.6) and (3.12) on the domain* E *satisfy the following conditions:*

(i) for fixed C *and* $\xi \in E$ *the map* $a \to L_\nu(C,a)\xi$ *is strongly continuous on* $\mathbb{R}$*;*

(ii) for fixed a *and* $\xi \in E$, $L_\nu(C,a)\xi$ *is strongly countably additive in* $C \in F_o$*;*

(iii) $L_\nu(C,a)$ *is a positive symmetric operator on the domain* E.

Proof: For $a,b \in \mathbb{R}$ we have from (3.11) and (3.12)

$$\langle L_\nu(C,a)\psi(u),\ L_\nu(C,b)\psi(u)\rangle$$

$$= (2\pi)^{-2}e^{||u||^2}\int_{\mathbb{R}^2\times C\times C} e^{i(ax-by)}\,\eta_u(s,t,x,y)dx\,dy\,\nu(ds)\nu(dt) \tag{3.17}$$

where η_u is defined by (3.9). Now (3.10) and the integrability condition in (3.5) imply that the right hand side of (3.17) is continuous in (a,b). This proves (i). Once again (3.17) and (3.10) imply

$$||L_\nu(C,a)\psi(u)||^2 \le (2\pi)^{-1} e^{||u||^2} \int_{C\times C} K_\phi(s,t)\nu(ds)\nu(dt) \quad . \qquad (3.18)$$

By definition $L_\nu(C,a)\psi(u)$ is finitely additive in $C \in F_O$. Hence strong countable additivity of $L_\nu(\cdot,a)\psi(u)$ in F_O follows from (3.18). This proves (ii). Since $L_\nu(C,a,\varepsilon)$ defined by (3.6) is a positive symmetric operator on H, (3.12) implies (iii).

□

Remark: We may call the operator valued measure $L_\nu(\cdot,a)$ on F_O the *local measure* at a for the process $\{P_\phi(t), t \in T\}$ in the *scale* ν. When T is an interval of the real line it may be called the *local time* at a in the scale ν.

Suppose the map $\phi : T \to h$ with which we began this section satisfies the condition that $<\phi(s), \phi(t)>$ is real for all $s,t \in T$. Then (3.2) implies that $\{P_\phi(t), t \in T\}$ is a commuting family of selfadjoint operators. We can find a real subspace h_O such that $\phi(t) \in h_O$ for all $t \in T$ and h is the complexification $h_O + ih_O$ of h. The family of self-adjoint operators $\{P(u), u \in h_O\}$ is commuting. In the Fock vacuum state $\{P(u), u \in h_O\}$ can be identified with a classical Gaussian field of random variables. Such an identification shows that $L_\nu(C,a)$ is a function of the Gaussian random variables $P_\phi(t)$, $t \in T$. This implies that $L_\nu(C,a)$ is essentially selfadjoint on the domain E. It is tempting to conjecture that $L_\nu(C,a)$ is essentially selfadjoint on the domain E without any hypothesis on ϕ. This would mean that local measure can always be interpreted as an observable concerning the process $\{P_\phi(t), t \in T\}$. When $T = \mathbb{R}_+$ and ν is Lebesque measure we may then interpret $L([o,t],a)$ as a local time observable in the quantum sense.

We shall now examine a few examples of maps ϕ and Radom measures ν such that $\nu \in R_\phi$ in the sense of (3.5).

Example 1: Let $T = \mathbb{R}_+$, ν = Lebesque measure, $h = L_2(\mathbb{R}_+)$ and $\phi(t) = \chi_{[0,t]}$, the indicator of the interval [0,t]. Then $t \to \phi(t)$ is a continuous map from $\mathbb{R}_+$ into h and

$$<\phi(s), \phi(t)> = s \wedge t,$$

$$K_\phi(s,t) = [st(1 - st^{-1})]^{-1/2} \quad \text{if } s < t$$

where $s \wedge t$ denotes the minimum of s and t. For any $a > 0$

$$\int_0^a \int_0^a K_\phi(s,t)\,ds\,dt = 2 \int_{0<s<t<a} [st(1-st^{-1})]^{-1/2} ds\,dt = 2a\pi. \tag{3.19}$$

Thus $\nu \in R_\phi$ in the sense of (3.5). We write

$$L(t,a) = L_\nu([0,t],a), \tag{3.20}$$

$$P(t) = P_\phi(t) \tag{3.21}$$

In view of the Remark after Theorem 3.5 the closure of $L(t,a)$ is self-adjoint. In the Fock vacuum state $\{P(t),\ t \geq 0\}$ is a standard Brownian motion and the closure of $L(t,a)$ is nothing but the classical local time of the Brownian motion at a in the sense of P. Levy. By Proposition 3.4

$$\langle\psi(u),\ L(t,a)\psi(v)\rangle = (2\pi)^{-1/2}\, e^{\langle u,v\rangle} \int_0^t s^{-1/2}\, e^{-\frac{1}{2s}(F(s)+a)^2} ds \tag{3.22}$$

where

$$F(t) = i\int_0^t (\bar{u}(s) - v(s))ds \quad . \tag{3.23}$$

<u>Example 2:</u> Let $T = \mathbb{R}_+^n$, ν = Lebesque measure,

$$h = L_2(\mathbb{R}_+^n),\quad \underline{t} = (t_1,t_2,\ldots,t_n),\quad t_j \geq 0,\quad j = 1,2,\ldots,n,$$

$$V(\underline{t}) = [0,t_1] \times [0,t_2] \times \ldots \times [0,t_n] \subset \mathbb{R}_+^n \quad ,$$

$$\phi(\underline{t}) = \chi_{V(\underline{t})} \quad .$$

Then $\underline{t} \to \phi(\underline{t})$ is a continuous map from T into h and

$$\langle\phi(\underline{s}),\ \phi(\underline{t})\rangle = \prod_{j=1}^n s_j \wedge t_j,$$

$$K_\phi(\underline{s},\ \underline{t}) = \{s_1 s_2 \ldots s_n t_1 \ldots t_n - (\prod_{j=1}^n s_j \wedge t_j)^2\}^{-1/2} \quad ,$$

$$\int_{V(\underline{a})\times V(\underline{a})} K_{\phi}(\underline{s},\underline{t})\, ds_1 \ldots ds_n\, dt_1 \ldots dt_n$$

$$= 2^n a_1 a_2 \ldots a_n \int_0^1 \ldots \int_0^1 [x_1 x_2 \ldots x_n (1 - x_1 x_2 \ldots x_n)]^{-1/2}\, dx_1 dx_2 \ldots dx_n < \infty .$$

In other words $\nu \in R_{\phi}$ in the sense of (3.5). Once again by the Remark after Theorem 3.5 the closures of $L_{\nu}(C,\alpha)$, $C \in F_o$, $\alpha \in \mathbb{R}$ constitute a commuting family of selfadjoint operators with domains including E. We note that in the Fock vacuum state $\{P_{\phi}(\underline{t}),\ \underline{t} \in T\}$ is a classical n-parameter Brownian motion.

<u>Example 3</u>: Let $T = \mathbb{R}$, ν = Lebesque measure and let $t \to U_t$ a strongly continuous one parameter group of unitary operators in h. Suppose $\phi(t) = U_t u$ where u is a unit vector in h. Then

$$\langle \phi(s), \phi(t) \rangle = \langle u, U_{t-s} u \rangle$$

$$= \rho(t-s), \text{ say.}$$

Suppose that

$$\int_0^a [1 - \operatorname{Re} \rho(s)]^{-1/2}\, ds < \infty \quad \text{for all } a > 0. \tag{3.24}$$

Since $\rho(0) = 1$ and $K_{\phi}(s,t) = \{1 - [\operatorname{Re} \rho(t-s)]^2\}^{-1/2}$ it follows from (3.24) that

$$\int_{-a}^{a} \int_{-a}^{a} K_{\phi}(s,t)\, ds\, dt < \infty \quad \text{for all } 0 < a < \infty.$$

Thus $\nu \in R_{\phi}$ and the operators $\{L_{\nu}(C,a)\}$ are defined on the domain E. However, we do not know whether $L_{\nu}(C,a)$ is essentially selfadjoint on E. If $\rho(t) = \rho(-t)$ for all t or equivalently the measure $\langle u, E(\cdot)u \rangle$ is symmetric, where $U_t = \int_{\mathbb{R}} e^{itx} E(dx)$ is the spectral representation of $\{U_t\}$ then the $\{P_{\phi}(t),\ t \in \mathbb{R}\}$ is a stationary Gaussian process with autocorrelation function $\rho(t)$ and by the Remark after Theorem 3.5, $\{L_{\nu}(C,a),\ C \in F_o,\ a \in \mathbb{R}\}$ is a commuting family of essentially selfadjoint operators. By Proposition 3.4

$$\langle\psi(v_1),\ L_\nu(C,a)\psi(v_2)\rangle = (2\pi)^{-1/2}\ e^{\langle v_1,v_2\rangle} \int_C \{\exp -\tfrac{1}{2}\ (G(t)+a)^2\}dt$$

where

$$G(t) = i\{\langle v_1,\ U_t u\rangle - \langle U_t u,\ v_2\rangle\}.$$

4. A Generalized Ito-Tanaka Formula

As an application of Proposition 3.4 we shall derive a generalized version of the classical Ito-Tanaka formula for standard Brownian motion (see Chapter 7, 8 in [1]). To this end we examine Example 1 in Section 3 with the same notation. In the language of quantum stochastic calculus in the Fock space $H = \Gamma(L_2(\mathbb{R}_+))$ the operator P(t) defined by (3.21) can be expressed as

$$P(t) = i[A(t) - A^+(t)] \text{ on } E \tag{4.1}$$

where A(t) and $A^+(t)$ are the annihilation and creation operators defined by $A(t) = a(\chi_{[0,t]}), A^{\dagger}(t) = a^{\dagger}(\chi_{[0,t]})$ respectively.

Proposition 4.1: Let μ be any totally finite signed measure on $\mathbb{R}$ and let the family $\{L(t,x),\ x \in \mathbb{R},\ t > 0\}$ of essentially selfadjoint operators on the domain E be defined by (3.20). Then for any $u \in h$ the integral

$$\int_{\mathbb{R}} \{L(t,x)\psi(u)\}\ \mu(dx) \tag{4.2}$$

is well defined in the sense of Bochner and

$$\left|\left|\int_{\mathbb{R}}\{L(t,x)\psi(u)\}\ \mu(dx)\right|\right| \le t^{1/2}\ ||\psi(u)||\ \ ||\mu|| \tag{4.3}$$

where $||\mu||$ denotes the total variation of μ.

Proof: By property (i) in Theorem 3.5, $L(t,x)\psi(u)$ is continuous in x. By (3.20), (3.18) and (3.19) we have

$$||L(t,x)\psi(u)|| \leq t^{1/2}\, e^{1/2||u||^2} ,$$

Hence (4.2) is well defined and (4.3) obtains.

□

Definition: The operator $\int L(t,x)\mu(dx)$ is defined on the domain E for every totally finite signed measure μ by putting

$$\int L(t,x)\mu(dx)\psi(u) = \int \{L(t,x)\psi(u)\}\mu(dx), \quad u \in h$$

using Proposition 4.1 and extending to E linearly.

Remark: By the Remark after Theorem 3.5 the closure of $\int L(t,x)\mu(dx)$ is a selfadjoint operator for every real totally finite signed measure μ.

Proposition 4.2: Let $u \in h$, $m(t) = \int_0^t u(s)ds$. *In the pure state determined by the unit vector* $\tilde{\psi}(u) = ||\psi(u)||^{-1}\psi(u)$ *the family* $\{P(t)-2 \operatorname{Im} m(t)\}$ *is a standard Brownian motion.*

Proof: For any real valued function $f \in h$ put

$$P_f(t) = P(f\chi_{[0,t]}) = \int_0^t f dP \quad \text{on } E$$

where the last expression denotes quantum stochastic integral. By (2.4), (2.1) and (2.2)

$$\langle\psi(u), e^{ixP_f(t)}\psi(u)\rangle = \langle\psi(u), W(xf\chi_{[0,t]})\psi(u)\rangle$$

$$= \exp\{||u||^2 - \tfrac{1}{2} x^2 \int_0^t f^2\, ds + 2ix \int_0^t f(s) \operatorname{Im} u(s)ds\} .$$

In other words, in the pure state $\tilde{\psi}(u)$ the observable $P_f(t)$ has distribution with mean $2\int_0^t f(s) \operatorname{Im} u(s)ds$ and variance $\int_0^t f^2(s)ds$. This is equivalent to saying that the commuting family of observables

$\{P(t)-2 \text{ Im } m(t)\}$ is a standard Brownian motion in the state $\tilde{\psi}(u)$.

□

Proposition 4.3: Let $f \in L_\infty(\mathbb{R}) + L_2(\mathbb{R})$. *Then the quantum stochastic integral* $\int_0^t f(P(s))dP(s)$ *is well defined on the domain* E.

Proof: By the definition of quantum stochastic integrals in [3] we have to only verify that

$$\int_0^t ||f(P(s))\psi(u)||^2 \, ds < \infty \quad \text{for all } u \in h,\ t > 0. \tag{4.4}$$

By Proposition 4.2 we have

$$||f(P(s))\psi(u)||^2 = ||\psi(u)||^2 < \tilde{\psi}(u), |f(P(s))|^2 \tilde{\psi}(u)>$$

$$= ||\psi(u)||^2 \int_{-\infty}^{\infty} |f(x)|^2 (2\pi \ s)^{-1/2}\{\exp . \frac{1}{2s}(x-2 \text{ Im } m(s))^2\}dx. \tag{4.5}$$

If $f \in L_\infty(\mathbb{R})$ it follows that

$$||f(P(s))\psi(u)||^2 \leq ||f||_\infty^2 ||\psi(u)||^2$$

and if $f \in L_2(\mathbb{R})$

$$||f(P(s))\psi(u)||^2 \leq (2\pi s)^{-1/2} \ ||f||_2^2 \ ||\psi(u)||^2 .$$

In either case we obtain (4.4).

□

Theorem 4.4: Let f *be an absolutely continuous function on* $\mathbb{R}$ *satisfying the following conditions:*

(i) $\int_{-\infty}^{\infty} |f(x)|^2 e^{-ax^2} dx < \infty$ *for all* $a > 0$

(ii) f *admits the first two derivatives* f', f'' *as tempered distributions where* $f' \in L_\infty(\mathbb{R}) + L_2(\mathbb{R})$ *and* $f'' = \mu$ *is a totally finite signed measure.*

Then

$$f(P(t)) = f(o) + \int_0^t f'(P(s))\, dP(s) + \frac{1}{2}\int_{-\infty}^{\infty} L(t,x)\mu(dx) \qquad (4.6)$$

on the domain E *where the operators* $L(t,x)$ *are defined by (3.20).*

Proof: Condition (i) and (4.5) imply that $\psi(u)$ is in the domain of $f(P(t))$ for all $t > C$. Let $\mathcal{S}(\mathbb{R})$ denote the Schwarz space of rapidly decreasing test functions on $\mathbb{R}$. Then condition (ii) implies

$$\int (f\phi' + f'\phi)dx = \int f'\phi' dx + \int \phi\, d\mu = 0 \text{ for all } \phi \in \mathcal{S}(\mathbb{R}). \qquad (4.7)$$

Let $u,v \in h$. By Proposition 2.1 the indicator function $\chi_{(a-\varepsilon,a+\varepsilon)}$ satisfies the relation

$$<\psi(u),\ \chi_{(a-\varepsilon,a+\varepsilon)}(P(t))\psi(v)>$$

$$= \lim_{n\to\infty} \pi^{-1}\int_{-n}^{n} e^{-iax}\,\frac{\sin\varepsilon x}{x} <\psi(u), W(x\chi_{[0,t]})\psi(v)>dx$$

$$= \pi^{-1}\int_{-\infty}^{\infty} e^{-iax}\,\frac{\sin\varepsilon x}{x}\exp\{-\tfrac{1}{2}tx^2 - ixF(t)\}dx$$

$$\text{for all } a \in \mathbb{R},\ \varepsilon > 0 \qquad (4.8)$$

where $F(t)$ is given by (3.23). Since the unitary Fourier transforms in $L_2(\mathbb{R})$ of the functions $\chi_{(a-\varepsilon,a+\varepsilon)}$ and $t^{-1/2}\exp -\frac{1}{2t}(x+F(t))^2$ are respectively $(2\pi^{-1})^{1/2}e^{iax}x^{-1}\sin\varepsilon x$ and $\exp\{-\frac{1}{2}tx^2 - ixF(t)\}$ for each $t > 0$ we obtain from (4.8) and the Parseval identity

$$<\psi(u),\ \chi_{(a-\varepsilon,a+\varepsilon)}(P(t))\psi(v)>$$

$$= (2\pi t)^{-1/2} e^{<u,v>}\int_{-\infty}^{\infty}\chi_{(a-\varepsilon,a+\varepsilon)}(x)\exp\{-\tfrac{1}{2t}(x+F(t))^2\}dx.$$

Now a routine approximation argument yields

$$\langle\psi(u),\ f(P(t))\psi(v)\rangle$$

$$= (2\pi t)^{-1/2}\ e^{\langle u,v\rangle} \int_{-\infty}^{\infty} f(x)\exp\{-\ \frac{1}{2t}\ (x+F(t))^2\}dx. \tag{4.9}$$

By (4.1), condition (ii) and Proposition 4.3 we have

$$\langle\psi(u),\ \int_0^t f'(P(s))dP(s)\psi(v)\rangle = \int_0^t i(\bar{u}(s)-v(s))\langle\psi(u),f'(P(s))\psi(v)\rangle ds.$$

Condition (ii) also implies that $\int_{-\infty}^{\infty} |f'(x)|^2\ e^{-ax^2}\ dx < \infty$ and hence (4.9) holds when f is replaced by f'. Thus

$$\langle\psi(u),\ \int_0^t f'(P(s))dP(s)\psi(v)\rangle$$

$$= e^{\langle u,v\rangle} \int_0^t \int_{-\infty}^{\infty} F'(s)f'(x)e^{-\frac{1}{2s}(x+F(S))^2}(2\pi s)^{-1/2}\ dx\ ds.$$

Since functions of the form exp $-\ \alpha(x+\beta)^2$ are in $\mathcal{S}$ for $\alpha > 0$, $\beta \in ¢$ we obtain from (4.7)

$$\langle\psi(u),\ \int_0^t f'(P(s))dP(s)\psi(v)\rangle$$

$$= -(2\pi)^{-1/2}\ e^{\langle u,v\rangle} \int_0^t \int_{\infty}^{\infty} s^{-\frac{3}{2}}F'(s)f(x)(x+F(s))e^{-\frac{1}{2s}(x+F(s))^2}dx\ ds.$$

By Proposition 3.4 and (4.7) we have

$$\frac{1}{2}\langle\psi(u),\ \int L(t,x)d\mu(x)\psi(v)\rangle$$

$$= (8\pi)^{-1/2}\ e^{\langle u,v\rangle} \int_0^t \int_{-\infty}^{\infty} s^{-1/2}\ e^{-\ \frac{1}{2s}(x+F(s))^2}\mu(dx)ds$$

$$= (8\pi)^{-1/2}\ e^{\langle u,v\rangle} \int_0^t \int_{-\infty}^{\infty} f(x)e^{-\ \frac{1}{2s}(x+F(s))^2}\{(\frac{x+F(s)}{x})^2 - \frac{1}{s}\}dx\ ds. \tag{4.11}$$

Straightforward differentiation with respect to t shows that

$$t^{-1/2} e^{-\frac{1}{2t}(x+F(t))^2} = -\int_0^t F'(s)s^{-\frac{3}{2}}(x+F(s))e^{-\frac{1}{2s}(x+F(s))^2} ds$$

$$+ \frac{1}{2}\int_0^t s^{-1/2}\{(\frac{x+F(s)}{s})^2 - \frac{1}{s}\} e^{-\frac{1}{2s}(x+F(s))^2} ds.$$

Now (4.9) - (4.11) and the above relation imply (4.6) on the domain E.

□

Remark: Put $f(x) = |x - a|$ in Theorem 4.4. Then conditions (i) and (ii) are fulfilled with

$$f'(x) = \text{sgn}(x - a), \quad f''(x) = 2\delta(x - a).$$

In other words f" is twice the Dirac measure at the point a. Thus we obtain the Ito-Tanaka formula (1.1) if we identify $\{P(t), t \geq 0\}$ with the standard Brownian notion in the Fock vacuum state.

References

[1] K.L. Chung and R.J. Williams: Introduction to Stochastic Integration, Birkhauser, Boston, Basel, Stuttgart 1983.

[2] H. Cramer: Mathematical Methods of Statistics, Princeton University Press, Princeton, 1946.

[3] R.L. Hudson and K.R. Parthasarathy: Quantum Ito's formula and stochastic evolutions, Comm. Math. Phys. 93, 301-323 (1984).

[4] K. Ito and H.P. Mckean: Diffusion Processes and their Sample Paths, Springer Verlag, Berlin Heidelberg New York, 1974.

[5] P.A. Meyer: Elements de probabilités quantiques in Séminaire de Probabilités XX 1984/85, 1204 Lecture Notes in Mathematics, Springer-Verlag, Berlin Heidelberg New York (Ed. J. Azema and M. Yor), pp. 186-312.

Representation of a Class of Quantum Martingales II

by

K.R. Parthasarathy and Kalyan B. Sinha

Indian Statistical Institute, New Delhi

1. Introduction

Here we extend the results obtained in [1] for the representation of a bounded regular martingale to the case where there is an underlying degeneracy space. This is necessary to deal with, for example, the uniqueness of the Fermion martingales where the test functions are in $L^2(\mathbb{R}^\nu)$, $\nu > 1$, as has been used in [2].

Due to technical difficulties, we are at present unable to do this in a basis independent way, and for this reason, we have to extend the stochastic integration theory of [3] to the present situation. This we do briefly in section 2 and then prove the representation theorem in section 3 with the section 4 dealing with applications, particularly that of Fermion martingales.

2. Stochastic Integral and Ito Formula

In [3] and [4], stochastic integrals of the type $\int_0^t \{E_1(s)d\Lambda(s) + E_2(s)dA(s) + E_3(s)dA^+(s) + E_4(s)ds\}$ were defined on the exponential domain for adapted processes $\{E_j(s)\}$ $(j = 1,2,3,4)$ satisfying

$$\int_0^t \{|f(s)|^2 \, ||E_1(s)u \otimes \psi(f)||^2 + \sum_{j=2}^{4} ||E_j(s)u \otimes \psi(f)||^2\}ds<\infty, \qquad (2.1)$$

where $u \in h_0$, $f \in L^2(\mathbb{R}_+)$. Here we extend the theory to the case where $L^2(\mathbb{R}_+)$ is replaced by $L^2(\mathbb{R}_+, h_1)$ with h_1 a separable Hilbert space. As mentioned in the introduction, this we shall do only for a fixed orthonormal basis of h_1.

Let h_0 and h_1 be two separable Hilbert spaces, and we write $h \equiv L^2(\mathbb{R}_+) \otimes h_1 \simeq L^2(\mathbb{R}_+, h_1)$, $\tilde{H} \equiv h_0 \otimes H$, $\tilde{H}_t \equiv h_0 \otimes \Gamma(L^2([0,t],h_1))$, $H^t \equiv \Gamma(L^2([t,\infty), h_1))$, so that $\tilde{H} = \tilde{H}_t \otimes H^t$. Also let $\mathcal{D}_0$ and M be dense linear manifolds in h_0 and h respectively such that $\{\chi_{[0,t]}f\}$ belong to M for every $t \geq 0$ whenever f does, and let

$$M_t = \{\chi_{[0,t]}f \mid f \in M\}; \; M^t = \{\chi_{[t,\infty)}f \mid f \in M\} .$$

For $f \in h$, we define the exponential vector of f as

$$\psi(f) = 1 \oplus f \oplus \ldots \oplus f^{\otimes n}/\sqrt{n!} \oplus \ldots .$$

and by $E(\mathcal{D})$ we mean the linear manifold generated by $\{\psi(f) \mid f \in \mathcal{D} \subset h\}$. The *annihilation, creation* and *conservation* operators are then defined on $E(h)$ as follows:

$$\begin{aligned} a(f)\psi(g) &= \langle f,g\rangle \, \psi(g) \\ a^+(f)\psi(g) &= \frac{d}{d\varepsilon}\, \psi(g + \varepsilon f)\big|_{\varepsilon=0} \qquad (2.2) \\ \lambda(T)\psi(g) &= \frac{d}{d\varepsilon}\, \psi(e^{\varepsilon T} g)\big|_{\varepsilon=0} \quad , \end{aligned}$$

all the derivatives are in the strong sense and $f,g \in h$, $T \in \mathcal{B}(h)$. The operators $a(f)$ and $a^+(f)$ are adjoint to each other while $\lambda(T^+) = (\lambda(T))^+$ all on $E(h)$.

We also set

$$E = \mathcal{D}_0 \otimes E(M), \; E_t = \mathcal{D}_0 \otimes E(M_t), \; E^t = E(M^t),$$

where $\otimes$ denotes the algebraic tensor product. Then the definitions of adapted, bounded adapted and martingale processes with respect to the pair $(\mathcal{D}_0, M)$ remain unchanged from those in [1]. Now let $\{e_i\}$ be an orthonormal basis in h_1 and let $P_{ij} = |e_i\rangle\langle e_j|$. Next we define the three basis martingale families called *annihilation, creation,* and *conservation* respectively with respect to the pair (h_0, h) :

$$A_i(t) = I_0 \otimes a(\chi_{[0,t]} \otimes e_i)$$

$$A_i^+(t) = I_0 \otimes a^+(\chi_{[0,t]} \otimes e_i)$$

$$\Lambda_{ij}(t) = I_0 \otimes \lambda(\chi_{[0,t]} \otimes P_{ij}), \tag{2.3}$$

where I_0 is the identity in h_0; a, a^+, λ are defined in (2.2), $\chi_{[0,t]}$ is looked upon as a vector in $L^2(\mathbb{R}_+)$ in the first two equations while it is treated as a multiplication operator in $L^2(\mathbb{R}_+)$ in the last.

Let $\{E_{ij}^{(1)}\}$, $\{E_i^{(2)}\}$, $\{E_i^{(3)}\}$, $\{E^{(4)}\}$ be four sets of families of processes adapted with respect to the pair $(\mathcal{D}_0, M)$. We want to define stochastic integrals of the type $\int_0^t \{ \sum_{ij} E_{ij}^{(1)} d\Lambda_{ij} + \sum_i E_i^{(2)} dA_i + \sum_i E_i^{(3)} dA_i^+ + E^{(4)} ds\}$. As in [3] we proceed by first defining in integral for a finite family of simple adapted processes $\{E^{(k)}\}_{k = 1,2,3,4}$, establish quantum Ito's product formula and then extend the definition for a certain class of countable families $\{E^{(k)}\}$. We shall not give all the details since the calculations are essentially similar to those in [3].

The quantum Ito's formula is described in Theorem 4.5 of [3] and can be best summarized in the following table:

	$d\Lambda_{k\ell}$	dA_k	dA_k^+	dt
$d\Lambda_{ij}$	$\delta_{jk} d\Lambda_{i\ell}$	0	$\delta_{jk} dA_i^+$	0
dA_i	$\delta_{ik} dA_\ell$	0	$\delta_{ik} dt$	0
dA_i^+	0	0	0	0
dt	0	0	0	0

(2.4)

The next theorem constitutes the basic result in the definition of the stochastic integral in such a set up.

Theorem 2.1: Let $A_2(\mathcal{D}_0, M)$ *denote the class of all ordered quadruples of families of processes* $\mathbf{E} \equiv \{E_{ij}^{(1)}(t), E_i^{(2)}(t), E_i^{(3)}(t), E^{(4)}(t);$ $1 \leq i,j < \infty, t \geq 0\}$ *adapted with respect to the pair* $(\mathcal{D}_0, M)$ *satisfying:*

$$\int_0^t \{\sum_{i=1}^{\infty} || \sum_{j=1}^{\infty} f_j(s) E_{ij}^{(1)}(s) u \otimes \psi(f) ||^2 + \sum_{k=2}^{3} \sum_{i=1}^{\infty} ||E_i^{(k)}(s) u \otimes \psi(f)||^2$$

$$+ ||E^{(4)}(s) u \otimes \psi(f)||^2\} ds < \infty, \tag{2.5}$$

for all $t > 0$, $u \in \mathcal{D}_0$, $f \in M$. *Then the stochastic integral*

$$X(t) = \int_0^t \{ \sum_{i,j} E_{ij}^{(1)}(s) d\Lambda_{ij}(s) + \sum_i E_i^{(2)}(s) dA_i(s)$$

$$+ \sum_i E_i^{(3)}(s) dA_i^{+}(s) + E^{(4)}(s)\, ds\} \tag{2.6}$$

is well defined as an adapted process with respect to the pair $(\mathcal{D}_0, M)$. *Moreover one has for* $u, v \in \mathcal{D}_0$; $f, g \in M$:

$$\langle u \otimes \psi(f), X(t) v \otimes \psi(g)\rangle = \int_0^t \langle u \otimes \psi(f), \{\sum_{ij} \bar{f}_i(s) g_j(s) E_{ij}^{(1)}(s)$$

$$+ \sum_i g_i(s) E_i^{(2)}(s) + \sum \bar{f}_i(s) E_i^{(3)}(s)$$

$$+ E^{(4)}(s)\} v \otimes \psi(g)\rangle ds , \tag{2.7}$$

and an estimate:

$$||X(t) u \otimes \psi(f)||^2$$

$$\leq \int_0^t \exp\{t-s+3 \int_0^t ||f(\tau)||^2 d\tau\}\{3 \sum_i || \sum_j f_j(s) E_{ij}^{(1)}(s) u \otimes \psi(f)||^2$$

$$+ \sum_i ||E_i^{(2)}(s)u \otimes \psi(f)||^2 + 3 \sum_i ||E_i^{(3)}(s)u \otimes \psi(f)||^2 \; .$$

$$+ ||E^{(4)}(s)u \otimes \psi(f)||^2 \} ds < \infty. \tag{2.8}$$

Sketch of proof: Proceeding as in [3] and using the Ito formula (2.4), one gets after a simple but long computation that for simple quadruple **E** with finite number of non-zero components that

$$\frac{d}{dt} ||X(t)u \otimes \psi(f)||^2 = 2 \text{ Re } \{\sum_i \langle f_i(t)X(t)u \otimes \psi(f), \sum_j f_j(t)E_{ij}^{(1)}(t)u \otimes \psi(f)\rangle$$

$$+ \sum_i \langle \bar{f}_i(t)X(t)u \otimes \psi(f), E_i^{(2)}(t)u \otimes \psi(f)\rangle$$

$$+ \sum_i \langle f_i(t)X(t)u \otimes \psi(f), E_i^{(3)}(t)u \otimes \psi(f)\rangle$$

$$+ \sum_i \langle E_i^{(3)}(t)u \otimes \psi(f), \sum_j f_j(t)E_{ij}^{(1)}(t)u \otimes \psi(f)\rangle$$

$$+ \langle X(t)u \otimes \psi(f), E^{(4)}(t)u \otimes \psi(f)\rangle$$

$$+ \sum_i || \sum_j f_j(t)E_{ij}^{(1)}(t)u \otimes \psi(f)||^2$$

$$+ \sum_i ||E_i^{(3)}(t)u \otimes \psi(f)||^2 \; . \tag{2.9}$$

On using the inequality $2 \text{ Re } \bar{a}b \leq |a|^2 + |b|^2$ we have from (2.8) that

$$\frac{d}{dt} ||X(t)u \otimes \psi(f)||^2 \leq (1 + 3||f(t)||^2) \; X(t)u \otimes \psi(f)||^2$$

$$+ 3 \sum_i || \sum_j f_j(t)E_{ij}^{(1)}(t)u \otimes \psi(f)||^2$$

$$+ 3 \sum_i ||E_i^{(3)}(t)u \otimes \psi(f)||^2$$

$$+ \sum_i ||E_i^{(2)}(t)u \otimes \psi(f)||^2$$

$$+ ||E^{(4)}(t)u \otimes \psi(f)||^2.$$

The final estimate (2.8) follows from the above inequality and assumption (2.5) and also allows for the extension of the integral to the whole class $A_2(\mathcal{D}_0,M)$. Having done that, (2.9) extends to the whole class $A_2(\mathcal{D}_0,M)$.

□

3. Regular martingales

A bounded martingale X on H is said to be *regular* with respect to a Radon measure μ on $\mathbb{R}_+$ or simply regular if for all $t > a \geq 0$, $\tilde{u} \in \tilde{H}_a$

$$\max\{||[X(t) - X(a)]\tilde{u}||^2, ||[X^+(t) - X^+(a)]\tilde{u}||^2\} \leq ||\tilde{u}||^2 \mu([a,t]). \tag{3.1}$$

The necessity of (3.1) for a large class of bounded martingales admitting representation (2.6) with $E^{(4)} = 0$ is contained in the next lemma.

Lemma 3.1: Let X *be a bounded martingale with representation* $dX = \sum_{ij} M_{ij}d\Lambda_{ij} + \sum_i K_i^+ dA_i + \sum_i L_i dA_i^+$ *where the quadruples* $(M_{ij}, K_i^+, L_i, 0)$ *and* $(M_{ji}^+, L_i^+, K_i, 0)$ *belong to* $A_2(h_0,M)$*;* M_{ij}, K_i, L_i *are bounded adapted processes such that* $\sum_i L_i^+(s)L_i(s)$ *and* $\sum_i K_i^+(s)K_i(s)$ *converge weakly to (self-adjoint) bounded operators* L(s) *and* K(s) *respectively with* $||L(s)||$ *and* $||K(s)||$ *locally integrable. Then* X *is regular.*

Proof: It follows from (2.9) that for $\tilde{u} \in \tilde{E}_a$

$$||[X(t)-X(a)]\tilde{u}||^2 = ||X(t)\tilde{u}||^2 - ||X(a)\tilde{u}||^2 = \int_0^t \sum ||L_i(s)\tilde{u}||^2 \, ds$$

$\leq ||\tilde{u}||^2 \int_a^t ||L(s)||ds$, and similarly one has that $||[X^+(t)-X^+(a)]\tilde{u}||^2$

$\leq ||\tilde{u}||^2 \int_a^t ||K(s)||ds.$

□

Our aim is to prove the converse of Lemma 3.1 and for that we prepare ourselves with a series of lemmas beginning with an extension of classical representation theorem of Kumita-Watanabe for L^2-martingales adapted to one Brownian motion to an h_0-valued L^2-martingale adapted to a countable family of independent Brownian motions.

Proposition 3.2: Let $\{X(t)\}$ *be a square integrable martingale adapted to* $\{w_1(t), w_2(t),...\}$ *where* w_j, $j = 1,2,...$ *is an i.i.d. sequence of standard Brownian motions. Then*

$$X(t) = X(0) + \sum \int_0^t \xi_j \, dw_j$$

where $\{\xi_j(\cdot)\}$ *is a sequence of adapted processes satisfying*

$$\int_0^t \sum_j \mathbb{E} \, |\xi_j(s)|^2 \, ds < \infty \;\forall\; t.$$

Proof: Let P denote the probability measure of $\{w_1,w_2,...\}$. Identify $L_2(P)$ with $H = \Gamma(L_2(\mathbb{R}_+) \otimes \ell_2)$ by the correspondence

$$\psi(\underline{f}) \to \exp \sum_j (\int f_j \, dw_j - \tfrac{1}{2} \int f_j^2 \, dt)$$

where $\underline{f} = (f_1,f_2,...) \in \bigoplus_{j=1}^{\infty} L_2(\mathbb{R}_+) \simeq L_2(\mathbb{R}_+) \otimes \ell_2$. The set $E_0 = \{\psi(\underline{f}),\ \underline{f} = (f_1,f_2,...)$ where all but a finite number of f_j are $0\}$ is total in H. We have

$$\psi(\underline{f}\,\chi_{[0,t]}) - 1 = \sum_j \int_0^t f_j(s) \, \psi(\underline{f}\,\chi_{[0,s]})dw_j(s)$$

for $\psi(\underline{f}) \in E_0$. For any $a > 0$ consider $X(a)$. Then we can find

$\underline{f}_{nj}$, $j = 1,2,\ldots,k_n$ in $L_2(\mathbb{R}_+) \otimes \ell_2$ and scalars $a_{n1}, a_{n2},\ldots,a_{nk_n}$ such that $\psi(\underline{f}_{nj}) \in E_0$, $\lim_{n\to\infty} ||X(a)-X(0) - \sum_{j=1}^{k_n} a_{nj}[\psi(\underline{f}_{nj}\, \chi_{[0,a]})-1]|| = 0$.

Since projections are continuous it follows that

$$\lim_{n\to\infty} ||X(t) - X(0) - \sum_{j=1}^{k_n} a_{nj}\, [\psi(\underline{f}_{nj}\, \chi_{[0,t]}) - 1]|| = 0 \,.$$

We have

$$\sum_{j=1}^{k_n} a_{nj}[\psi(\underline{f}_{nj}\, \chi_{[0,t]}) - 1] = \int_0^t \sum_{n,j} a_{nj}\, f_{njk}(s)\psi(\underline{f}_{nj}\, \chi_{[0,s]})dw_k(s),$$

$$0 \leq t \leq a.$$

Put

$$\xi_{nk}(s) = \sum_j a_{nj}\, f_{njk}(s)\psi(\underline{f}_{nj}\, \chi_{[0,s]}),\ 0 \leq s \leq a.$$

Then

$$\lim_{n\to\infty} ||X(t) - X(0) - \sum_k \int_0^t \xi_{nk}(s)dw_k(s)||^2 = 0.$$

In particular

$$\lim_{m,n\to\infty} \sum_k \int_0^a \mathbb{E}|\xi_{nk}(s) - \xi_{mk}(s)|^2\, ds = 0 \quad .$$

Thus there exist adapted processes $\xi_1(s)$, $\xi_2(s)$, ... in $[0,a]$ such that

$$\lim_{n\to\infty} \sum_k \int_0^a \mathbb{E}|\xi_{nk}(s) - \xi_k(s)|^2\, ds = 0 \quad .$$

Hence

$$X(t) = X(0) + \sum_k \int_0^t \xi_k(s)dw_k(s), \quad 0 \leq t \leq a \quad .$$

Once again by isometry of Ito integrals it follows that

$$X(t) = X(0) + \sum_k \int_0^t \xi_k(s)dw_k(s) \quad \text{for } 0 \le t < \infty$$

where

$$\sum_k \int_0^t \mathbb{E}|\xi_k(s)|^2 \, ds < \infty \;\forall\, t \ge 0.$$

□

Lemma 3.3: Let X be a bounded martingale regular with respect to a Radon measure μ *on* $\mathbb{R}_+$. *Then (i)* $||X(t)||$ *is non-decreasing, (ii)* μ *can be replaced by its absolutely continuous part, (iii) there exist two countable families of bounded adapted processes* $\{L_i(t)\}$ *and* $\{K_i(t)\}$ *such that for* $\tilde{u} \in \tilde{H}_a$, $t > a \ge 0$,

$$[X(t) - X(a)]\tilde{u} = \int_a^t \sum_i L_i(s)\tilde{u}\, dw_i(s), \quad \textit{and}$$

$$[X^+(t) - X^+(a)]\tilde{u} = \int_a^t \sum_i K_i(s)\tilde{u}\, dw_i(s),$$

where $\{w_i(s)\}$ *is the countable family of Brownian motions in Proposition 3.2, (iv) furthermore,* $\sum_i L_i^+(s)L_i(s)$ *and* $\sum_i K_i^+(s)K_i(s)$ *converge strongly to self-adjoint bounded adapted processes, say* L(s) *and* K(s) *respectively with the property that* $||L(s)||$ *and* $||K(s)||$ *are locally integrable.*

Proof: The proof of (i) is same as in [1]. For (ii) we observe that for fixed $\tilde{u} \in \tilde{H}_a$, $\{X(t)\tilde{u};\ t \ge a\}$ and $\{X^+(t)\tilde{u};\ t \ge a\}$ are classical h_0-valued square integrable martingales in $[a,\infty)$ adapted to the countable family $\{w_i(s)\}$ of independent Brownian motions and therefore by Proposition 3.2 there exist two countable families of h_0-valued adapted square integrable Brownian functionals $\{\xi_i(t,\tilde{u})\}$ and $\{\eta_i(t,\tilde{u})\}$ such that

$$X(t)\tilde{u} - X(a)\tilde{u} = \int_a^t \sum_i \xi_i(s,\tilde{u})dw_i(s),$$

$$X^+(t)\tilde{u} - X^+(a)\tilde{u} = \int_a^t \sum \eta_i(s,\tilde{u})dw_i(s) \ . \tag{3.2}$$

By the isometry of classical Ito integrals (see Proposition 3.2) and the regularity assumption (3.1), we have for all $0 \leq a < b < t < \infty$

$$\max\{ \int_a^t \mathbb{E} \sum_i ||\xi_i(s,\tilde{u})||^2 \, ds, \int_b^t \mathbb{E} \sum_i ||\eta_i(s,\tilde{u})||^2 \, ds\}$$

$$= \max\{||[X(t) - X(b)]\tilde{u}||^2, \ ||[X^+(t)-X^+(b)]\tilde{u}||^2\}$$

$$\leq ||\tilde{u}||^2 \ \mu([b,t]), \tag{3.3}$$

which proves (ii).

Because of (ii) we can assume in the sequel without loss of generality that μ is absolutely continuous. As in the proof of the Proposition 3.4 in [1] we can see that $\{\xi_i(t,\tilde{u})\}$ and $\{\eta_i(t,\tilde{u})\}$ do not depend on the end point a and we set

$$L_i(s)u \otimes \psi(f\chi_{[0,a]}) \equiv \xi_i(s,u \otimes \psi(f\chi_{[0,a]})) \quad \text{a.e.} \quad s.$$

The measure theoretic considerations for the definition of the operator family $\{L_i(s)\}$ go through as in the proof of the Proposition 3.4 of [1] since we have a countable family involved. Thus by (3.3) we have

$$\int_b^t \sum_i ||L_i(s)\tilde{u}||^2 \, ds = \int_b^t \mathbb{E} \sum_i ||\xi_i(s,\tilde{u})||^2 \, ds \leq ||\tilde{u}||^2 \ \mu([b,t])$$

leading to $\sum_i ||L_i(s)\tilde{u}||^2 \leq \mu'(s)||\tilde{u}||^2$ for all s. (3.4)

This shows that each $L_i(s)$ is a bounded adapted process and that $\sum_i L_i^+(s)L_i(s)$ converges strongly to a bounded selfadjoint operator process which is locally norm integrable. In fact, we see that $\int_a^t || \sum_i L_i^+(s)L_i(s)||ds \leq \mu'(s)$ a.e. It is also clear that an identical construction and conclusion is valid for $X^+(t)$ and $\{K_i(t)\}$ replacing $X(t)$ and $\{L_i(t)\}$.

□

Now define $M_0 \equiv \{f \in h | f_i = 0 \in L^2(\mathbb{R}_+)$ for all but a finite number of i's$\}$ and $\tilde{E}_0 = h_0 \otimes E(M_0)$. Also set

$$S(t) = \int_0^t \sum_j (L_j(s)dA_j^+(s) + K_j^+(s)dA_j(s)),$$

$$S^+(t) = \int_0^t \sum_j (K_j(s)dA_j^+(s) + L_j^+(s)dA_j(s)),$$

$$Z(t) = X(t) - S(t) \quad \text{and} \quad Z^+(t) = X^+(t) - S^+(t) . \tag{3.5}$$

Lemma 3.4: Let $u \in h_0$ *and* $f \in M_0$. *Then (i) the processes* $\{S,S^+\}$ *and* $\{Z,Z^+\}$ *are disjoint pairs well defined on* $\tilde{E}_0$, *(ii)* $\{Z(t)u \otimes \psi(f\chi_{[0,t]})\}$ *and* $\{Z^+(t)u \otimes \psi(f\chi_{[0,t]})\}$ *are classical* h_0*-valued martingales adapted to Brownian motion family* $\underline{w}$, *(iii) there exist two* h_0*-valued square summable classical processes* $\{\xi_i(\cdot,u,f)\}$ *and* $\{\eta_i(\cdot,u,f)\}$ *such that*

$$Z(t)u \otimes \psi(f\chi_{[0,t]}) = X(0)u \otimes \Omega + \int_0^t \sum_i \xi_i(s,u,f)dw_i(s),$$

$$Z^+(t)u \otimes \psi(f\chi_{[0,t]}) = X^+(0)u \otimes \Omega + \int_0^t \sum_i \eta_i(s,u,f)dw_i(s);$$

and (iv)

$$Z(t)u \otimes \psi(f\chi_{[0,a]}) = Z(a)u \otimes \psi(f\chi_{[0,a]}),$$

$$Z^+(t)u \otimes \psi(f\chi_{[0,a]}) = Z^+(a)u \otimes \psi(f\chi_{[0,a]}) \;\; \forall\, t > a.$$

<u>Proof:</u> (i) $\int_0^t \sum_i L_i^+(s)dA_i^+$ and $\int_0^t \sum_i K_i(s)dA_i^+$ are well defined on $\tilde{E}$ with $M = h$ by (2.5) and (3.4). However, this is not true in general for the other two integrals e.g. $\int_0^t \sum_i K_i^+(s)dA_i(s)$ since we do not have any estimate for $\sum_i ||K_i^+(s)u||^2$. If we consider them on E_0, the infinite sum reduces to a finite one and an estimate like (3.4) for each i then

allows to define the other two stochastic integrals by (2.5). That the two pairs are adjoint to each other is immediate from (2.7) and the definitions.

The proof of (ii) and (iv) proceeds along lines similar to those in the proof of Proposition 3.6 except for the necessary restriction of f,g appearing in the exponential vectors to the set M_0. The part (iii) then follows from the Proposition 3.2 and the remarks following it.

□

Next our aim is to show that Z is a stochastic integral with respect to the martingale family $\{\Lambda_{ij}\}$. For this, as in [1], we use a special martingale $U^{(i)}$, related to the Weyl representation.

Lemma 3.5: Consider for each i *the unique bounded martingale* $U^{(i)}$ *given by:*

$$dU^{(i)} = (dA_i^+ - dA_i)U^{(i)}, \quad U^{(i)}(0) = I. \tag{3.6}$$

Then (i) $e^{-t/2}\, U^{(i)}(t) = I_0 \otimes W(e_i\chi_{[0,t]}, I)$, *where* I_0 *is the identity in* $B(h_0)$ *and* W *is the Weyl representation defined in* [3], *(iii)* $U^{(i)}(t)$ *leaves* $\tilde{E}_0$ *invariant.*

The proof of these statements follows from the properties of the Weyl representation and are omitted.

Lemma 3.6: Let X *be a bounded regular martingale in* H *and let* $\{L_i\}$, $\{K_i\}$ *be the associated bounded adapted families defined in Lemma 3.3. Set*

$$Y^{(i)}(t) = X(t)\, U^{(i)}(t) - \int_0^t K_i^+(s)U^{(i)}(s)ds , \tag{3.7}$$

where $U^{(i)}$ *is the martingale defined by (3.6). Then (i) for each* i, $Y^{(i)}$ *is a bounded regular martingale, (ii) there exist a unique family* $\{L_j^{(i)}\}$ *of bounded adapted processes such that for all* $t > a > 0$, $u \in \tilde{H}_a$

$$[Y^{(i)}(t) - Y^{(i)}(a)]u = \int_a^t \sum_j L_j^{(i)}(s)u \, dw_j(s) \tag{3.8}$$

Proof: (i) It is clear that $X(t)\, U^{(i)}(t)$ is bounded. We also have from the inequality similar to (3.4) with K_i replacing L_i that for every $t > 0$, $\int_0^t ||K_i(s)U^{(i)}(s)||ds \leq \int_0^t e^{s/2}\, (\mu'(s))^{1/2}\, ds < \infty$, showing that $Y^{(i)}(t)$ is bounded.

Note that since $U^{(i)}(t)$ leaves $\tilde{E}_0$ invariant, the relation:

$$Y^{(i)}(t) = Z(t)U^{(i)}(t) + S(t)U^{(i)}(t) - \int_0^t K_i^+(s)U^{(i)}(s)ds \tag{3.9}$$

holds on $\tilde{E}_0$. Then by Lemma 3.4 (iv) and the martingale property of $U^{(i)}$, one has that $\{Z(t)U^{(i)}(t)\}$ is a martingale with respect to the pair $\{h_0, M_0\}$. Now for $f,g \in M_0$ and $u,v \in h_0$ we apply Ito's product formula (2.4) to $S(t)U^{(i)}(t)$ to get

$$\frac{d}{dt} \langle S^+(t)u \otimes \psi(f),\, U^{(i)}(t)v \otimes \psi(g)\rangle$$

$$= \int_j \bar{f}_j(t)\langle\{L_j^+(t) + S^+(t)\delta_{ij}\}u \otimes \psi(f),\, U^{(i)}(t)v \otimes \psi(g)\rangle$$

$$+ \sum_j g_j(t)\langle\{K_j(t) - S^+(t)\delta_{ij}\}u \otimes \psi(f),\, U^{(i)}(t)v \otimes \psi(g)\rangle$$

$$+ \langle K_i(t)u \otimes \psi(f),\, U^{(i)}(t)v \otimes \psi(g)\rangle\ . \tag{3.10}$$

Writing (3.10) in integral form and then multiplying f and g by $\chi_{[0,a]}$ with $a < t$, we easily conclude that $S(t)U^{(i)}(t) - \int_0^t K_i^+(s)U^{(i)}(s)ds$ is a martingale.

The proof of regularity is similar to that in [1] and simple estimates show that for $t > a > 0$, $\tilde{u} \in \tilde{H}_a$,

$$||[Y^{(i)}(t) - Y^{(i)}(a)]\tilde{u}||^2$$

$$\leq ||\tilde{u}||^2\{4(e^t - e^a)||X(t)||^2 + 4e^a\, \mu([a,t]) + 2(e^t - e^a)\mu([a,t])\}$$

$$\leq 4||\tilde{u}||^2\{e^t[||X(t)||^2 + \mu([0,t])] - e^a[||X(a)||^2 + \mu([0,a])]\}$$

$$\equiv ||\tilde{u}||^2 \nu([a,t]). \tag{3.11}$$

Since $||X(t)||$ is non-decreasing by Lemma 3.3 (i), ν given by (3.11) is a measure on $\mathbb{R}_+$ and we have (i). The part (ii) follows from Lemma 3.3 (iii) being applied to the bounded regular martingale $\{Y^{(i)}(t)\}$.

□

Define $M_{00} = \{f \in M_0 \subset h | \ ||f(t)|| \text{ is a locally bounded function of } t\}$ and $\tilde{E}_{00} = h_0 \underline{\otimes} E(M_{00})$.

Lemma 3.7: Let X *be a bounded regular martingale and let* $\{L_i\}$, $\{K_i\}$, $\{S, S^+\}$, $\{Z, Z^+\}$, $U^{(i)}$ *and* $\{L_j^{(i)}\}$ *be as defined in lemmas 3.3 - 3.6. Set*

$$M_{ij}(t) = L_i^{(j)}(t)\, U^{(j)}(t)^{-1} - X(t)\delta_{ij} - L_i(t). \tag{3.12}$$

Then the vector processes $\{\eta_i(t,u,f)\}$ *defined in Lemma 3.4 (iii) satisfies the relation for every* $u \in h_0$, $f \in M_0$;

$$\eta_i(t,u,f) = \sum_j f_j(t)\{M_{ji}^+(t) + Z^+(t)\delta_{ij}\}u \otimes \psi(f\chi_{[0,t]}) \text{ a.e.t.} \tag{3.13}$$

Furthermore one has

$$Z(t) = \int_0^t \sum_{i,j} M_{ij}(s)\,d\Lambda_{ij}(s),$$

defined on $\tilde{E}_{00}$.

<u>Proof</u>: Let $u,v \in h_0$, $f,g \in M_0$, $t > a$. We have from Lemma 3.5 (ii) that

$$U^{(i)}(t)v \otimes \psi(g\chi_{[0,a]}) = \exp(-\int_0^a g_i(\tau)d\tau)v \otimes \psi(g\chi_{[0,a]} + e_i\, \chi_{[0,t]}).$$

Thus by Lemma 3.4 (iii) and the isometry of Ito integrals, we have

$$\frac{d}{dt} \langle Z^+(t)u \otimes \psi(f\chi_{[0,t]}),\ U^{(i)}(t)v \otimes \psi(g\chi_{[0,a]})\rangle$$

$$= e^{-\int_0^a g_i(\tau)d\tau} \frac{d}{dt} \int_0^t \sum_j (g_i(s)\chi_{[0,a]}(s)+\delta_{ij})\langle\eta_j(s,u,f),v \otimes \psi(g\chi_{[0,a]}+e_i\chi_{[0,a]})\rangle$$

$$= \langle\eta_i(t,u,f),\ U^{(i)}(t)v \otimes \psi(g\chi_{[0,a]})\rangle\ , \quad \text{a.e. } t > a. \tag{3.14}$$

On the other hand from (3.9), (3.8) isometry of Ito integrals and (3.10) we have that for $t > a$,

$$\langle u \otimes \psi(f\chi_{[0,t]}),\ [Z(t)U^{(i)}(t) - Z(a)U^{(i)}(a)]v \otimes \psi(g\chi_{[0,a]})\rangle$$

$$= \langle u \otimes \psi(f\chi_{[0,t]}),\ [Y^{(i)}(t) - Y^{(i)}(a)]v \otimes \psi(g\chi_{[0,a]})\rangle$$

$$- \langle u \otimes \psi(f\chi_{[0,t]}),\{S(t)U^{(i)}(t)-S(a)U^{(i)}(a)-\int_a^t K_i^+(s)U^{(i)}(s)ds\}v\otimes\psi(g\chi_{[0,a]})\rangle$$

$$= \int_a^t \sum_j \bar{f}_j(s)\langle u \otimes \psi(f\chi_{[0,s]}),\ L_j^{(i)}(s)v \otimes \psi(g\chi_{[0,a]})\rangle ds$$

$$- \int_a^t \sum_j \bar{f}_j(s)\langle u \otimes \psi(f\chi_{[0,s]}),[L_i(s)+S(s)\delta_{ij}]U^{(i)}(s)v \otimes \psi(g\chi_{[0,a]})\rangle ds$$

$$= \int_a^t \sum_j \bar{f}_j(s)\langle u \otimes \psi(f\chi_{[0,s]}),(M_{ij}(s) + Z(s)\delta_{ij})U^{(i)}(s)v \otimes \psi(g\chi_{[0,a]})\rangle ds \tag{3.15}$$

Comparing (3.14) and (3.15) and using the bounded invertibility of $U^{(i)}(t)$ and the totality of the set $\{v \otimes \psi(g\chi_{[0,a]}) | v \in h_0,\ g \in M_0,\ 0 < a < t\}$ in $\tilde{H}_t$, we get (3.13). Thus by Lemma 3.4 (iii) and (3.13)

$$Z^+(t)u \otimes \psi(f\chi_{[0,t]})$$

$$= X^+(0)u \otimes \Omega + \int_0^t \sum_{ij} f_j(s)\{M_{ji}^+(s)+Z^+(s)\delta_{ij}\}u \otimes \psi(f\chi_{[0,s]})dw_i(s) \tag{3.16}$$

It is easy to see that $\{M_{ij}(t)\}$ defined by (3.12) are adapted bounded processes and a simple estimate shows that the integral

$\int_0^t \sum_{i,j} M_{ij}(s)d\Lambda_{ij}(s)$ is well defined on $\tilde{E}_{00}$ since the integrability condition (2.5) is satisfied when $f \in M_{00}$. Now by (3.16) and the isometry of Ito integral, we have for $f,g \in M_{00}$,

$$<u \otimes \psi(f),\ Z(t)v \otimes \psi(g)>$$

$$= <Z^+(t)u \otimes \psi(f\chi_{[0,t]}), v \otimes \psi(g\chi_{[0,t]})>e^{\int_t^\infty <f(s),g(s)>ds}$$

$$= \exp(\int_t^\infty <f(s),g(s)>ds)\ \{<u \otimes \Omega,\ X(0)v \otimes \Omega>$$

$$+ \int_0^t ds \sum_{ij} \bar{f}_j(s)g_i(s)<u \otimes \psi(f\chi_{[0,s]}),(M_{ji}(s)+Z(s)\delta_{ij})v \otimes \psi(g\chi_{[0,s]})>\}$$

Finally a differentiation shows that

$$\frac{d}{dt} <u \otimes \psi(f),\ Z(t)v \otimes \psi(g)> = \sum_{ij} \bar{f}_j(t)g_i(t)<u \otimes \psi(f),M_{ji}(t)v \otimes \psi(g)>$$

which by (2.7) proves that $Z(t) = \int_0^t \sum_{ij} M_{ij}(s)d\Lambda_{ij}(s)$ on $\tilde{E}_{00}$.

□

Combining the Lemma 3.7 with definitions (3.5) we have

Theorem 3.8: Let X *be a bounded martingale in* $H = h_0 \otimes \Gamma(L^2(\mathbb{R}_+,h_1))$ *which is regular with respect to a Radon measure* μ *on* $\mathbb{R}_+$*, and let* M_{00} *and* $\tilde{E}_{00}$ *be the manifolds defined in Lemma 3.7. Then there exist three unique bounded families of adapted processes* $\{M_{ij}\}$, $\{K_i\}$, $\{L_i\}$ *such that*

$$dX = \sum_{ij} M_{ij}d\Lambda_{ij} + \sum_i K_i^+ dA_i + \sum_i L_i\ dA_i^+$$

with respect to the pair (h_0,M_{00})*. Furthermore,* $\sum_i K_i^+(s)K_i(s)$ *and* $\sum_i L_i^+(s)L_i(s)$ *converge strongly to* K(s) *and* L(s) *respectively with the property:* $\max(||K(s)||,\ ||L(s)|| \leq \mu'_{ac}(s)$ *for all* s*, where* μ_{ac} *denotes the absolutely continuous part of* μ.

4. Applications

(i) *Unitary martingales*: As in [1], we can readily deduce that every unitary martingale U is regular with respect to the zero measure and hence $L_i = K_i = 0$, $\forall$ i. Then by theorem 3.7, $dU = \sum_{ij} M_{ij}\, d\Lambda_{ij}$. An application of Ito formula (2.3) to $d(U^+U) = d(UU^+) = 0$ leads to

$$U^+M_{ij} + M^+_{ji}U + \sum_k M^+_{ki}\, M_{kj} = M_{ij}U^+ + UM^+_{ji} + \sum_k M_{ki}M^+_{kj} = 0.$$

From this it follows that

$$\sum_k (U^+\delta_{ki} + M^+_{ki})(U\delta_{kj} + M_{kj}) = \sum (U\delta_{ki} + M_{ki})(U^+\delta_{kj} + M^+_{kj}) = \delta_{ij}.$$

Setting $U\delta_{ij} + M_{ij} = W_{ij}U$, we have that

$$\sum_k W^+_{ki}\, W_{kj} = \sum_k W_{ki}\, W^+_{kj} = \delta_{ij} \tag{4.1}$$

or equivalently $W \equiv \{W_{ij}\}$ looked upon as an operator in $h_1 \otimes H$ is unitary.

(ii) *Fermion martingales*: Here $h_0 = ¢$ while $h_1 = L^2(\mathbb{R}_+^{\nu-1})\nu \geq 2$. Define reflection-valued process J(s) on $H = \Gamma(L^2(\mathbb{R}_+, h_1))$

$$J(s)\psi(f) = \psi(f\chi_{[0,s]} + f\chi_{[s,\infty]})$$

where $(f\chi_{[0,s]})(\tau) = f(\tau)\ \chi_{[0,s]}(\tau) \in h_1$, and for $\phi \in L^2(\mathbb{R}_+, h_1)$ $\simeq\ L^2(\mathbb{R}_+^{\nu})$,

$$F_\phi(t) = \int_0^t J(s)dA_\phi(s),\ F^+_\phi(t) = \int_0^t J(s)\ dA^+_\phi(s) \quad ,$$

where $A_\phi(t)\psi(f) = (\int_0^t \langle\phi(s), f(s)\rangle_1\ ds)\psi(f)$ and $A^+_\phi(t)$ is the adjoint of $A_\phi(t)$. Then as in [2] one verifies easily that $\{F_\phi, F^+_\psi\}$ verify CAR i.e.

$$[F_\phi(t),F_\psi(t)]_+ = 0, [F_\phi(t),F_\psi^+(t)]_+ = \int_0^t <\phi(s),\psi(s)>_1 ds. \qquad (4.2)$$

Also one observes that the algebra generated by $\{F_\phi(t), F_\phi^+(t); \phi,\psi \in L^2(\mathbb{R}_+, h_1)\}$ is irreducible in H and $F_\phi(t)\Omega = 0$. Now question arises whether there are any other representation of CAR in H and the answer is given in the next theorem.

Theorem 4.1: *Let $h_0 = \mathbb{C}$, $\nu \geq 2$, and $\phi \in h = L^2(\mathbb{R}_+,h_1)$. Also let (i) $\{X_\phi(t),\ X_\phi^+(t)\}$ be an adjoint pair of martingales w.r.t. $L^2(\mathbb{R}_+,h_1))$ satisfying CAR (4.2), (ii) the representation be factorizable i.e. $X_{\phi_\tau}(t) = X_\phi(\tau)$ for every $0 < \tau < t$ where $\phi_a = \phi\chi_{[0,a]}$, (iii) $X_\phi(t)\Omega = 0$, (iv) the algebra A_t generated by $\{X_\phi(t),\ X_\psi^+(t),\ \phi,\psi \in h\}$ is irreducible in H_t. Then there exists a measurable map V from $\mathbb{R}_+$ taking values in unitaries on h_1 such that*

$$X_\phi(t) = F_{V\phi}(t),\ \forall\ \phi \in h \quad \text{where } (V\phi)(\tau) = V(\tau)\phi(\tau).$$

Proof: As in [1] and [4] we conclude that X_ϕ, X_ϕ^+ are bounded adapted regular martingales. Also we note that the map $\phi \rightarrow X_\phi(t)$ is anti-linear in ϕ. Also as in the proof of Theorem 4.5 in [1] we see that there exist a unique unitary martingale U such that $X_\phi(t) = U(t)F_\phi(t)U^+(t)$, $U(t)\Omega_t = \Omega_t$. By preceeding discussion on unitary martingales, there exist a family of bounded adapted processes $\{W_{ij}\}$ satisfying (4.1) such that $dU = \sum_{ij} (W_{ij} - \delta_{ij})U\, d\Lambda_{ij}$. By Ito formula (2.4) we have

$$dX_\phi(t) = \sum_{ij} (\sum_\ell W_{i\ell}(t)X_\phi(t)W_{j\ell}^+(t) - \delta_{ij}X_\phi(t)d\Lambda_{ij}$$

$$+ U(t)J(t)U^+(t) \sum_{ij} \bar{\phi}_i(t)W_{ji}^+\, dA_j. \qquad (4.3)$$

For $0 < a < t$, choose ϕ to have support in $[0,a]$ so that $\phi = \phi_a$. Then by factorizability assumption, $dX_{\phi_a}(t) = 0 = \phi_{ai}(t)$. Thus (4.3) leads to the equation

$$\sum_{\ell} W_{i\ell}(t)\ X_{\phi_a}(t)\ W^{+}_{j\ell}(t) = \delta_{ij}\ X_{\phi_a}(t)$$

which on using (4.1) reduces to

$$W_{ij}(t)X_{\phi_a}(t) = X_{\phi_a}(t)W_{ij}(t)\ \forall\ i,j. \tag{4.4}$$

Looking upon $\{W_{ij}\}$ as a unitary operator and $X_{\phi_a}(t)$ as $1 \otimes X_{\phi_a}(t)$ in $h_1 \otimes H_t$ we can rewrite (4.4) as $W(t)X_{\phi_a}(t) = X_{\phi_a}(t)W(t)$. Using the antilinearity and L^2-continuity of $X_\phi(t)$ in ϕ we obtain that $W(t)$ commutes with $X_\phi(t)$ and $X^+_\phi(t)$ in $h_1 \otimes H_t$ for every ϕ. Thus by the assumption of irreducibility of A_t in H_t, it follows that $W(t) = V(t) \otimes I$ on $h_1 \otimes H_t$ where $V(t)$ is a unitary operator in h_1 for every $t > 0$. Thus U must be the second quantization of multiplication by $V\chi_{[0,t]} + \chi_{[t,\infty)}$. Therefore

$$X_\phi(t) = U(t)\ F_\phi(t)U(t)^+ = F_{V\phi}(t)\ \forall\ \phi \in h.$$

□

Remark: There was an error in the proof of Theorem 4.5 of [1] in the part where the coefficient of d was shown to be 0. One needs the assumption of factorizability of the CAR representation as has been added here and proof completed as above.

References

[1] Parthasarathy, K.R., Sinha, K.B., Stochastic integral representation of bounded quantum martingales in Fock space, Jour. Func. Anal. 67, 126-151 (1986).

[2] Parthasarathy, K.R., Sinha, K.B., Boson-fermion relations in several dimensions, Pramana - J. Phys. 27, 105-116 (1986).

[3] Hudson, R.L., Parthasarathy, K.R., Quantum Ito's formula and stochastic evolutions, Comm. Math. Phys. 93, 301-323 (1984).

[4] Hudson, R.L., Parthasarathy, K.R., Unification of fermion and boson stochastic calculus, Comm. Math. Phys. 104, 457-470 (1986).

CONDITIONAL EXPECTATION IN QUANTUM PROBABILITY

Dènes Petz
Mathematical Institute HAS,
H-1364 Budapest, PF. 127, Hungary

Let (Ω,S,P) be a probability space and S_0 be a sub-σ-algebra of S. In 1954 Moy observed that the property

$$E(fE(g)) = E(f)E(g) \quad (f,g \text{ are bounded})$$

plays a crucial role in the characterization of the conditional expectation operator $E:L^1(S,P) \to L^1(S_0,P)$ ([15]). More or less at the same time Nakamura, Turumaru and Umegaki created the algebraic form of the conditional expectation ([16], [24])). Before saying their definition we fix the main notations of the lecture.

On operator algebras our basic reference is [4]. H always stands for a Hilbert space and M denotes a von Neumann algebra. By a state of M we mean a linear functional $\varphi \in M^*$ such that

(1) $\varphi(x) \geq 0$ for all $x \in M_+$ and the equality holds in the only case $x = 0$,

(2) $\varphi(I)=1$,

(3) φ is w-continuous.

(In other words, φ is a faithful normal state.) If in addition

(4) $\varphi(xy) = \varphi(yx)$ for every $x,y \in M$,

then φ is called tracial state. When M is finite dimensional (and so it is a direct summe of some full matrix algebras), then it always possesses a tracial state τ. In this case every state φ has the form $\varphi(a) = \tau(\rho_\varphi a)$ $(a \in M)$ with a unique $\rho_\varphi \in M_+$, called the density of φ. Due to the lack of unbounded operators the finite dimensional case is rather easily computable.

Let φ be a fixed state on M and $E:M \to M$ a linear mapping so that

(5) $E(x) \geq 0$ if $x \geq 0$,

(6) $E(I) = I$,

(7) $\varphi(E(x)) = \varphi(x)$ for all $x \in M$,

(8) $E(xE(y)) = E(x)E(y)$ for every $x,y \in M$.

Property (7) implies that E is w-continuous and hence $M_0 = \{ a \in M : E(a) = a \}$ is w-closed. It follows also easily that M_0 is an algebra and the range of E. Such an E is called φ-preserving conditional expectation (in the sense of Umegaki) onto M_0.

THEOREM 1. ([25]) Let τ be a tracial state on M and a von Neumann subalgebra of M. Then there exists a unique conditional expectation preserving τ and mapping M onto M_0.

$\langle a,b\rangle = \tau(b^*a)$ defines an inner product on M (Hilbert-Schmidt inner product). To avoid the completion proccess assume that M is finite dimensional and so it is a Hilbert space. M_0 is a closed subspace of M and we consider the orthogonal projection $E:M \to M_0$. The properties (5)-(8) are verified straightforwardly.

The case of tracial τ reminds very much the commutative situation. From this point of view the following example may be surprising. If $M=M_2(\mathbb{C})\otimes M_2(\mathbb{C}) \simeq M_4(\mathbb{C})$ and $M_0=M_2(\mathbb{C})\otimes\mathbb{C} \simeq M_2(\mathbb{C})$, then for a state φ on M the conditional expectation $E:M \to M_0$ (preserving φ) exists if and only if $\varphi = \varphi_1\otimes\varphi_2$. The Takesaki theorem clarifies the existence of the conditional expectation by means of the modular group of the state.

THEOREM 2. ([23]) Let M_0 be a von Neumann subalgebra of M and φ a state on M. Then the φ-preserving conditional expectation from M onto M_0 exists if and only if M_0 is stable under the modular group of φ. If the conditional expectation exists then it is unique.

If we try to apply the argument which lead to success for a tracial state then we arrive at troubles with the positivity condition (5). Accardi and Cecchini have defined a mapping $E_\varphi:M \to M_0$, which is completely positive, always exists and (therefore) possesses only a weaker form of (8). E_φ, called φ-conditional expectation, does not map M onto M_0. In fact, they did not consider postulates like (5)-(8) but extended to von Neumann algebras the construction of measure theoretic conditional expectation. If g is a bounded S-measurable function, then $F(f)=\int fg\,dP$ is a functional on $L^1(S,P)$ and its restriction determines an S_0-measurable function $E(g)$ such that $\int E(g)h\,dP=\int gh\,dP$ for all $h\in L^1(S_0,P)$. So in this construction the emphasis is on the duality $(L^1)^*=L^\infty$.

The following result is standard in the Tomita-Takesaki theory.

LEMMA 3. If φ is a state on M and $a,b\in M$, then there exists a function $F:\{z\in\mathbb{C}:\ 0 \le \operatorname{Im} z \le 1\} \to \mathbb{C}$ so that

(1) F is bounded continuous and it is analytic on the interior of its domain.

(2) $F(t) = \varphi(\sigma_t(b)a)$ for all $t\in\mathbb{R}$.

(3) $F(t+i) = \varphi(a\sigma_t(b))$ for all $t\in\mathbb{R}$.

if σ_t stands for the modular group of φ.

Instead of $F(z)$ we shall write simply $\varphi(\sigma_z(b)a)$.

THEOREM 4. ([1]) Let $M_0\subset M$ be von Neumann algebras and φ a state on M. If σ_t ($\tilde{\sigma}_t$) is the modular group of φ ($\tilde{\varphi} = \varphi|M_0$) then the formula

$$\varphi(\sigma_{i/2}(b_0)a) = \varphi(\tilde{\sigma}_{i/2}(b_0)E_\varphi(a))$$

determines a completely positive mapping $E_\varphi : M \to M_0$. For $a_0\in M_0$

$$E_\varphi(aa_0) = E_\varphi(a)a_0 \quad \text{for every } a\in M$$

holds if and only if $\sigma_t(a_0)\in M_0$ for every $t\in\mathbb{R}$.

This result is a localized generalization of Takesaki theorem. In fact, $\varphi(\sigma_{i/2}(b)a^*) = \langle b,a\rangle$ is an inner product on M (the selfpolar form of Connes [7]). M_0 endowed with the corresponding inner product is not a subspace of M any more. Nevertheless the inclusion has an adjoint and this is E_φ.

We sketch the proof of the first part of Theorem 4. Assume that the algebra acts on a Hilbert space H and the state φ is determined by the cyclic and separating vector $\Phi\in H$. Let p' be the projection onto the closure H_0 of $M_0\Phi$. Then p' belongs to M_0'. The closure of the (closable) operator $S_0 : a\Phi \to a^*\Phi$ ($S_0|M_0\Phi$) possesses a polar decomposition $J\Delta^{1/2}$ ($J_0(\Delta_0)^{1/2}$). In this new language we have

$$\varphi(\sigma_{i/2}(b^*)a^*)=\langle Ja\Phi,\ b\Phi\rangle.$$

The operator $E(a) = J_0p'JaJp'J_0$ satisfies

$$\langle Ja\Phi,\ b_0\Phi\rangle = \langle J_0E(a)\Phi,\ b_0\Phi\rangle$$

for every $b_0\in M_0$. Due to Tomita's theorem $JaJ\in M'\subset M_0'$. Hence $p'JaJp'\in M_0'$. Refering to Tomita's theorem again we obtain $J_0p'JaJp'J_0\in M_0$.

We note that the above construction of the φ-conditiuonal expectation may be generalized in two different directions. Using the selfpolar form $\varphi(\sigma_{i/2}(b)a^*)$ $=\langle b,a\rangle$ one can find a dual or adjoint for any positive mapping between von Neumann algebras (see below and [18]). On the other hand, for every $\lambda\in[0,1]$

$(b,a)\to\varphi(\sigma_{\lambda i}(b)a^*)$ is a scalar product $\langle .,.\rangle_\lambda$ and many mappings admit an adjoint with respect to $\langle .,.\rangle_\lambda$ (see [13]).

The subset $\mathbf{F}(E_\varphi) = \{ a\in M : E_\varphi(a)=a \}$ is a von Neumann subalgebra of M_0. By the mean ergodic theorem $n^{-1}\Sigma_{k=0}(E_\varphi)^k$ converges strongly to the conditional expectation onto $\mathbf{F}(E_\varphi)$. The fixed point algebra $\mathbf{F}(E_\varphi)$ is the set of elements behaving regularly with respect to E_φ, $\mathbf{F}(E_\varphi) = \{ a_0\in M_0 : E_\varphi(aa_0) = E_\varphi(a)a_0$ for every $a\in M \}$. Unfortunately, $\mathbf{F}(E_\varphi)$ can be rather small in M_0. For example, if M_0 is commutative, then $\mathbf{F}(E_\varphi) = \{ a_0\in M_0 : \varphi(aa_0) = \varphi(a_0a)$ for every $a\in M \}$.

Let us see some examples. If M is finite dimensional with a tracial state τ, then φ ($\varphi|M_0$) has a density ρ (ν) with respect to τ ($\tau|M_0$) and there is an Umegaki conditional expectation $E:M \to M_0$ preserving τ. Then E_φ is of the form

$$E_\varphi(a) = \nu^{-1/2}E(\rho^{1/2}a\rho^{1/2})\nu^{-1/2}.$$

Another example is treated by Frigerio in [9].

A special φ-conditional expectation has great inportance in quantum field theory. If $M_0\subset M\subset B(H)$ and Φ is a separating vector for M and it is cyclic for M_0, then the φ conditional expectation corresponding to the vector state $\varphi(.) = \langle .\Phi,\Phi\rangle$ is a an algebraic isomorphism onto a proper subalgebra of M_0 (see [1], p.259 and [8]). In this case, $(E_\varphi)^k$ converges to the conditional expectation onto $\mathbf{F}(E_\varphi)$, which may be trivial ([14]).

The latest example shows that the range of E_φ can be smaller than M_0. However, it can not be arbitrarily small. Keeping the notation of the proof of the previous theorem we prove the following

LEMMA 6. The real linear subspace $E_\varphi(M^{sa})\Phi$ is dense in $[M_0^{sa}]$.

Proof. Since M^{sa} (M_0^{sa}) is stable both under S_0 ($S_0|M_0^{sa}$) and Δ^{it} (Δ_0^{it}) the closed real linear subspaces $[M^{sa}]$ and $[M_0^{sa}]$ are stable under the modular conjugations J and J_0, respectively. It follows that $J_0p'JM^{sa}\Phi = E_\varphi(M^{sa})\Phi$ is dense in $[M_0^{sa}]$.

THEOREM 7. The φ-conditional expectation mapping $E_\varphi:M \to M_0$ determines M_0 uniquely.

Proof. The operator $T: E_\varphi(a)\Phi \to p'\Delta^{1/2}a\Phi$ ($a\in M$) is densely defined on H_0 and the restriction of $\Delta_0^{1/2}$ to $E_\varphi(M)\Phi$.

$$\Delta_0^{1/2} E_\varphi(a)\Phi = J_0 E_\varphi(a)^*\Phi = J_0 J_0 p' J a^*\Phi = p' J J \Delta^{1/2} a\Phi$$

We show that if $a=a^*\in M_0$ then $a\Phi$ belongs to the domain of the Friedrichs extension $\tilde{T}$ of T (see [22] for the Friedrichs extension). According to Lemma 6 there is a sequence $(a_n)\subset M^{sa}$ so that $E_\varphi(a_n) \to a\Phi$. We can establish

$$\langle T E_\varphi(a_n-a_m)\Phi, E_\varphi(a_n-a_m)\Phi\rangle \to 0$$

since $TE_\varphi(a_n)\Phi = J_0 E_\varphi(a_n)\Phi$.

$M_0\Phi$ is a core for $\Delta_0^{1/2}$ and $M_0\Phi\subset D(\tilde{T})$. Hence $\Delta_0^{1/2} = \tilde{T}$ must be. Knowing $D(\Delta_0^{1/2})$ we may identify the algebra M_0 as follows.

$$p'M_0p' = \{p'a : a\in M,\ ap'=p'a,\ a\Phi\in D(\Delta_0^{1/2})\}$$

THEOREM 8. ([12], [18]) Let $M_1\subset M_2\subset M_3\subset\ldots\subset M$ be von Neumann algebras and assume that $\cup M_n$ is w-dense in M. If φ is state on M and $E_n: M \to M_n$ is the φ-conditional expectation, then $E_n(a)\to a$ in the strong operator topology as $n\to\infty$ for every $a\in M$.

Assume that $M\subset B(H)$ and $\varphi(a)=\langle a\Phi, a\Phi\rangle$ $(a\in M)$. Let p_n be the projection of H onto $H_n=[M_n\Phi]$. Then $p_n \to I$ strongly. Let $S: a\Phi \to a^*\Phi$ $(a\in M)$ and take $\bar{S}_n$ ($\bar{S}$) as the closure of $S|M_n\Phi$ (S). Since $\cup M_n\Phi$ is a core for S, if $S=J\Delta^{1/2}$ and $S_n=J_n\Delta_n^{1/2}$ are the polar decompositions, we obtain that $J_np_n \to J$ (and also $\Delta_n^{it}p_n \to \Delta^{it}$) strongly. Now it is sufficient to remember that $p_nE_n(a)p_n= J_np_nJaJp_nJ_n$.

This Theorem was used in [5] to prove the convergence also in L^p-norm. It is not known if $E_n(a)\to a$ in a stronger sense (for example, almost uniformly) as in the the case of Umegaki conditional expectation (see [17] for a review). It is worthwile pointing out that the decreasing version of Theorem 7 is not true.

Let φ and ω be states of the von Neumann algebra M. The von Neumann subalgebra M_0 of M is called sufficient with respect to (φ,ω) if the φ-conditional expectation into is the same as the ω-conditional expectation. We note that this definition is weaker than the one first appeared in [25]. (See [10] and [11] for a more modern treatment.) There it was required that the conditional expectations involved are projections. Let $M=M_0\otimes N$ and $\varphi=\varphi_1\otimes\varphi_2$ and $\omega=\omega_1\otimes\omega_2$ be states on M. Then $E_\varphi(a\otimes b)=\varphi_2(b)a$ and $E_\omega(a\otimes b)=\omega_2(b)a$ for $a\in M_0$ and $b\in N$. (Actually, they are

Umegaki conditional expectations.) Therefore, M_0 is sufficient for φ and ω if and only if $\varphi_2=\omega_2$.

In the characterization of sufficient subalgebras the quantity $P_A(\varphi,\omega)$ has been proved to be useful. $P_A(\varphi,\omega)$ is a kind of mutual information (,interpreted also as a transition probability) and it generalizes the affinity of measures ([21]). If Φ and Ω are the vector representatives of φ and ω in a natural positive cone, then

$$P_A(\varphi,\omega) = \langle\Phi, \Omega\rangle.$$

(In the finite dimensional case this reduces to $\tau(\rho^{1/2}\nu^{1/2})$ if ρ and ν are the densities.)

THEOREM 8. ([19], [20]) If $M_0\subset M$ and φ,ω are states on M, then the following conditions are equivalent.

(i) M_0 is sufficient with respect to φ and ω.

(ii) $\varphi\circ E_\omega=\varphi$.

(iii) There is a completely positive mapping $\alpha:M\to M_0$ such that $\varphi\circ\alpha=\varphi$ and $\omega\circ\alpha=\omega$.

(iv) $P_A(\varphi,\omega) = P_A(\varphi|M_0,\omega|M_0)$.

(v) $[D\varphi,D\omega]_t= [D(\varphi|M_0),D(\omega|M_0)]_t$ for all $t\in\mathbb{R}$.

(vi) $[D\varphi,D\omega]_t\in M_0$ for all $t\in\mathbb{R}$.

(vii) $[D\varphi,D\omega]_t\in\mathbf{F}(E_\varphi)$ for all $t\in\mathbb{R}$.

Proof. The implications (i)→(ii)→(iii) are obvious. If β is a completely positive mapping then $P_A(\psi_1,\psi_2)\le P_A(\psi_1\circ\beta,\psi_2\circ\beta)$ ([21]). Applying this twice both to the embedding $M_0\to M$ and to $\alpha:M\to M_0$ we obtain (iii)→(iv).

In the proof of (iv)→(v) we use the relative modular operators $\Delta=\Delta(\varphi,\omega)$ and $\Delta_0=\Delta(\varphi|M_0,\omega|M_0)$ (see [4]). If Φ (Φ_0) and Ω (Ω_0) are the vector representatives for φ ($\varphi|M_0$) and ω ($\omega|M_0$) in a natural positive cone then $\Phi=\Delta^{1/2}\Omega$ and $\Phi_0=\Delta_0{}^{1/2}\Omega$. We identify $a\Omega$ and $a\Omega_0$ for all $a\in M_0$ and denote by P the projection of $H=[M\Omega]$ onto $[M_0\Omega]$. Then

$$P_A(\varphi,\omega) = \langle\Delta^{1/2}\Omega, \Omega\rangle = \pi^{-1}\int_0^\infty\lambda^{-1/2}\langle\Delta(\Delta+\lambda)^{-1}\Omega, \Omega\rangle d\lambda$$

and

$$P_A(\varphi|M_0,\omega|M_0) = \langle\Delta_0{}^{1/2}\Omega, \Omega\rangle = \pi^{-1}\int_0^\infty\lambda^{-1/2}\langle\Delta_0(\Delta_0+\lambda)^{-1}\Omega, \Omega\rangle d\lambda.$$

Here

$$\langle\Delta(\Delta+\lambda)^{-1}\Omega,\Omega\rangle = 1-\lambda\langle(\Delta+\lambda)^{-1}\Omega,\Omega\rangle = 1-\lambda\langle P(\Delta+\lambda)^{-1}P\Omega,\Omega\rangle \le$$
$$\le 1-\lambda\langle P(\Delta_0+\lambda)^{-1}P\Omega,\Omega\rangle = \langle\Delta_0(\Delta_0+\lambda)^{-1}\Omega,\Omega\rangle$$

and

$$\langle(\Delta+\lambda)^{-1}\Omega,\Omega\rangle = \langle(\Delta_0+\lambda)^{-1}\Omega,\Omega\rangle \text{ for all } \lambda\in\mathbb{C} \text{ with } \mathrm{Re}\lambda>0.$$

From that one can deduce that $\Delta^{it}\Omega=\Delta_0{}^{it}\Omega$ for all $t\in\mathbb{R}$. Since $\Delta^{it}\Omega = [D\varphi,D\omega]_t\Omega$ and $\Delta_0{}^{it}\Omega = [D(\varphi|M_0),D(\omega|M_0)]_t\Omega$ the cyclicity of Ω guarantees (v).

(v)→(vi) is trivial. Due to the cocycle property

$$\sigma^{\omega}{}_s([D\varphi,D\omega]_t) = [D\varphi,D\omega]_s{}^*[D\varphi,D\omega]_{t+s}$$

(vi) gives that the orbit of $[D\varphi,D\omega]_t$ under σ^{ω} remains in M_0, therefore (vii) follows.

There exists the φ-preserving Umegaki conditional expectation E of M onto $\mathbf{F}(E_\varphi)$. It is well-known that (vii) is equivalent to $\omega\circ E=\omega$. In order to show (vii) implies (iv) we establish a sequence of inequalities.

$$P_A(\varphi,\omega) \le P_A(\varphi|M_0,\omega|M_0) \le P_A(\varphi|\mathbf{F}(E_\varphi),\omega|\mathbf{F}(E_\varphi)) \le P_A(\varphi,\omega)$$

Finally we note that (v) implies $E_\varphi=E_\omega$ may be proved by analytic continuation (see [12]).

COROLLARY 9. There exists a smallest sufficient subalgebra M_1. It is generated by $\{ [D\varphi,D\omega]_t : t\in\mathbb{R} \}$ and admits and Umegaki conditional expectation (preserving both φ and ω).

Proof. $\{ [D\varphi,D\omega]_t : t\in\mathbb{R} \}$ is contained in any sufficient subalgebra and clearly it generates the smallest one. The cocycle property gives that M_1 is stable under the modular group σ^{ω}. Therefore, the ω-conditional expectation $E_\omega: M\to M_1$ is a projection.

P_A has the monotonicity property: $P_A(\varphi,\omega) \le P_A(\varphi|M_0,\omega|M_0)$. If φ does not commute with ω (that is, $[D\varphi,D\omega]_t$ is not a group) and M_0 is commutative then

$$P_A(\varphi,\omega) < P_A(\varphi|M_0,\omega|M_0).$$

We would expect that

$$P_A(\varphi,\omega) < \inf \{ P_A(\varphi|M_0,\omega|M_0) : M_0\subset M \text{ is commutative}\}.$$

In particular, we would like to have an estimate for the right hand side in terms of the (modular objects related to) φ and ω.

Instead of taking a pair of states we may consider a family of states Θ and we speak of sufficient subalgebras for Θ. Contrary to classical statistics, sufficiency is the same as pairwaise sufficiency.

Now we turn to the connection between sufficiency and inner perturbations of states. If φ is a state on M and $h=h^*\in M$, then the perturbed state φ^h is defined by the formula

$$[D\varphi^h, D\varphi]_t = \exp(itL+ith)\exp(-itL)$$

where L stands for the logarithm of the modular operator of φ ([3], [4]).

THEOREM 10. ([19]) M_0 is sufficient for the states φ and φ^h if and only if $h\in \mathbf{F}(E_\varphi)$.

Proof. We should establish that $[D\varphi^h, D\varphi]_t \in \mathbf{F}(E_\varphi)$ for all $t\in\mathbb{R}$ if and only if $h\in\mathbf{F}(E_\varphi)$. On the one hand the derivative of $\sigma^\varphi_s([D\varphi^h, D\varphi]_t)$ at $t=0$ is $-i\sigma^\varphi_s(h)$, hence $\sigma^\varphi_s([D\varphi^h, D\varphi]_t)\in\mathbf{F}(E_\varphi)$ implies $\sigma^\varphi_s(h)\in\mathbf{F}(E_\varphi)$. Conversely, one can refer to the expansion

$$[D\varphi^h, D\varphi]_t = \Sigma_{n=0}^{\infty} i^n \,{}_0\!\int^t dt_1 \,{}_0\!\int^{t_1} dt_2 \ldots {}_0\!\int^{t_{n-1}} dt_n \,\sigma^\varphi_{t_n}(h)\ldots\sigma^\varphi_{t_1}(h).$$

COROLLARY 11. Let $h=h^*\in M$ and φ a state on M. Then the commutative subalgebra M_0 generated by h is sufficient for φ and φ^h if and only if $\varphi(ah)=\varphi(ha)$ for every $a\in M$

Proof. We saw above that for a commutative subalgebra M_0 the fixed point algebra is the intersection of M_0 and the centralizer of φ, i.e. $\mathbf{F}(E_\varphi) = \{a_0\in M_0 : \varphi(aa_0) = \varphi(a_0a)$ for every $a\in M\}$.

Until now M_0 always was a subalgebra of M. Most of the above results have an extension to the case when N and M are quite different algebras and they are connected by a completely positive unital w-continuous mapping $\beta: N\to M$.

THEOREM 12. ([20]) Let $\beta: N\to M$ be a completely positive mapping between the von Neumann algebras N and M and assume that φ is a state on M and $\varphi\circ\beta$ is a state on N. Then there exists a completely positive mapping $\tilde\beta_\varphi$ characterized as

$$\varphi(\sigma_{1/2}(\beta(b))a) = \varphi\circ\beta(\tilde\sigma_{1/2}(b)\tilde\beta_\varphi(a))$$

for every $b\in N$ and $a\in M$. (Here σ and $\tilde\sigma$ stand for the modular group of φ and $\varphi\circ\beta$, respectively.) Furthermore, for $b\in N$ the following two conditions are equivalent.

(i) $\tilde\beta_\varphi\circ\beta(b) = b$.

(ii) $\beta(b^*b) = \beta(b)^*\beta(b)$ and $\beta(\tilde{\sigma}_t(b)) = \sigma_t(\beta(b))$ for all $t\in\mathbb{R}$.

This result is the extension of Theorem 4. A triple (N, M, β) is called channel and in [20] its sufficiency is treated with respect to a family Θ of states of M. We say that (N, M, β) is sufficient for Θ if there is a channel (M, N, γ) such that $\varphi\circ\beta\circ\gamma=\varphi$ for all $\varphi\in\Theta$. Sufficiency of (N, M, β) for Θ has several characterization corresponding to Theorem 8.

Finally we mention that some recent results on conditional expectations are reviewed also in the paper [6].

REFERENCES

1. L.Accardi, C.Cecchini, Conditinal expectations in von Neumann algebras and a theorem of Takesaki, J. Funct. Anal. ,45(1982), 245-273.
2. L.Accardi, C.Cecchini, Surjectivity of the conditional expectation on the L^1-spaces, Lecture Notes in Math., 992, 436-442, Springer, 1983.
3. H.Araki, Relative Hamiltonian for faithful normal states of a von Neumann algebra, Publ. Res. Inst. Math. Sci., 9(1973), 165-209.
4. O.Bratteli, D.W.Robinson, Operator algebras and quantum statistical mechanics I,II, Springer 1979 and 1981.
5. C.Cecchini, D.Petz, Norm convergence of generalized martingales in L^p-spaces over von Neumann algebras, Acta Sci. Math., 48(1985), 55-63.
6. C.Cecchini, Some non-commutative Radon-Nikodym theorems, in this volume.
7. A.Connes, Caracterization des espaces vectoriels ordonnées sousjacents aux algebres de von Neumann, Ann. Inst. Fourier, 24(1974), 121-155.
8. S.Doplicher, R.Longo, Standard and split inclusions of von Neumann algebras, Invent. Math., 75(1984), 493-536.
9. A.Frigerio, Duality of completely positive quasi-free maps and a theorem of L.Accardi and C.Cecchini, Boll. Un. Mat. Ital., 6(1983), 269-281.
10. F.Hiai, M.Ohya, M.Tsukada, Sufficiency, KMS condition and relative entropy in von Neumann algebras, Pacific J. Math., 96(1981), 99-109.
11. F.Hiai, M.Ohya, M.Tsukada, Sufficiency and relative entropy in *-algebras with applications to quantum systems, Pacific J. Math., 107(1983), 117-140.
12. F.Hiai, M.Tsukada, Strong martingale convergence of generalized conditional expectations on von Neumann algebras, Trans. Amer. Math. Soc., 282(1984), 791-798.
13. B.Kümmerer, Adjoints of operators on W^*-algebras, Preprint, Tübingen, 1985.

14. R.Longo, Solution of the factorial Stone-Weierstrass conjecture. An application of the theory of standard split inclusions, Invent. Math., 76(1984), 145-155.

15. S.Ch.Moy, Characterization of conditional expectation as a transform of function spaces, Pacific J. Math., 4(1954), 47-63.

16. M.Nakamura, T.Turumaru, Expectations in an operator algebra, Tohoku J. Math., 6(1954), 189-204;

17. D.Petz, Quantum ergodic theorems, Proceedings of the Workshop on Quantum Probability, Lecture Notes in Math., 1055, 289-300, Springer,1984.

18. D.Petz, A dual in von Neumann algebras, Quart. J. Math. Oxford, 35(1984), 475-483.

19. D.Petz, Sufficient subalgebras and the relative entropy of states of a von Neumann algebra, Comm. Math. Phys., 105(1986), 123-131.

20. D.Petz, Sufficiency of channels over von Neumann algebras, Quart. J. Math. Oxford, to appear.

21. G.A.Raggio, Comparison of Uhlmann's transition probability with the one induced by the natural positive cone of a von Neumann algebra, Lett. Math. Phys., 6(1982), 233-236.

22. F.Riesz, B.Sz.-Nagy, Lectures on functional analysis, Ungar, New York, 1955.

23. M.Takesaki, Conditional expectations in von Neumann algebras, J.Funct. Analysis, 9(1972), 306-321.

24. H.Umegaki, Conditional expectations in an operator algebra, Tohoku J. Math., 6(1954),177-181.

25. H.Umegaki, Conditional expectations in an operator algebra IV (entropy and information), Kodai Math. Sem. Rep., 14(1962), 59-85.

MUTUAL QUADRATIC VARIATION AND ITO'S TABLE IN QUANTUM STOCHASTIC CALCULUS

Johan Quaegebeur
Departement Wiskunde, Katholieke Universiteit Leuven
Celestijnenlaan 200B, B-3030 LEUVEN

ABSTRACT

We discuss the relation between the mutual quadratic variation of two measures and the Ito table in quantum stochastic calculus. We prove the existence of the mutual quadratic variation of field- and gauge-measures in a quasi-free representation of the CCR in the topology of strong convergence on an appropriate invariant domain. We characterise the Fock state in terms of the mutual quadratic variation of the field- and gauge-measures. The higher order quadratic variations of these measures are computed in a general quasi-free representation.

1) Relation between mutual quadratic variation and Ito's table

In general, quantum stochastic calculus uses the following ingredients:

- a (non commutative) topological algebra $\mathcal{A}$ having a filtration $\mathcal{A}_{t]}$, $t \in \mathbb{R}^+$, i.e. a family of subalgebras $\mathcal{A}_{t]}$ of $\mathcal{A}$ s.t. $\mathcal{A}_{s]} \subset \mathcal{A}_{t]}$ if $s < t$.
- adapted processes i.e. functions $F: \mathbb{R}^+ \to \mathcal{A}$ s.t. $F(t) \in \mathcal{A}_{t]}$.
- $\mathcal{A}$-valued measures X on $\mathbb{R}^+$ s.t. $X([s,t)) \in \mathcal{A}_{t]}$.

A stochastic integration theory is then constructed roughly as follows. One starts off by defining in an obvious way the integral $\int F\,dX$ of an adapted $\mathcal{A}$-valued stepfunction F on $\mathbb{R}^+$ with respect to an $\mathcal{A}$-valued measure X on $\mathbb{R}^+$. Integrals of more general processes F are then obtained by considering limits of integrals of approximating stepfunctions for F i.e.

$$\int F\,dX \quad := \lim_{\sup|t_{k+1}-t_k| \to 0} \sum_k F(t_k)\, X([t_k, t_{k+1}))$$

Ito's formula arises if one wants to express the product of two stochastic integrals as a linear combination of stochastic integrals. More explicitly, let F_j $(j = 1,2)$ be adapted stepfunctions and let $I_j := \int F_j\,dX_j$ and $I_j(t) := \int F_j \chi_{[0,t]}\,dX_j$. Then for any partition $0=t_0 < t_1 < t_2 < \ldots < t_n$ in $\mathbb{R}^+$ which is finer than the partition given by the discontinuity points of F_1 and F_2 we have that

$$I_1 I_2 = \sum_{k,\ell=0}^{n} F_1(t_k)\, X_1([t_k, t_{k+1}))\, F_2(t_\ell)\, X_2([t_\ell, t_{\ell+1}))$$

We split this double sum in three parts: the off diagonal parts (a):= $\sum_{k<\ell}$ and (b):= $\sum_{k>\ell}$, and the diagonal part (c):= $\sum_{k=\ell}$ and we consider the limit of these parts as $\sup|t_{j+1}-t_j|$

tends to zero. Clearly we have that

$$\lim (a) = \lim \sum_{\ell=0}^{n} I_1(t_\ell)\, F_2(t_\ell)\, X_2([t_\ell, t_{\ell+1})) = \int I_1(t)\, F_2(t)\, dX_2(t)$$

In order to compute the limit of part (b) one needs a commutation rule between the "past" algebra $\mathcal{A}_{s]}$ and the measure $X_1([s,t))$ for $s<t$, e.g. $[\mathcal{A}_{s]}, X_1([s,t))] = 0$. A commutation rule of this type is always satisfied in some form or other in all examples of quantum stochastic calculus ([1] → [4]). Using this we find that

$$\lim (b) = \lim \sum_{k=0}^{n} F_1(t_k)\, I_2(t_k)\, X_1([t_k, t_{k+1})) = \int F_1(t)\, I_2(t)\, dX_1(t)$$

Finally, the diagonal part (c) becomes

$$\lim (c) = \lim \sum_{k=0}^{n} F_1(t_k)\, F_2(t_k)\, X_1([t_k, t_{k+1}))\, X_2([t_k, t_{k+1}))$$

$$= \int F_1(t)\, F_2(t)\, d[\![X_1, X_2]\!](t)$$

provided

$$[\![X_1, X_2]\!]([s,t)) := \lim_{\substack{s=s_0<\ldots<s_n=t \\ \sup|s_{j+1}-s_j|\to 0}} \sum_{k=1}^{n} X_1([s_k, s_{k+1}))\, X_2([s_k, s_{k+1})) \tag{1}$$

exists. We call the measure $[\![X_1, X_2]\!]$ the mutual quadratic variation of X_1 and X_2.

Hence, assuming the existence of $[\![X_1, X_2]\!]$, we have that

$$\int F_1(t)\, dX_1(t) \cdot \int F_2(t)\, dX_2(t) =$$

$$\int I_1(t)\, F_2(t)\, dX_2(t) + \int F_1(t)\, I_2(t)\, dX_1(t) + \int F_1(t)\, F_2(t)\, d[\![X_1, X_2]\!](t)$$

from which we read the Ito formula

$$dX_1\, dX_2 = d[\![X_1, X_2]\!] \tag{2}$$

Therefore we conclude that if a "decent" integration theory for the measures X_1 and X_2 can be constructed (i.e. if the integration theory is rich enough to make the limit procedures used in the discussion above rigorous), then the existence of the mutual quadratic variation $[\![X_1, X_2]\!]$ implies the Ito formula (2).

In this contribution we will concentrate upon the notion of mutual quadratic variation. The formulation of this part of the theory does not depend on any particular filtration of the algebra $\mathcal{A}$, nor on the fact that the measures we consider are defined on $\mathbb{R}^+$. Therefore we generalise $\mathbb{R}^+$ by an arbitrary measurable space.

Let $(T, \mathfrak{T}_0)$ be a measurable space and let $\mathfrak{T}$ be a subset of $\mathfrak{T}_0$ such that $\emptyset \in \mathfrak{T}$ and $\mathfrak{T}$ is closed for finite unions and intersections. For every $I \in \mathfrak{T}$ we denote by $\mathcal{P}(I)$ the set of all finite $\mathfrak{T}$-partitions $(I_1, I_2, \ldots, I_n)$ of I. Clearly $\mathcal{P}(I)$ is a directed set for the order relation "is finer then", and hence $\mathcal{P}(I)$ can be used to index nets.

Now we make the following definition.

DEFINITION 1

Let $(T,\mathcal{T})$ be as above and let $\mathfrak{A}$ be an algebra

A (finitely additive) $\mathfrak{A}$-valued measure X on $(T,\mathcal{T})$ is a mapping $X: \mathcal{T} \to \mathfrak{A}$ such that $X(I_1 \cap I_2) = X(I_1) + X(I_2)$ for all disjoint $I_1, I_2 \in \mathcal{T}$.

The space of all $\mathfrak{A}$-valued measures on $(T,\mathcal{T})$ is denoted by $\mathcal{M}(T,\mathcal{T};\mathfrak{A})$.

If $\mathfrak{A}$ is a topological algebra and $X_1, X_2 \in \mathcal{M}(T,\mathcal{T};\mathfrak{A})$, then we say that the mutual quadratic variation $[\![X_1,X_2]\!]$ of X_1 and X_2 exists (in the topology of $\mathfrak{A}$) if for all $I \in \mathcal{T}$

$$[\![X_1,X_2]\!](I) \quad := \quad \lim_{(I_k)_{k=1\ldots n} \in \mathcal{P}(I)} \quad \sum_{k=1}^{n} X_1(I_k)\, X_2(I_k)$$

exists. (Remark that $[\![X_1,X_2]\!]$ is then again an $\mathfrak{A}$-valued measure.)

2) Field- and gauge-measures in quasi-free representations of the CCR

In this section we will "second quantize" measures taking values in the bounded operators on some Hilbertspace H.

First we introduce the CCR C^*-algebra. Let H be a complex Hilbertspace. Then the CCR algebra $\mathcal{C}(H)$ is a C^*-algebra which is characterised by the following property:

there is a mapping $W: H \to \mathcal{C}(H): f \to W(f)$ such that

- $W(f)^* = W(-f)$ and $W(f)W(g) = W(f+g) \exp -i \operatorname{Im}\langle f|g\rangle$ for all $f,g \in H$,
- $\operatorname{span}\{W(f) \mid f \in H\}$ is dense in $\mathcal{C}(H)$

Consider now a quasi-free state ω_Q on $\mathcal{C}(H)$ given by

$$\omega_Q(W(f)) \quad = \quad \exp -\frac{1}{2}\langle f|Qf\rangle \tag{3}$$

where $Q \in \mathcal{B}(H)$ and $Q \geq \mathbb{1}$. Notice that ω_Q is the Fock state if $Q = \mathbb{1}$. Denote by $(\mathcal{H},\pi,\Omega)$ the GNS triplet for ω_Q. It is well known (e.g. [5]) that $\lambda \in \mathbb{R} \to \pi(W(\lambda f))$ is a strongly continuous unitary group for all $f \in H$, and that its generator $B(f)$ (this is the Boson field with testfunction f) has all the vectors of the form $B(f_1)\ldots B(f_n)\pi(W(g))\Omega$, $n \in \mathbb{N}$, $f_j, g \in H$, in its domain. Denote by $\mathcal{D}$ the linear span of all those vectors. For convenience we will suppress the "π" in the notation if no confusion can arise. Notice that $B(f)\mathcal{D} \subset \mathcal{D}$. Also remark that the mapping $f \in H \to B(f)$ is real linear (if $B(f)$ is restricted to the invariant domain $\mathcal{D}$), and $[B(f),B(g)] \subset 2i \operatorname{Im}\langle f|g\rangle \mathbb{1}$ and $[B(f),W(g)] \subset 2 \operatorname{Im}\langle g|f\rangle W(g)$.

Let now X be a bounded self adjoint operator on H such that $[X,Q] = 0$, and consider the socalled gauge-automorphism group $(\alpha_\theta^X)_{\theta \in \mathbb{R}}$ of $\mathcal{C}(H)$ associated with X given by

$$\alpha_\theta^X(W(f)) \quad = \quad W(e^{i\theta X} f)$$

Then $\omega_Q \circ \alpha_\theta^X = \omega_Q$, hence α_θ^X can be implemented in the GNS representation of ω_Q, i.e. there exists a unitary group $(U_\theta^X)_{\theta \in \mathbb{R}}$ acting on $\mathcal{H}$ such that

$$\pi \circ \alpha_\theta^X(\cdot) \quad = \quad U_\theta^X\, \pi(\cdot)\, U_{-\theta}^X \qquad \text{and} \qquad U_\theta^X\, \Omega \quad = \quad \Omega\,.$$

We now have the following result.

PROPOSITION 1

The unitary group $(U_\theta^X)_{\theta \in \mathbb{R}}$ as defined above, is strongly continuous. Its generator λ_X has $\mathfrak{D}$ in its domain; moreover we have that

$$\lambda_X\, B(f_1)\ldots B(f_n)W(g)\Omega \;=\; -i \sum_{j=1}^{n} B(f_1)\ldots B(iXf_j)\ldots B(f_n)W(g)\Omega$$
$$+\; B(f_1)\ldots B(f_n)[\, B(iXg) - \langle g|Xg\rangle]\, W(g)\Omega$$

Proof

Strong continuity of $(U_\theta^X)_{\theta \in \mathbb{R}}$ follows from a density argument and the continuity of $\theta \in \mathbb{R} \to \omega_Q(W(f_1)\alpha_\theta^X(W(f_2))W(f_3))$. One can check in a standard way that

$$U_\theta^X\, B(f_1)\ldots B(f_n)W(g)\Omega \;=\; B(e^{i\theta X}f_1)\ldots B(e^{i\theta X}f_n)W(e^{i\theta X}g)\Omega$$

So the proposition is proved if we show the existence of and compute the value of the following limit

$$\lim_{\theta \to 0} \frac{1}{i\theta}[\; B(e^{i\theta X}f_1)\ldots B(e^{i\theta X}f_n)W(e^{i\theta X}g)\Omega \;-\; B(f_1)\ldots B(f_n)W(g)\Omega\;]$$

For simplicity we only treat here the case n=0. The general case involves some combinatorial arguments.

Now we have that

$$\left\|\frac{1}{i\theta}[\; W(e^{i\theta X}g)\Omega - W(g)\Omega\;] - [\;B(iXg) - \langle g|Xg\rangle]\,W(g)\Omega\,\right\|$$
$$= \left\|W(g)\,\{\frac{1}{i\theta}[\; W(e^{i\theta X}g - g)\,\exp i\,\mathrm{Im}\langle g|e^{i\theta X}g\rangle \;-\; 1] - B(iXg) - \langle g|Xg\rangle\,\}\Omega\,\right\|$$
$$\leqslant\; (a) + (b) + (c)$$

where

$$(a) := \left\|\exp i\,\mathrm{Im}\langle g|e^{i\theta X}g\rangle\,\{\frac{1}{i\theta}[\; W(e^{i\theta X}g - g) \;-\; 1] \;-\; B(iXg)\}\;\Omega\,\right\|$$
$$(b) := \left|\exp i\,\mathrm{Im}\langle g|e^{i\theta X}g\rangle \;-\; 1\,\right|\,\left\|B(iXg)\Omega\right\|$$
$$(c) := \left|\frac{1}{i\theta}[\,\exp i\,\mathrm{Im}\langle g|e^{i\theta X}g\rangle - 1\,] - \langle g|Xg\rangle\,\right|$$

Clearly $\lim_{\theta\to 0}(b) = \lim_{\theta\to 0}(c) = 0$, whereas

$$(a) := \left\|\{\frac{1}{i\theta}[\; W(e^{i\theta X}g - g) \;-\; 1\,] - B(iXg)\}\;\Omega\,\right\|$$
$$\leqslant \left\|\frac{1}{i\theta}[\; W(e^{i\theta X}g - g) \;-\; W(i\theta Xg)]\,\Omega\,\right\| \;+\; \left\|\{\frac{1}{i\theta}[\; W(i\theta Xg) - 1\,] \;-\; B(iXg)\}\;\Omega\,\right\|$$

The second term tends to zero by definition of $B(iXg)$ and using the explicit form (3) for ω_Q one checks that also the first term vanishes as θ tends to zero. ∎

If $X \in \mathcal{B}(H)$, not necessarily self adjoint, such that $[X,Q] = 0$, we define $\lambda_X = \lambda_{X_1} + i\lambda_{X_2}$ where $X_1 = \frac{1}{2}(X + X^*)$ and $X_2 = \frac{1}{2i}(X - X^*)$. Thereby $X \in \mathcal{B}(H) \to \lambda_X$ becomes a complex linear map (if λ_X is restricted to the invariant domain $\mathfrak{D}$).

This linearity together with the real linearity of $X \in \mathcal{B}(H) \to B(Xf)$ allows us now to state the definition of the field- and gauge-measures. Let us first introduce the following notation: $\mathcal{B}_Q(H)$ is the algebra of bounded operators on H which commute with Q, and $\mathcal{L}_{\mathfrak{D}}(\mathcal{H})$ is the algebra of (possibly) unbounded operators on the GNS space $\mathcal{H}$ of ω_Q which have $\mathfrak{D}$ as an invariant domain.

DEFINITION 2

Let $(T, \mathcal{T})$ be as above and let $X \in \mathcal{M}(T, \mathcal{T}; \mathcal{B}_Q(H))$ and $f \in H$.
Then we define $\mathcal{L}_{\mathfrak{D}}(\mathcal{H})$-valued measures on $(T, \mathcal{T})$ by

$$\mathbf{B}_X^f(I) := B(X(I)f) \quad \text{and} \quad \Lambda_X(I) := \lambda_{X(I)} \quad , I \in \mathcal{T}$$

The measures $\mathbf{B}_X^f$ and Λ_X obtained in this way are called the field- and gauge-measure associated with X.

In the sequel it will be convenient to consider special linear combinations of the field measures, namely: $A_X^f := \frac{1}{2}(\mathbf{B}_X^{if} - i\,\mathbf{B}_X^f)$ (the annihilation measure) and $A_X^{f+} := \frac{1}{2}(\mathbf{B}_X^{if} + i\,\mathbf{B}_X^f)$ (the creation measure). Sometimes we will use the notation $A_X^{f\#}$ to indicate either A_X^f or A_X^{f+}.

3) Mutual quadratic variation of the field- and gauge measures.

In this section we will prove the existence of the mutual quadratic variation of the field- and gauge measures $\mathbf{B}_X^f$ and Λ_Y in the topology of strong convergence on the invariant domain $\mathfrak{D}$. In order to achieve this we will need some regularity conditions on the measures X and Y which roughly express the non atomicity of the measures.

DEFINITION 3

Let $X \in \mathcal{M}(T, \mathcal{T}; \mathcal{B}(H))$. We say that X is regular if the following conditions are satisfied:

- $\forall I \in \mathcal{T},\ \forall f,g \in H : \quad \lim_{(I_k)_{k=1\ldots n} \in \mathcal{P}(I)} \sum_{k=1}^{n} |\langle f | X(I_k) g\rangle|^2 = 0$
- $\forall I \in \mathcal{T},\ \exists M \in \mathbb{R}^+,\ \forall (I_k)_{k=1\ldots n} \in \mathcal{P}(I),\ \forall f \in H : \quad \sum_{k=1}^{n} \|X(I_k)f\|^2 \leq M\|f\|^2$
- $\forall I \in \mathcal{T},\ \forall f,g \in H : \quad \lim_{(I_k)_{k=1\ldots n} \in \mathcal{P}(I)} \sum_{k,\ell=1}^{n} |\langle X(I_\ell)f | X(I_k)g\rangle|^2 = 0$

If $X, Y \in \mathcal{M}(T, \mathcal{T}; \mathcal{B}(H))$ we say that X and Y are jointly regular if both X and Y are regular and

$$\forall I \in \mathcal{T},\ \forall f,g \in H : \quad \lim_{(I_k)_{k=1\ldots n} \in \mathcal{P}(I)} \sum_{k,\ell=1}^{n} |\langle X(I_\ell)f | Y(I_k)g\rangle|^2 = 0$$

THEOREM 1

Let $X,Y \in \mathcal{M}(T,\mathcal{T}\,;\mathcal{B}_Q(H))$ be self adjoint and jointly regular measures. If $[\![X,Y]\!]$ exists in the strong operator topology on $\mathcal{B}_Q(H)$, then $[\![\mathbf{B}^f_X, \mathbf{B}^g_Y]\!]$, $[\![\Lambda_X,\Lambda_Y]\!]$, $[\![\mathbf{B}^f_X,\Lambda_Y]\!]$ and $[\![\Lambda_X, \mathbf{B}^f_Y]\!]$ exist in the topology of strong convergence on the invariant domain $\mathfrak{D}$ for all $f,g \in H$.

Moreover we have for all $I \in \mathcal{T}$:

(i) $[\![\mathbf{B}^f_X, \mathbf{B}^g_Y]\!](I) \subset [\,\mathrm{Re}\langle Qf|[\![X,Y]\!](I)g\rangle + i\,\mathrm{Im}\langle f|[\![X,Y]\!](I)g\rangle]\,\mathbb{1}$

(ii)
$$[\![\Lambda_X,\Lambda_Y]\!](I)\, B(g_1)\ldots B(g_n)W(h)\Omega =$$
$$- \sum_{1\leq k<\ell\leq n} B(g_1)\ldots\widehat{B(g_k)}\ldots\widehat{B(g_\ell)}\ldots B(g_n)W(h)\Omega\,[\,\mathrm{Re}\langle Qg_k|[\![X,Y]\!](I)g_\ell\rangle + i\,\mathrm{Im}\langle g_k|[\![X,Y]\!](I)g_\ell\rangle + \mathrm{Re}\langle [\![X,Y]\!](I)g_k|Qg_\ell\rangle + i\,\mathrm{Im}\langle [\![X,Y]\!](I)g_k|g_\ell\rangle\,]$$
$$-i\sum_{k=1}^{n} B(g_1)\ldots\widehat{B(g_k)}\ldots B(g_n)W(h)\Omega\,[\,\mathrm{Re}\langle Qg_k|[\![X,Y]\!](I)h\rangle + i\,\mathrm{Im}\langle g_k|[\![X,Y]\!](I)h\rangle + \mathrm{Re}\langle [\![X,Y]\!](I)g_k|Qh\rangle + i\,\mathrm{Im}\langle [\![X,Y]\!](I)g_k|h\rangle\,]$$
$$+\sum_{k=1}^{n} B(g_1)\ldots B([\![X,Y]\!](I)g_k)\ldots B(g_n)W(h)\Omega$$
$$+\, i\, B(g_1)\ldots B(g_n)B([\![X,Y]\!](I)h)W(h)\Omega$$
$$+\, B(g_1)\ldots B(g_n)\,[\,\mathrm{Re}\langle [\![X,Y]\!](I)h|Qh\rangle + i\,\mathrm{Im}\langle [\![X,Y]\!](I)h|h\rangle\,]W(h)\Omega$$

(iii)
$$[\![\Lambda_X,A^f_Y]\!](I)\, B(g_1)\ldots B(g_n)W(h)\Omega =$$
$$[\,- A^f_{[\![X,Y]\!]}(I) + \langle [\![X,Y]\!](I)f|\tfrac{1}{2}(Q+\mathbb{1})h\rangle\,]\, B(g_1)\ldots B(g_n)W(h)\Omega$$
$$-i\sum_{k=1}^{n} B(g_1)\ldots\widehat{B(g_k)}\ldots B(g_n)\,\langle [\![X,Y]\!](I)f|\tfrac{1}{2}(Q+\mathbb{1})g_k\rangle\, W(h)\Omega$$

(iv)
$$[\![\Lambda_X,A^{f+}_Y]\!](I)\, B(g_1)\ldots B(g_n)W(h)\Omega =$$
$$[\,A^{f+}_{[\![X,Y]\!]}(I) + \langle \tfrac{1}{2}(Q-\mathbb{1})h|[\![X,Y]\!](I)f\rangle\,]\, B(g_1)\ldots B(g_n)W(h)\Omega$$
$$-i\sum_{k=1}^{n} B(g_1)\ldots\widehat{B(g_k)}\ldots B(g_n)\,\langle \tfrac{1}{2}(Q-\mathbb{1})g_k|[\![X,Y]\!](I)f\rangle\, W(h)\Omega$$

(v)
$$[\![A^f_X,\Lambda_Y]\!](I)\, B(g_1)\ldots B(g_n)W(h)\Omega =$$
$$\langle \tfrac{1}{2}(Q+\mathbb{1})f|[\![X,Y]\!](I)h\rangle\, B(g_1)\ldots B(g_n)W(h)\Omega$$
$$-i\sum_{k=1}^{n} B(g_1)\ldots\widehat{B(g_k)}\ldots B(g_n)\,\langle \tfrac{1}{2}(Q+\mathbb{1})f|[\![X,Y]\!](I)g_k\rangle\, W(h)\Omega$$

(vi)
$$[\![A^{f+}_X,\Lambda_Y]\!](I)\, B(g_1)\ldots B(g_n)W(h)\Omega =$$
$$\langle [\![X,Y]\!](I)h|\tfrac{1}{2}(Q-\mathbb{1})f\rangle\, B(g_1)\ldots B(g_n)W(h)\Omega$$
$$-i\sum_{k=1}^{n} B(g_1)\ldots\widehat{B(g_k)}\ldots B(g_n)\,\langle [\![X,Y]\!](I)g_k|\tfrac{1}{2}(Q-\mathbb{1})f\rangle\, W(h)\Omega$$

Remark: The theorem has been formulated for self adjoint measures X and Y. The non self adjoint case follows in an obvious way from this theorem.

Proof:

(i) Let X and Y be as stated in the theorem. Then, using the shorthand notation X_k and Y_k instead of $X(I_k)$ and $Y(I_k)$, we have to prove that

$$\lim_{(I_k)_{k=1\dots n}\in \mathcal{P}(I)} \sum_{k=1}^{n} B(X_k f)\, B(Y_k g)\, B(g_1)\dots B(g_n)W(h)\Omega = z\, B(g_1)\dots B(g_n)W(h)\Omega$$

where $z := \operatorname{Re}\langle Qf|[\![X,Y]\!](I)g\rangle + i \operatorname{Im}\langle f|[\![X,Y]\!](I)g\rangle$.

For simplicity we only compute here this limit for n=0. The general case involves some combinatorial arguments.

Using the definition of the fields B(f) and the explicit expression (3) for ω_Q one can derive that

$$\begin{aligned}(a) := {}& \|[\sum_{k=1}^{n} B(X_k f)\, B(Y_k g) - z]\, W(h)\Omega\|^2 \\ = {}& \sum_{k,\ell=1}^{n} (y_k x_k x_\ell y_\ell + y_k y_\ell s_{k,\ell} + x_k y_\ell \overline{r_{\ell,k}} + x_\ell y_\ell \overline{r_{k,k}} + x_k x_\ell t_{k,\ell} \\ & + y_k x_\ell r_{k,\ell} + y_k x_k r_{\ell,\ell} + s_{k,\ell} t_{k,\ell} + r_{k,\ell}\overline{r_{\ell,k}} + r_{\ell,\ell}\overline{r_{k,k}}) \\ & - z \sum_{k=1}^{n} (x_k y_k + \overline{r_{k,k}}) - \bar{z} \sum_{\ell=1}^{n} (x_\ell y_\ell + r_{\ell,\ell}) + |z|^2 \end{aligned} \tag{4}$$

where

$$x_k := 2 \operatorname{Im}\langle h|X_k f\rangle \quad , \quad y_k := 2 \operatorname{Im}\langle h|Y_k g\rangle$$

$$r_{i,j} := \operatorname{Re}\langle QX_i f|Y_j g\rangle + i \operatorname{Im}\langle X_i f|Y_j g\rangle$$

$$s_{i,j} := \operatorname{Re}\langle QX_i f|X_j f\rangle + i \operatorname{Im}\langle X_i f|X_j f\rangle$$

$$t_{i,j} := \operatorname{Re}\langle QY_i g|Y_j g\rangle + i \operatorname{Im}\langle Y_i g|Y_j g\rangle$$

From the assumed existence of $[\![X,Y]\!]$ we get $\lim \sum_{k=1}^{n} r_{k,k} = z$. Futhermore, by the joint regularity of X and Y we have that

$$|\sum_k x_k y_k| \leq 4 \sum_k |\langle h|X_k f\rangle|\,|\langle h|Y_k g\rangle| \leq 4\, (\sum_k |\langle h|X_k f\rangle|^2)^{1/2} (\sum_k |\langle h|Y_k f\rangle|^2)^{1/2} \to 0$$

and

$$\begin{aligned} & |\sum_{k,\ell} y_k y_\ell s_{k,\ell}| \\ & \leq 4 \sum_{k,\ell} |\langle h|Y_k g\rangle|\,|\langle h|Y_\ell g\rangle|\, (\|X_\ell Qf\| + \|X_\ell f\|)\, \|Y_k g\| \\ & \leq 8\, (\sum_k |\langle h|Y_k g\rangle|^2)^{1/2} (\sum_k |\langle h|Y_\ell g\rangle|^2)^{1/2} (\sum_\ell \|X_\ell Qf\|^2 + \|X_\ell f\|^2)^{1/2} (\sum_k \|Y_k g\|^2)^{1/2} \\ & \to 0 \end{aligned}$$

and analogously all the terms but the last in the double sum of (4) vanish in the limit of infinitely fine partitions.

Combining these results we get that lim (a) = 0 which proves statement (i).

(ii) Also here we will for convienience restrict ourselves to compute the limit of $\sum_k \Lambda_X(I_k)\ \Lambda_Y(I_k)\ W(h)\Omega$. By proposition 1 we know that

$$\sum_{k=1}^{n} \Lambda_X(I_k)\ \Lambda_Y(I_k)\ W(h)\Omega$$

$$= \sum_{k=1}^{n} \Lambda_X(I_k)[\ B(iY_k h) \ - \langle h|Y_k h\rangle\ ]\, W(h)\Omega$$

$$= \sum_{k=1}^{n} i\ B(X_k Y_k h)\ W(h)\Omega \ + \ \sum_{k=1}^{n} B(iY_k h)\ B(iX_k h)\ W(h)\Omega$$

$$- \sum_{k=1}^{n} B(iY_k h)\ \langle h|X_k h\rangle\ W(h)\Omega \ - \ \sum_{k=1}^{n} \langle h|Y_k h\rangle\ B(iX_k h)\ W(h)\Omega$$

$$+ \sum_{k=1}^{n} \langle h|Y_k h\rangle \langle h|X_k h\rangle\ W(h)\Omega$$

Using regularity of X and Y one can check that the last three terms vanish in the limit. From the existence of $[\![X,Y]\!]$ and the explicit form (3) for ω_Q it follows that the first term tends to $i\ B([\![X,Y]\!](I)h)$. Finally, by the commutation relations for the fields and (i) we find that the second term tends to $[\ \mathrm{Re}\langle [\![X,Y]\!](I)h|Qh\rangle \ + \ i\ \mathrm{Im}\langle [\![X,Y]\!](I)h|h\rangle\ ]\ W(h)\Omega$. This proves statement (ii).

The proofs of (iii) → (vi) can be done along the same lines.

∎

4) Characterisation of the Fock state

In this section we will characterise the Fock state in terms of the mutual quadratic variations of the field and the gauge measures.

DEFINITION 4

We call $\mathcal{M}(T,\mathcal{I}\ ;\ \mathcal{B}_Q(H))$ full if the linear span of the set

$\{\ [\![X,Y]\!](I)f \mid f\in H,\ I\in\mathcal{I}\ ,\ X,Y\in\mathcal{M}(T,\mathcal{I}\ ;\ \mathcal{B}_Q(H))$ such that X and Y are self adjoint jointly regular and $[\![X,Y]\!]$ exists $\}$

is dense in H.

THEOREM 2

Consider the following statements:

(i) ω_Q is the Fock state (i.e. $Q=\mathbb{1}$)

(ii) $[\![A_X^f, \Lambda_Y]\!] = A^f_{[\![X,Y]\!]^*}$ for all $f\in H$ and all X and Y

(iii) $[\![A_X^{f+}, \Lambda_Y]\!] = 0$ for all $f\in H$ and all X and Y

(iv) $[\![\Lambda_X, A_Y^f]\!] = 0$ for all $f\in H$ and all X and Y

(v) $[\![\Lambda_X, A_Y^{f+}]\!] = A^{f+}_{[\![X,Y]\!]}$ for all $f\in H$ and all X and Y

(vi) $[\![\Lambda_X, \Lambda_Y]\!] = \Lambda_{[\![X,Y]\!]}$ for all X and Y

(The phrase "for all X and Y" should be understood as "for all jointly regular self adjoint X and Y such that $[\![X,Y]\!]$ exists".)

Then (i) implies each of the other statements.

If $\mathcal{M}(T,\mathcal{I};\mathcal{B}_Q(H))$ is full, then all statements are equivalent.

Remark: All equalities in statements (ii) → (vi) hold on $\mathcal{D}$.

Proof

For $Q = \mathbb{1}$ the following holds:

$$A^f_X(I)\, B(g_1)\ldots B(g_n) W(h)\Omega$$
$$= -i \sum_{j=1}^{n} B(g_1)\ldots \widehat{B(g_j)}\ldots B(g_n)\, \langle X(I)f|g_j\rangle\, W(h)\Omega$$
$$+ \langle X(I)f|h\rangle\, B(g_1)\ldots B(g_n)W(h)\Omega$$

This fact together with theorem 1 yields immediately that (i) implies (ii) → (v).

Now we prove that (i) implies (vi). In principle we have to check that (vi) holds on all vectors of $\mathcal{D}$. Again for simplicity we only show the equality on vectors of the form $W(h)\Omega$.

Thus, by proposition 1 and theorem 1, we have to prove that for all $I \in \mathcal{I}$:

$$i\, B(Zh)\, W(h)\Omega + \langle Zh|h\rangle\, W(h)\Omega$$
$$= B(iZ_s h)\, W(h)\Omega + i\, B(iZ_a h)\, W(h)\Omega - \langle h|Zh\rangle\, W(h)\Omega \qquad (5)$$

where $Z := [\![X,Y]\!](I)$, $Z_s := \frac{1}{2}(Z + Z^*)$ and $Z_a := \frac{1}{2i}(Z - Z^*)$.

But if $Q = \mathbb{1}$ one can straightforwardly verify that $B(ig)\Omega = i\, B(g)\Omega$ for all $g \in H$. Using this together with the commutation relations we have that

$$B(iZ_s h)\, W(h)\Omega = W(h)\, B(iZ_s h)\Omega + 2\, \mathrm{Im}\langle h|iZ_s h\rangle\, W(h)\Omega$$
$$= i\, W(h)\, B(Z_s h)\Omega + 2\langle h|Z_s h\rangle\, W(h)\Omega$$
$$= i\, B(Z_s h)\, W(h)\Omega + 2\langle h|Z_s h\rangle\, W(h)\Omega$$

Hence the righthand side of (5) can be rewritten as

$$i\, B(Z_s h)\, W(h)\Omega + i\, B(iZ_a h)\, W(h)\Omega + (\, 2\langle h|Z_s h\rangle - \langle h|Zh\rangle\,)\, W(h)\Omega$$
$$= i\, B(Zh)\, W(h)\Omega + \langle Zh|h\rangle\, W(h)\Omega$$

which proves the validity of (5).

Suppose now that $\mathcal{M}(T,\mathcal{I};\mathcal{B}_Q(H))$ is full. We prove that then (ii) implies (i). Take any pair of jointly regular selfadjoint measures X,Y such that $[\![X,Y]\!]$ exists. Choose any $I \in \mathcal{I}$ and put $Z := [\![X,Y]\!](I)$. We know from theorem 1 that for all $f,h \in H$:

$$[\![A^f_X, A_Y]\!](I)\, W(h)\Omega = \langle \tfrac{1}{2}(Q+1)f|Zh\rangle\, W(h)\Omega$$

Hence by (ii) we have that

$$\langle\Omega|W(h)\Omega\rangle\langle \tfrac{1}{2}(Q+\mathbb{1})f|Zh\rangle = \langle\Omega|A^f_{Z^*}\, W(h)\Omega\rangle = \langle\Omega|W(h)\Omega\rangle\langle f|Zh\rangle$$

so $\langle \frac{1}{2}(\mathbb{1}-Q)f|Zh\rangle = 0$ for all $f,h \in H$, thus $Q = \mathbb{1}$. The other implications are proven in the same way.

∎

Theorem 2 tells us that in the Fock case the scalar-, field- and gauge measures are closed for the operation of taking the mutual quadratic variation. This inspires us to introduce the following notion.

DEFINITION 5

Let $\mathcal{A}$ be a topological algebra and let $\mathcal{M}(T,\mathcal{T};\mathcal{A})$ be as in definition 1.
We say that a linear subspace $\mathcal{X}$ of $\mathcal{M}(T,\mathcal{T};\mathcal{A})$ is an Ito algebra if for all X and Y in $\mathcal{X}$ $[\![X,Y]\!]$ exists and belongs to $\mathcal{X}$.

Now we can rephrase part of theorem 2 as follows:
Let $\mathcal{X}\subset\mathcal{M}(T,\mathcal{T};\mathcal{B}_Q(H))$ be a regular Ito algebra (i.e. all X and Y in $\mathcal{X}$ are also assumed to be jointly regular). Then the linear span of the set

$$\{\langle f|X(\cdot)g\rangle\,\mathbb{1}\ ,\ A_X^f\ ,\ A_X^{f+}\ ,\ \Lambda_X\ |\ X\in\mathcal{X}\ ,\ f,g\in H\}$$

forms an Ito algebra in $\mathcal{M}(T,\mathcal{T};\mathcal{L}_{\mathcal{D}}(\mathcal{H}))$ if $Q=\mathbb{1}$. The mapping $\Lambda : X\to\Lambda_X$ is then a homomorphism.

Example 1

Take $H=L^2(\mathbb{R}^+)$, $Q=\mathbb{1}$ and let $(T,\mathcal{T})$ be $\mathbb{R}^+$ with its usual Borel σ-algebra. Define a $\mathcal{B}(H)$-valued measure X by $X(I)$:= multiplication by the characteristic function χ_I of the Borelset $I\subset\mathbb{R}^+$. It is clear that X is regular and that $[\![X,X]\!]=X$ (in the strong operator topology on $\mathcal{B}(H)$). So $\mathcal{X}:=\mathbb{C}X$ is a regular Ito algebra. Denote $\Lambda:=\Lambda_X$, $A^f:=A_X^f$ and $A^{f+}:=A_X^{f+}$. These measures together with the scalar measures $\bar{f}g\,dt$ span an Ito algebra of $\mathcal{L}_{\mathcal{D}}(\mathcal{H})$-valued measures. Using the formal notation (2) motivated in section 1, its multiplication table can, by theorem 2, be written as

	dt	dA^f	dA^{g+}	$d\Lambda$
dt	0	0	0	0
dA^f	0	0	$\bar{f}g\,dt$	dA^f
dA^{g+}	0	0	0	0
$d\Lambda$	0	0	dA^{g+}	$d\Lambda$

(compare with [2])

5) The Ito algebra for a non-Fock quasi-free representation

If $Q\neq\mathbb{1}$ the measures $[\![\Lambda_X,\Lambda_Y]\!]$, $[\![\Lambda_X,A_Y^f]\!]$, etc., are new objects, i.e. they are not linear combinations of the scalar-, field- and gauge measures we started from. So it is a natural question to ask whether we keep on generating new objects if we compute higher order quadratic variations such as $[\![A_X^f,[\![\Lambda_Y,\Lambda_Z]\!]]\!]$. In this section we will show that, under some commutativity condition on the measures X,Y and Z, no more new objects are generated after the second order.

DEFINITION 6

Let $X,Y\in\mathcal{M}(T,\mathcal{T};\mathcal{A})$. We say that X and Y commute infinitesimally if both $[\![X,Y]\!]$ and $[\![Y,X]\!]$ exist and $[\![X,Y]\!]=[\![Y,X]\!]$.

Remark that commuting measures X and Y (i.e. $X(I)\,Y(I) = Y(I)\,X(I)$ for all $I \in \mathcal{I}$) also commute infinitesimally. The converse however is not true.

THEOREM 3

Let $\mathcal{X}$ be an infinitesimally commutative, self adjoint and regular Ito algebra in $\mathcal{M}(T,\mathcal{I}\,;\,\mathcal{B}_Q(H))$. Then we have that

(i) $[\![\Lambda_X, A^f_Y]\!] = -A^f_{[\![X,Y]\!]} + [\![A^f_Y,\Lambda_X]\!]$ and $[\![\Lambda_X, A^{f+}_Y]\!] = A^{f+}_{[\![X,Y]\!]} + [\![A^{f+}_Y,\Lambda_X]\!]$ for all $X,Y \in \mathcal{X}$ and $f \in H$.

(ii) The linear span of the set
$\{\langle f|X(\cdot)g\rangle\ ,\ A^{f\#}_X\ ,\ \Lambda_X\ ,\ [\![A^{f\#}_X,\Lambda_Y]\!]\ ,\ [\![\Lambda_X,\Lambda_Y]\!]\ |\ X,Y \in \mathcal{X}\ ,\ f \in H\ \}$
forms an Ito algebra in $\mathcal{M}(T,\mathcal{I}\,;\,\mathcal{L}_{\mathcal{D}}(\mathcal{H}))$.
The multiplication in this Ito algebra is given by theorem 1 and the following relations:

$$[\![A^{f\#}_X, [\![A^{g\#}_Y,\Lambda_Z]\!]]\!] = [\![[\![A^{f\#}_X,A^{g\#}_Y]\!],\Lambda_Z]\!] = 0$$

$$[\![A^f_X,[\![\Lambda_Y,\Lambda_Z]\!]]\!] = [\![A^f_X,\Lambda_{[\![Y,Z]\!]}]\!]\ ,\quad [\![[\![A^f_X,\Lambda_Y]\!],\Lambda_Z]\!] = [\![A^f_{[\![X,Y]\!]},\Lambda_Z]\!]$$

$$[\![A^{f+}_X,[\![\Lambda_Y,\Lambda_Z]\!]]\!] = -[\![A^{f+}_X,\Lambda_{[\![Y,Z]\!]}]\!]\ ,\quad [\![[\![A^{f+}_X,\Lambda_Y]\!],\Lambda_Z]\!] = -[\![A^{f+}_{[\![X,Y]\!]},\Lambda_Z]\!]$$

$$[\![\Lambda_X,[\![A^{f\#}_Y,\Lambda_Z]\!]]\!] = [\![[\![\Lambda_X,A^{f\#}_Y]\!],\Lambda_Z]\!] = 0$$

$$[\![\Lambda_X,[\![\Lambda_Y,\Lambda_Z]\!]]\!] = \Lambda_{[\![X,[\![Y,Z]\!]]\!]}\ ,\quad [\![[\![\Lambda_X,\Lambda_Y]\!],\Lambda_Z]\!] = \Lambda_{[\![[\![X,Y]\!],Z]\!]}$$

$$[\![[\![A^f_X,\Lambda_Y]\!],A^{g+}_Z]\!] = \langle \tfrac{1}{2}(Q+\mathbf{1})f\,|\,[\![[\![X,Y]\!],Z]\!]g\rangle\ \mathbf{1}$$

$$[\![[\![A^{f+}_X,\Lambda_Y]\!],A^g_Z]\!] = \langle [\![[\![X,Y]\!],Z]\!]g\,|\,\tfrac{1}{2}(\mathbf{1}-Q)f\rangle\ \mathbf{1}$$

$$[\![[\![A^{f+}_X,\Lambda_Y]\!],A^{g+}_Z]\!] = [\![[\![A^f_X,\Lambda_Y]\!],A^g_Z]\!] = 0$$

$$[\![[\![A^{f\#}_X,\Lambda_Y]\!],[\![A^{g\#}_Z,\Lambda_V]\!]]\!] = 0$$

$$[\![[\![A^f_X,\Lambda_Y]\!],[\![\Lambda_Z,\Lambda_V]\!]]\!] = [\![A^f_{[\![X,Y]\!]},\Lambda_{[\![Z,V]\!]}]\!]$$

$$[\![[\![A^{f+}_X,\Lambda_Y]\!],[\![\Lambda_Z,\Lambda_V]\!]]\!] = -[\![A^{f+}_{[\![X,Y]\!]},\Lambda_{[\![Z,V]\!]}]\!]$$

$$[\![[\![\Lambda_X,\Lambda_Y]\!],A^f_Z]\!] = -[\![\Lambda_{[\![X,Y]\!]},A^f_Z]\!]\ ,\quad [\![[\![\Lambda_X,\Lambda_Y]\!],A^{f+}_Z]\!] = [\![\Lambda_{[\![X,Y]\!]},A^{f+}_Z]\!]$$

$$[\![[\![\Lambda_X,\Lambda_Y]\!],[\![A^{f\#}_Z,\Lambda_V]\!]]\!] = 0$$

$$[\![[\![\Lambda_X,\Lambda_Y]\!],[\![\Lambda_Z,\Lambda_V]\!]]\!] = [\![\Lambda_{[\![X,Y]\!]},\Lambda_{[\![Z,V]\!]}]\!]$$

Remark: All quadratic variations listed above exist in the topology of strong convergence on the invariant domain $\mathcal{D}$. The relations hold on $\mathcal{D}$.

Proof

(i) follows immediately from theorem 1 and the fact that $[\![X,Y]\!]$ is self adjoint if X and Y are self adjoint and commute infinitesimally.

(ii) follows from the explicit actions of the measures on $\mathcal{D}$ given in proposition 1 and theorem 1. The verifications of the relations listed in the theorem are all done along the same lines. As an illustration we show that for all $I \in \mathcal{I}$

$$[\![\Lambda_X,[\![\Lambda_Y,\Lambda_Z]\!]]\!](I)\ W(h)\Omega \ = \ \Lambda_{[\![X,[\![Y,Z]\!]]\!]}(I)\ W(h)\Omega$$

So we have to compute for all $I \in \mathcal{Z}$

$$\lim_{(I_k)_{k=1\ldots n} \in \mathcal{A}(I)} \sum_{k=1}^{n} \Lambda_X(I_k)\ [\![\Lambda_Y,\Lambda_Z]\!](I_k)\ W(h)\Omega$$

Using the shorthand notation X_k and $[\![Y,Z]\!]_k$ instead of $X(I_k)$ and $[\![Y,Z]\!](I_k)$, we have by theorem 1 and proposition1 that

$$\begin{aligned}
&\Lambda_X(I_k)\ [\![\Lambda_Y,\Lambda_Z]\!](I_k)\ W(h)\Omega \\
&= \Lambda_X(I_k)\ i\ B([\![Y,Z]\!]_k\ h)\ W(h)\Omega \\
&\quad + [\ \mathrm{Re}\langle [\![Y,Z]\!]_k\ h|Qh\rangle \ + \ i\ \mathrm{Im}\langle [\![Y,Z]\!]_k\ h|h\rangle]\ \Lambda_X(I_k)\ W(h)\Omega \\
&= B(i\ X_k\ [\![Y,Z]\!]_k\ h)\ W(h)\Omega \\
&\quad + \ i\ B([\![Y,Z]\!]_k\ h)\ B(iX_k\ h)\ W(h)\Omega \qquad (6)\\
&\quad - \ B([\![Y,Z]\!]_k\ h)\ \langle h|X_k\ h\rangle \ W(h)\Omega \\
&\quad + \ [\mathrm{Re}\langle [\![Y,Z]\!]_k\ h|Qh\rangle \ + \ i\ \mathrm{Im}\langle [\![Y,Z]\!]_k\ h|h\rangle]\ \ [\ B(iX_k\ h) - \langle h|X_k\ h\rangle]\ W(h)\Omega
\end{aligned}$$

Using the explicit form (3) for ω_Q , one can check that by the regularity of the measures, the third and the fourth terms vanish when they are summed over k and when the limit is taken. On the other hand, one can verify that the first term will give a contribution

$$B(i\ [\![X,[\![Y,Z]\!]]\!](I)\ h)\ W(h)\Omega \qquad (7)$$

and the second term becomes by theorem 1(i)

$$i[\ \mathrm{Re}\langle Qh|[\![[\![Y,Z]\!],X]\!](I)\ ih\rangle \ + \ i\ \mathrm{Im}\langle h|[\![[\![Y,Z]\!],X]\!](I)\ ih\rangle]\ \ W(h)\Omega$$

But due to the self adjointness and the infinitesimal commutativity, this is equal to

$$-\ \langle h|[\![X,[\![Y,Z]\!]]\!](I)\ h\rangle \ \ W(h)\Omega \qquad (8)$$

So summing (6) over k and taking the limit yields (7)+(8) and comparing this with $\Lambda_{[\![X,[\![Y,Z]\!]]\!]}(I)\ W(h)\Omega$ (cfr. proposition 1) gives the result.

∎

Example 2

Take $(T,\mathcal{Z})$, H and X as in example 1. Let Q be multiplication in $L^2(\mathbb{R}^+)$ by some measurable function $q(t)$ (with $1 \leqslant q(t) \leqslant M$). Take $f = \chi_{[0,S]}$ for some $S \in \mathbb{R}^+$. Denote $A := A_X^f$, $A^+ := A_X^{f+}$ and $\Lambda := \Lambda_X$.

Then the measures A , A^+ , Λ , $[\![A,\Lambda]\!]$, $[\![A^+,\Lambda]\!]$ and $[\![\Lambda,\Lambda]\!]$ together with the scalar measures which are absolutely continous with respect to the Lebesgue measure dt , span an Ito algebra whose multiplication table can, using the formal notation (2), be written as

	dt	dA	dA^+	$d\Lambda$	$dA\,d\Lambda$	$dA^+d\Lambda$	$d\Lambda\,d\Lambda$
dt	0	0	0	0	0	0	0
dA	0	0	r dt	$dA\,d\Lambda$	0	0	$dA\,d\Lambda$
dA^+	0	s dt	0	$dA^+d\Lambda$	0	0	$-dA^+d\Lambda$
$d\Lambda$	0	$dA\,d\Lambda - dA$	$dA^+d\Lambda + dA^+$	$d\Lambda\,d\Lambda$	0	0	$d\Lambda$
$dA\,d\Lambda$	0	0	r dt	$dA\,d\Lambda$	0	0	$dA\,d\Lambda$
$dA^+d\Lambda$	0	-s dt	0	$-dA^+d\Lambda$	0	0	$-dA^+d\Lambda$
$d\Lambda\,d\Lambda$	0	$dA - dA\,d\Lambda$	$dA^+ + dA^+d\Lambda$	$d\Lambda$	0	0	$d\Lambda\,d\Lambda$

where $r(t) := \frac{1}{2}(q(t)+1)\,\chi_{[0,S]}$ and $s(t) := \frac{1}{2}(q(t)-1)\,\chi_{[0,S]}$.

6) Mutual quadratic variation of the field measures in a non quasi-free representation

The existence of the boson fields and the generators of the unitary groups implementing the gauge automorphisms is of course not a privilege of the quasi-free states. Indeed, for all states ω on $\mathcal{C}(H)$ which are sufficiently regular, these objects are well defined and have a nice domain. So, one may wonder whether the mutual quadratic variations of the field- and gauge measures exist and under what conditions they generate an Ito algebra. We could not solve this problem yet without imposing unelegant and hardly verifiable conditions on ω .

However, the fact that one can go beyond the quasi-free case may be illustrated by the following example. Consider a state ω on the CCR-algebra $\mathcal{C}(H)$ given by

$$\omega(W(f)) = \exp \psi(f)$$

with

$$\psi(f) = -\frac{1}{2}\langle f|Qf\rangle + i\,\mathrm{Re}\,\langle f_0|f\rangle + \int_H \left[\exp i\,\mathrm{Re}\,\langle f'|f\rangle - 1 - \frac{i\,\mathrm{Re}\,\langle f'|f\rangle}{1+\|f'\|^2}\right] \frac{1+\|f'\|^2}{\|f'\|^2}\, d\mu(f')$$

where $Q\in\mathcal{B}(H)$ and $Q\geq 1$, $f_0\in H$ and μ is a finite Borelmeasure on H without mass at zero. Notice that ω is infinitely divisible in the sense that for all $n\in\mathbb{N}_0$ there exists a state ω_n on $\mathcal{C}(H)$ such that $(\omega_n(W(f/\sqrt{n})))^n = \omega(W(f))$ for all $f\in H$. Denote as before the GNS triplet of ω by $(\mathcal{H},\pi,\Omega)$. We will assume futhermore that $\int_H \|g\|^k\, d\mu(g) < \infty$ for all $k\in\mathbb{N}_0$. Thereby the map $\lambda\in\mathbb{R}\to\omega(W(\lambda f+g))$ is C^∞ for all $f,g\in H$, and one can verify that therefore $\lambda\in\mathbb{R}\to\pi(W(\lambda f))$ is a strongly continuous unitary group and that its generator, B(f), has all vectors of the form $B(g_1)\ldots B(g_n)\,W(h)\Omega$ in its domain. Denote again by $\mathcal{D}$ the span of those vectors in .

Let $(T,\mathcal{T})$ be as before and define for $X\in\mathcal{M}(T,\mathcal{T},\mathcal{A}_0(H))$ and for $f\in H$ the field measure $\mathbb{B}^f_X$ by $\mathbb{B}^f_X(I) := B(X(I)f)$, $I\in\mathcal{T}$. Suppose now that X and Y are self adjoint jointly regular measures in $\mathcal{M}(T, \mathcal{Z}, \mathcal{A}_Q(H))$ such that $[\![X,Y]\!]$ exists in the strong operator topology on $\mathcal{A}_Q(H)$. We claim that then $[\![\mathbb{B}_X^f, \mathbb{B}_Y^g]\!]$ exists strongly on $\mathfrak{D}$ and equals $[\operatorname{Re}\langle Qf|[\![X,Y]\!](\cdot)\, g\rangle + i \operatorname{Im}\langle f|[\![X,Y]\!](\cdot)\, g\rangle]\, \mathbb{1}$.

For simplicity we will only show here the existence of $[\![\mathbb{B}_X^f, \mathbb{B}_Y^g]\!]$ weakly on vectors of the form $W(h)\Omega$. The stronger result requires some extra combinatorial arguments.

Let $I \in \mathcal{Z}$ and let $(I_k)_{k=1\ldots n} \in \mathcal{P}(I)$. Then, using the same notations as before, we have that

$$\langle W(h)\Omega \mid \sum_k \mathbb{B}_X^f(I_k)\, \mathbb{B}_Y^g(I_k)\, W(h')\Omega\rangle = \langle W(h)\Omega | W(h')\Omega\rangle \sum_k (\, s(X_k f)\, s(Y_k g) + t_k) \tag{9}$$

where

$$s(\cdot) := i \operatorname{Re}\langle Q(h'-h)|\cdot\rangle + \operatorname{Im}\langle h'+h|\cdot\rangle + \operatorname{Re}\langle f_0|\cdot\rangle$$
$$+ \int_H [\exp i \operatorname{Re}\langle f'|h'-h\rangle - \frac{1}{1+\|f'\|^2}]\ \operatorname{Re}\langle f'|\cdot\rangle\ \frac{1+\|f'\|^2}{\|f'\|^2}\ d\mu(f')$$

and

$$t_k := \operatorname{Re}\langle QX_k f|Y_k g\rangle + i \operatorname{Im}\langle X_k f|Y_k g\rangle$$

Hence, by joint regularity of X and Y and the dominated convergence theorem, we find that (9) becomes in the limit of infinitely fine partitions of I

$$\langle W(h)\Omega|W(h')\Omega\rangle\ [\operatorname{Re}\langle Qf|[\![X,Y]\!](I)\, g\rangle + i \operatorname{Im}\langle f|[\![X,Y]\!](I)\, g\rangle]$$

which gives the result.

REFERENCES

[1] C. Barnett, R.F. Streater, I.F. Wilde; The Ito Clifford integral I, J. Funct. Anal. 48, 172-212 (1982).

[2] R.L. Hudson, K.R. Parthasarathy; Quantum Ito's formula and stochastic evolutions, Commun. Math. Phys. 93, 301-323 (1984)

[3] D.B. Applebaum, R.L. Hudson; Fermion Ito's Formula and Stochastic Evolutions, Commun. Math. Phys. 96, 473-496 (1984)

[4] R.L. Hudson, J.M. Lindsay; Stochastic integration and a martingale representation theorem for non-Fock quantum Brownian motion, J. Funct. Anal. 61, 202-221 (1985)

[5] O. Bratteli, D.W. Robinson; Operator Algebras and Quantum Statistical Mechanics II, Springer-Verlag New-York, Heidelberg, Berlin, 1979.

ON MIXING PROPERTIES OF AUTOMORPHISMS OF VON NEUMANN ALGEBRAS RELATED TO MEASURE SPACE TRANSFORMATIONS*

Uwe Quasthoff

Karl-Marx-Universität Leipzig

Sektion Mathematik

DDR-7010 Leipzig

Abstract

The mixing properties of shift-automorphisms of an infinite tensor product of matrix algebras with a shift invariant (but, in general, non-product) state are studied. It is shown that if the centralizer of the von Neumann algebra is a type II_1 factor, then the mixing properties of the shift automorphism correspond to the mixing properties of the underlying measure space shift.

1. Non-Commutative Mixing Properties

Let M be a von Neumann algebra of dimension greater than 1, ω a state on M and α an automorphism. First we want to recall mixing properties and the relations between them.

Definition 1.1

i) α is called ergodic if $\alpha(a) = a$ implies $a = \lambda 1$ for some complex number λ.

ii) α is called weakly mixing with respect to ω if for $a,b \in M$

$$\lim_{n\to\infty} \frac{1}{n} \sum_{k=0}^{n-1} |\omega(a\cdot\alpha^k(b)) - \omega(a)\,\omega(b)| = 0 .$$

*) Work finished during a visit at the Institut für Theoretische Physik der Universität Wien.

iii) α is called (strongly) mixing with respect to ω if for $a,b \in M$

$$\lim_{n\to\infty} \omega(a\cdot\alpha^n(b)) = \omega(a)\,\omega(b) .$$

There is the following characterization of ergodicity similar to the definition of mixing and weak mixing [8].

Proposition 1.2

An automorphism α is ergodic if and only if for any α-invariant faithful state ω the following holds:

$$\lim_{n\to\infty} \frac{1}{n} \sum_{k=0}^{n-1} \omega(a\cdot\alpha^k(b)) = \omega(a)\,\omega(b) .$$

Corollary

Mixing implies weak mixing and weak mixing implies ergodicity.

Proof

Let $\{a_n\}$ be a sequence of real numbers. Then $\lim a_n = 0$ implies $\lim_{n\to\infty} \frac{1}{n} \sum_{k=0}^{n-1} |a_k| = 0$ and $\lim_{n\to\infty} \frac{1}{n} \sum_{k=0}^{n-1} |a_k| = 0$ implies $\lim_{n\to\infty} \frac{1}{n} \sum_{k=0}^{n-1} a_k = 0.$ □

An automorphism α is called inner if there is a unitary $u \in M$ such that $\alpha(x) = u^* xu$ for all x. Let $p(\alpha)$ be the largest projection e satisfying $\alpha(e) = e$ and $\alpha_{|eMe}$ is inner. Then α is called properly outer if $p(\alpha) = 0$.

Proposition 1.3

Any ergodic automorphism is properly outer.

Proof

By [1] we have $p(\alpha) \in M$. The invariance of $p(\alpha)$ under α implies $p(\alpha) = 0$ or $p(\alpha) = 1$, i.e. α is either inner or properly outer. Assume that α is inner, i.e. $\alpha = \mathrm{Ad}\, u$ for some unitary $u \in M$. Now $u \neq \lambda 1$ and $\alpha(u) = u$ in contradiction to the ergodicity. □

Now we want to introduce K-automorphisms.

Definition 1.4

Let α be an automorphism of the von Neumann algebra M and N be a subalgebra of M such that

i) $N \subset \alpha(N)$,

ii) $(\bigcup_{n=0}^{\infty} \alpha^n(N))'' = M$ and

iii) $\bigcap_{n=0}^{\infty} \alpha^{-n}(N) = \mathbb{C}$.

Then the triple (M,N,α) is called a K-system. If there is a faithful normal conditional expectation $E: M \to N$ then the K-system is said to satisfy the expectation property. α is called a K-automorphism if there is a subalgebra N such that (M,N,α) is a K-system.

K-automorphisms have the strongest mixing properties so far [9]:

Proposition 1.5

K-automorphisms are strongly mixing.

2. Automorphisms Related to Measure Space Shifts and Their Mixing Properties

The aim of this chapter is to start with a shift transformation of a measure space and to construct a corresponding shift automorphism on an infinite tensor product von Neumann algebra. If the measure space shift satisfies some mixing conditions the von Neumann algebra will be a factor containing a shift-invariant subalgebra isomorphic to the hyperfinite factor of type II_1.

Assume that T is a shift transformation on a measure space X, i.e. $X = Y^{\mathbb{Z}}$ for some state space Y which will be assumed as finite here and T acts as follows: An element $x \in X$ is a doubly infinite sequence of elements of Y. If $x = (\ldots,x_{-1},x_0,x_1,\ldots)$, then $Tx = y$ with $y_n = x_{n+1}$.

A subset of X of the form $\{x: x_\ell = a_0, x_{\ell+1} = a_1,\ldots,x_{\ell+k} = a_k\}$ is called a cylinder set and denoted by $[a_0,a_1,\ldots,a_k]_\ell$. The measure μ on X will always be assumed as shift invariant. For a general description of shift invariant measures, see [3]. Here we only want to present two examples.

Example 2.1

Bernoulli shifts. Here we choose a probability measure p on Y and let μ be the product measure, i.e. $\mu([a_0,\dots,a_k]) = p(a_0)\cdot\ldots\cdot p(a_k)$. (Because of the shift invariance we can drop the index ℓ at the cylinder.)

Example 2.2

Markov shifts. Let $P = (p_{ij})$ be a matrix of transition probabilities (i.e. $\sum_i p_{ij} = 1$ for all j and $p_{ij} \geq 0$). If the matrix P is irreducible, then there is a unique positive left eigenvector $p = (p_i)$ to the eigenvalue 1 with $\sum_i p_i = 1$. Define μ by $\mu([a_0,a_1,\dots,a_k]) = p_{a_0}\cdot p_{a_1,a_2}\cdots$ $\cdots p_{a_{k-1}a_k}$.

Having a shift measure space we can introduce the so-called locally finite dimensional transformations. A transformation S is called locally finite dimensional if for a.e. $x \in X$ the two points x and Sx differ in at most finitely many coordinates. These transformations form a group G. Let G_0 be the subgroup of G consisting only of measure preserving transformations of G.

Now we replace Y by a finite dimensional matrix algebra M_0 where the dimension of M_0 equals the cardinality of Y, the cartesian product by a tensor product and lift the measure μ to the tensor product.

So let $M_{n+1} = M_0 \otimes M_n \otimes M_0$ with canonical embedding $M_n \hookrightarrow M_{n+1}$. Let $a_1,a_2,\dots$ be a set of minimal projections of M_0 with $\sum_i a_i = 1$. Then define the state ω_n on M_n with the following density matrix ρ_n:

$$\rho_n = \sum_{(x_{-n},\dots,x_n)\in Y^{2n+1}} \mu([x_{-n},\dots,x_n])\cdot a_{x_{-n}} \otimes \dots \otimes a_{x_n},$$

i.e. $\omega(a) = \mathrm{Tr}\, \rho a$ for arbitrary $a \in M$. Because of the properties of the measure μ we have $\omega_{n+1}|_{M_n} = \omega_n$. So we can perform the inductive limit $\varinjlim_{n\to\infty} (M_n,\omega_n)$ and after a GNS representation we get a von Neumann algebra M together with a faithful state ω.

To describe the von Neumann algebra M we can use the groups G and G_0.

Proposition 2.3

M is a factor if and only if G acts ergodically on X. If M is a factor and $G = G_0$ then M is of type II_1, otherwise it is of type III.

Sketch of Proof

G can be generated by a single transformation [4]. Then M is isomorphic to the crossed product of $L^\infty(X,\mu)$ with this transformation. See [7] for details.

In the type III situation, one can often find a shift invariant subalgebra isomorphic to the hyperfinite factor of type II_1.

Proposition 2.4

If G_0 acts ergodically on X, then the centralizer $Z_\omega(M)$ is isomorphic to the hyperfinite factor of type II_1.

Proof

As in the above proof $Z_\omega(M)$ is isomorphic to the crossed product of $L^\infty(X,\mu)$ with a transformation generating G_0. □

Now we want to assume that the assumptions of proposition 2.4 are fulfilled and study the mixing properties.

Note that Bernoulli and mixing Markov shifts have ergodic groups of measure preserving locally finite dimensional transformations by [5].

Proposition 2.5

Assume that T is a shift transformation on a measure space X having an ergodic group G_0 of measure preserving locally finite dimensional transformations. Then T is ergodic.

Proof

Assume that T is not ergodic. Then there is a set A with $0 < \mu(A) < 1$ and $TA = A$. We show that no cylinder set is included in A which contradicts $\mu(A) > 0$. There is a locally finite dimensional transformation $U \in G_0$ with $\mu(UA \cap \bar{A}) > 0$. Now let $B = A \cap U^{-1}\bar{A}$. Then B is a subset of A which is mapped into $\bar{A}$ and one can find a subset C of B of positive measure such that U restricted to C changes no more than, let's say, N coordinates. By taking a larger N if necessary we can assume that C is a cylinder of the form $[x_{-N},\dots,x_N]_{-N}$. Now consider the sets $C_k = T^{(2N+1)k} C$, $k \in Z$. They cannot be all disjoint, let $\mu(C_k \cap C_{k'}) > 0$. Now $T^{(2N+1)k}\, U\, T^{-(2N+1)k}(C_k \cap C_{k'}) \subset C_{k'}$ because only C_k-coordinates are changed. On the other hand, $T^{(2N+1)k}\, U\, T^{-(2N+1)k}\, C_k \subset \bar{A}$, a contradiction to $\mu(C_k \cap C_{k'}) > 0$. □

Proposition 2.6

Assume T satisfies the assumptions of proposition 2.4. Then T has positive entropy.

Sketch of Proof

As in the proof of Prop. 2.5 we choose a set $A \subset X$ and a transformation $U \in G_0$ such that for $x \in A$ the points x and Ux differ only in coordinates between 0 and N. If T would have zero entropy then the knowledge of the coordinates x_k for all $k < 0$ of some point $x \in A$ is with probability 1 enough to determine any finite number of coordinates $x_0, x_1, \dots$. Then U had to be the identity, a contradiction. □

Connes and Størmer introduced in [2] entropy for automorphisms of type II_1 von Neumann algebras. It was shown in [6] that in the above construction, the Connes-Størmer entropy equals the classical Kolmogorov-Sinai entropy of the corresponding shift transformation T. So the above proposition says that the shift automorphisms considered here have positive entropy.

Now we want to compare mixing properties.

Theorem 2.7

Assume T satisfies the assumptions of proposition 2.4. Then the shift automorphism α is ergodic, weakly mixing or mixing if and only if T has the same property.

Proof

We will prove the theorem only in the case of ergodicity, the other proofs are similar. Assume that T is ergodic. We show that

$$\lim_{n\to\infty} \frac{1}{n} \sum_{k=0}^{n-1} \tau(a\alpha^k(b)) = \tau(a)\,\tau(b) \quad \text{for } \tau = \text{trace and arbitrary } a,b \in Z_\omega(M).$$

Choose sequences $a_k, b_k \in M_k$ such that $a_k \to a$ and $b_k \to b$. Take a small $\varepsilon > 0$, then there is a k such that

$$|\tau(a\,{}^{n}(b)) - \tau(a_k\alpha^n(b_k))| < \varepsilon \quad \text{for all } n$$

and

$$|\tau(a)\tau(b) - \tau(a_k)\tau(b_k)| < \varepsilon .$$

The first inequality implies

$$\left|\frac{1}{N}\sum_{n=0}^{N-1} \tau(a\alpha^n(b)) - \frac{1}{N}\sum_{n=0}^{N-1} \tau(a_k\alpha^n(b_k))\right| < \varepsilon .$$

Now,

$$\left|\frac{1}{N}\sum_{n=0}^{N-1} \tau(a\alpha^n(b)) - \tau(a)\tau(b)\right| \leq$$

$$\leq \left|\frac{1}{N}\sum_{n=0}^{N-1} \tau(a\alpha^n(b)) - \frac{1}{N}\sum_{n=0}^{N-1} \tau(a_k\alpha^n(b_k))\right| + \left|\frac{1}{N}\sum_{n=0}^{N-1} \tau(a_k\alpha^n(b_k)) - \tau(a_k)\tau(b_k)\right|$$

$$+ |\tau(a_k)\tau(b_k) - \tau(a)\tau(b)| .$$

The first and the last term of the right hand side of the above inequality are smaller than ε, and the sbcond term tends to zero as follows: Let A be the abelian subalgebra of $Z_\omega(M)$ generated by the projections $a_{x_{-n}} \otimes \ldots \otimes a_{x_n} \in M_n$ used to define the density matrices ρ_n. (This A is canonically isomorphic to $L^\infty(X,\mu)$.) Let E be the faithful conditional expectation E: $Z_\omega(M) \to A$. Then $E(a_k\alpha^n(b_k)) = E(a_k)\ E(\alpha^n(b_k))$ for $n > 2k$ and so $\tau(a_k\alpha^n(b_k)) = \tau(a_k)\tau(b_k)$ for $n > 2k$.

For the converse, assume that α is ergodic. Restrict α to $A \simeq L^\infty(X,\mu)$. Then α is induced by the point transformation T and the only invariant projections of A are 0 and 1, so the T-invariant sets of X have measure 0 or 1 and consequently T is ergodic. □

Corollary

The shift automorphism α is ergodic and properly outer.

Proof

Combine theorem 2.7 and the propositions 2.5 and 1.2. □

For K-systems, we have a version of theorem 2.7 only if the subalgebra N of M has some special form.

Theorem 2.8

Let C be the σ-algebra of X generated by cylinders of the form $[a_o,\ldots,a_n]_o$. Similarly let N be the von Neumann subalgebra of M generated by the operators $x_{-n} \otimes x_{-n+1} \otimes \ldots \otimes x_n \in M_n$ with $x_{-n} = \ldots = x_{-1} = 1$. Then (X,C,T) is a classical K-system if and only if (M,N,α) is a K-system. Any such K-system satisfies the expectation property.

Proof

Assume (X,C,T) is a K-system. The only nontrivial property to show is $\bigcap_{n=0}^{\infty} \alpha^{-n}(N) = \mathbf{C}$. Note that the subalgebra $A \subset M$ is maximal abelian in M, and $A \cap N$ is maximal abelian in N. So $\bigcap_{n=0}^{\infty} \alpha^{-n}(A \cap N)$ is maximal abelian in $\bigcap_{n=0}^{\infty} \alpha^{-n}(N)$, but $\bigcap_{n=0}^{\infty} \alpha^{-n}(A \cap N) = \mathbf{C}$ because (X,C,T) was a K-system. So (M,N,α) is a K-system.

For the converse, assume that (M,N,α) is a K-system. Apply the conditional expectation $E: M \to A$ to see that (X,C,T) is a K system. A conditional expectation $E_N: M \to N$ always exists and can be given for the generating operators as follows: Let $x = x_{-n} \otimes \ldots \otimes x_n \in M_n$. Then $x = \bar{x}_N \cdot x_N$ with $\bar{x}_N = x_{-n} \otimes \ldots \otimes x_{-1} \otimes 1 \otimes \ldots \otimes 1$ and $x_N = 1 \otimes \ldots \otimes 1 \otimes \otimes x_o \otimes \ldots \otimes x_n$. Then

$$E_N(x) = \begin{cases} x_N \dfrac{\omega(x)}{\omega(x_N)} & \text{if } \omega(x_N) \neq 0 \\ x_N\, \omega(\bar{x}_N) & \text{if } \omega(x_N) = 0 . \end{cases}$$ □

3. Examples

The first two examples shall demonstrate the different mixing properties. In the third example, the automorphism will not be a shift but still induced by a measure space transformation and it is shown that in this case theorem 2.7 is no longer true.

Example 1

Mixing Markov shifts. A Markov shift is mixing if there is a natural number n such that the n-th power of the transition matrix has only non-zero entries. Such a mixing Markov shift is a K-transformation and the σ-algebra C can be chosen as in theorem 2.8 [10]. So every mixing Markov shift gives rise to a K-automorphism satisfying the expectation property.

Example 2

Ergodic, but non-mixing automorphisms. We can choose a four point state space and the transition matrix as

$$P = \begin{pmatrix} 0 & 1/2 & 0 & 1/2 \\ 1/2 & 0 & 1/2 & 0 \\ 0 & 1/2 & 0 & 1/2 \\ 1/2 & 0 & 1/2 & 0 \end{pmatrix}$$

This gives rise to an ergodic, but not mixing automorphism.

Example 3

An ergodic transformation inducing a non-ergodic automorphism. Take $Y = \{0,1\}$ with symmetric probability measure and $X = Y^{\mathbb{Z}}$ with product measure. Note that for a.e. x and $y \in X$ the element $x + y$ (where + denotes coordinatewise addition with carrying) is well defined. Fix $z \in X$ such that the points z, $z + z$, $z + z + z$, ... form a dense set in X in the natural topology [3]. Then the transformation T_z defined by $T_z x = x + z$ is ergodic and gives rise to an automorphism α_z on the corresponding von Neumann algebra. Now consider the transformation T_0 defined by $T_0 x = x + z_0$ with $z_0 = (\ldots 0,0,1,0,0,\ldots)$. Then T_0 is a locally finite dimensional transformation and induces an inner automorphism Ad u on the von Neumann algebra. But, T_z and T_0 commute, so $\alpha_z(u) = u$ and α_z is not ergodic.

References

[1] Connes, A.: Une classification des facteurs de type III, Ann. Sci. Ec. Norm. Sup. 6 (1973) p. 133. 252.

[2] Connes, A. and Størmer, E.: Entropy for automorphisms of II_1 von Neumann Algebras, Acta Math. 134 (1975) p. 289 - 306.

[3] Denker, M., Grillenberger, Ch. and Sigmund, K.: Ergodic Theory on Compact Sets, Springer Lecture Notes in Math. 527, 1976.

[4] Krieger, W.: On non-singular transformations of a measure space I, Zeitschrift für Wahrscheinlichkeitstheorie u. verw. Geb. 11 (1969) p. 83 - 97.

[5] Krieger, W.: On the finitary isomorphisms of Markov shifts that have finite expected coding time, Zeitschrift für Wahrscheinlichkeitstheorie u. verw. Geb. 65 (1983) p. 323 - 328.

[6] Quasthoff, U.: Shift automorphisms of the hyperfinite factor, Math. Nachrichten 131 (1987) p. 101 - 106.

[7] Quasthoff, U.: On autmorphisms of factors related to measure space transformations, in preparation.

[8] Ruelle, D.: Statistical Mechanics, Benjamin, New York, 1969.

[9] Schröder, W.: W^*-K-Systems, Thesis, Tübingen 1983.

[10] Walters, P.: An Introduction to Ergodic Theory, Springer 1982.

FIRST EXIT TIME:
A THEORY OF STOPPING TIMES IN QUANTUM PROCESSES

Jean-Luc Sauvageot
CNRS, Laboratoire de Probabilités
tour 56, 4 place Jussieu, F-75230 Paris Cedex 05

INTRODUCTION.

This paper is the second step of a threefold work, whose ultimate aim is to solve the Dirichlet problem in C^*-algebras, by means of quantum processes methods.

The first step [6] was dilation theory: how to associate a non commutative Markov process to a Markov semigroup of completely positive maps on a C^*-algebra. This one is devoted to the powerful tool which allows to link the (non commutative Borel) structure provided by the quantum process, with the (topological) structure given by the C^*-algebra: the so-called "first exit time" from an open set (that is: from a closed ideal).

We do not claim that we have developped the most general theory of stopping times; other papers in those Proceedings provide theories of their own, in some way different and as convincing (cf. the contribution of K.R.Parthasarathy and Sinha). Ours is practical, and its stopping times are fitted to their job, which is to stop time evolutions in stochastic processes.

The first three section are a presentation of the theory: definition of a stopping time, conditional expectation associated with it, and which kind of time evolutions it is able to stop.

The two last ones are devoted to the specific example of the "first exit time". the data are:- a C^*-algebra A,

- a closed ideal I in A,
- a pointwise norm continuous semigroup $(\phi_t)_{t\geq 0}$ of (Markov) completely positive maps from A into itself, with infinitesimal generator Δ.

Section 4 is the construction of a (W^*-) covariant Markov stochastic process which dilates the semigroup, and satisfies some centrality properties which allow us to build the "first exit time". This construction is an adaptation (and a complication) of the one made in [6], and we give only a sketch of the proof. The last section is the study of the properties of the time evolution of the elements of A, when it is stopped by the "first exit time", and of what you get when you read it at time zero.

The main result (proposition 5.5) must be understood as the construction of a "Dirichlet kernel": under a rather weak locality assumption, one obtains a canonical completely positive map Λ from the quotient algebra A/I into the envelopping von Neumann algebra I^{**} of the ideal I, and this map is a nice substitute of the one introduced by the classical theory as a candidate for the solution of the Dirichlet problem (cf. [2], chapters XII and XIII).

In other words, we have made half of the way: to any element α of the quotient algebra A/I ("boundary values") is associated a "non commutative Borel function" $\Lambda(\alpha)$ in I^{**}. Forthcoming work will be for answering those two questions :

1/ Under which conditions is $\Lambda(\alpha)$ "harmonic", that is a solution of the differential equation "$\Delta(x) = 0$" localized on I ?

2/ When do α and $\Lambda(\alpha)$ glue together in order to recover an element of A ? i.e. when does there exist a completely positive lifting from A/I into A which solves the Dirichlet problem ?

1.A DEFINITION OF STOPPING TIMES.

1.1.Notations.

We start with a quantum process $\{M,(M_t)_{t\geq 0},(E_t)_{t\geq 0}\}$ indexed by the half real line $\mathbb{R}_+$:

- M is a von Neumann algebra;
- for each $t\geq 0$, E_t is a conditional expectation of M onto M_t ($E_t(1_M)=1_M$);
- [filtration property] for $s\leq t$, one has $M_s \subset M_t$ and $E_s E_t = E_s$.

No continuity, nor even measurability condition is assumed; the conditional expectations are onto, but not necessarily faithful.

S_0 is the support of E_0, that is the smaller projection p in M satisfying $E_0(p) = 1_M$. An element x of M satisfies $E_0(x^*x) = 0$ if and only if $x.S_0 = 0$; such an element will be considered as null with respect to E_0 .

1.2.Definition.

A stopping time will be an increasing family $\{Q(t)\}_{t\geq 0}$ of projections in M, satisfying the property:

$$Q(t) \in M_t \ , \ \forall t\geq 0.$$

This definition is more general than the classical one, which requires left continuity of the family; but it is easy to associate to our stopping time a true spectral measure $\mathcal{Q}$ on the extended real line $\overline{\mathbb{R}}_+$, characterized by

$$\mathcal{Q}([0,t]) = Q^+(t) = \lim_{s\downarrow t} Q(s), \forall t\geq 0; \ \mathcal{Q}([0,t[) = Q^-(t) = \lim_{s\uparrow t} Q(s), \forall t>0.$$

The spectral measure can have non zero weight at infinity:

$$\check{Q}(\{+\infty\}) = 1 - Q(+\infty) \text{ , where } Q(+\infty) = \text{increasing limit}_{t\to+\infty} Q(t).$$

1.3. Algebras associated with a stopping time.

1.3.1. Put $M \wedge Q' = \{x \in M \,/\, xQ(t) = Q(t)x, \ \forall t \geq 0\}$

$M_Q = \{x \in M \wedge Q' \,/\, xQ(t) \in M_t, \ \forall t \geq 0\}$.

1.3.2. A subdivision of $\mathbb{R}_+$ will be an increasing sequence $\mathcal{S} = \{t_\circ = 0 < t_1 < \ldots < t_n < \ldots\}$ in $\mathbb{R}_+$ with $\lim_{n\to\infty} t_n = +\infty$.

To such a subdivision will be associated:

- the von Neumann algebra $M_Q^{\mathcal{S}} = \{ x \in M \wedge Q' \,/\, xQ(t) \in M_t, \ \forall t \in \mathcal{S} \}$
- the conditional expectation $E_Q^{\mathcal{S}}$ from $M \wedge Q'$ onto $M_Q^{\mathcal{S}}$ defined by, $\forall\, x \in M \wedge Q'$:

$$E_Q^{\mathcal{S}}(x) = E_\circ(x)Q(0) + \sum_{n\geq 0} E_{t_{n+1}}(x)[Q(t_{n+1}) - Q(t_n)] + x[1 - Q(+\infty)]$$

$$= E_\circ[xQ(0)] + \sum_{n\geq 0} E_{t_{n+1}}[x(Q(t_{n+1}) - Q(t_n))] + x[1 - Q(+\infty)] .$$

The following lemma has obvious proof:

1.4. Lemma.

With the above notations, one has

(i) $M_Q \subset M_Q^{\mathcal{S}}$ for any subdivision $\mathcal{S}$, and $M_Q = \cap \{M_Q^{\mathcal{S}} \,/\, \mathcal{S} \text{ subdivision of } \mathbb{R}_+\}$;

(ii) for any subdivision $\mathcal{S}$, any x in $M \wedge Q'$,

$$E_\circ(E_Q^{\mathcal{S}}(x)) = E_\circ(x) \ ;$$

(iii) if $\mathcal{S}'$ is a refinement of $\mathcal{S}$ (i.e. $\mathcal{S} \subset \mathcal{S}'$), then $M_Q^{\mathcal{S}'} \subset M_Q^{\mathcal{S}}$ and

$$E_Q^{\mathcal{S}'} E_Q^{\mathcal{S}} = E_Q^{\mathcal{S}} E_Q^{\mathcal{S}'} = E_Q^{\mathcal{S}'} .$$

2. CONDITIONAL EXPECTATION ASSOCIATED WITH A STOPPING TIME.

2.1. Proposition.

There exists a projection Σ_Q in the center of M_Q such that:

(i) $E_\circ(\Sigma_Q) = 1_M$.

(ii) for any x in $M \wedge Q'$, the net $\{E_Q^{\mathcal{S}}(x).\Sigma_Q\}$ converges $*$-σ-strongly to a limit $E_Q(x)$ in M_Q, along the filter of all subdivisions of $\mathbb{R}_+$.

(iii) the map $x \to E_Q(x)$ is a conditional expectation from $M \wedge Q'$ into (and not necessarily onto) M_Q such that $E_Q(1_M) = \Sigma_Q$, and $E_\circ(E_Q(x)) = E_\circ(x)$, $\forall x \in M \wedge Q'$.

Proof.

By Stinespring's theory of completely positive maps [7], one can suppose that the von Neumann algebra M is realized in a Hilbert space $\mathcal{H}$ where exists an increasing family of projections $\{P_t\}_{t\geq 0}$ such that:

$$\forall\, t \geq 0,\ P_t \in M_t' \ ;\ \text{central support of } P_t \text{ in } M_t = 1_M \ ;\ P_t x P_t = E_t(x) P_t, \forall x \in M.$$

For $S = \{t_0 = 0 < t_1 < \ldots < t_n < \ldots\}$ a subdivision of $\mathbb{R}_+$, put

$$P_Q^S = P_0 Q(0) + \sum_{n\geq 0} P_{t_{n+1}} [Q(t_{n+1}) - Q(t)] + [1 - Q(+\infty)] \ .$$

Then P_Q^S is a projection in $\mathcal{H}$ which commutes with M_Q^S, and such that:

$$P_Q^S \times P_Q^S = E_Q^S(x)\, P_Q^S \ , \quad \forall\, x \in M \wedge Q' .$$

Moreover, one has $P_Q^{S'} \leq P_Q^S$ whenever S' is a refinement of S, so that the net $\{P_Q^S\}$ converges strongly, along the filter of all subdivisions of $\mathbb{R}_+$, towards a projection P_Q of $\mathcal{H}$ which, by 1.4.(i), commutes with M_Q.

Let Σ_Q be the central support of P_Q in M_Q.

(i) As $P_0 P_Q^S = P_0$ for any subdivision S, one has $E_0(\Sigma_Q) P_0 = P_0$; P_0 having central support 1_M in M_0, $E_0(\Sigma_Q)$ is necessarily equal to 1_M.

(ii) For x in $M \wedge Q'$, let y be any weak accumulation point of the net $\{E_Q^S(x)\}$ of elements of M: then y belongs to M_Q, and $y.P_Q = \lim_S E_Q^S(x).P_Q^S$

$$= \lim_S \ P_Q^S \times P_Q^S = P_Q \times P_Q \ .$$

Thus $y.\Sigma_Q$ is uniquely determined by x, and the net

$$\{E_Q^S(x).\Sigma_Q\}$$

converges σ-weakly, along the filter of all subdivisions of $\mathbb{R}_+$, towards a limit denoted by $E_Q(x)$.

By the same argument, the net $\{E_Q^S(x)^* E_Q^S(x) \Sigma_Q\}$ converges σ-weakly to $E_Q(x)^* E_Q(x)$: hence (ii) is proved.

(iii) is an immediate consequence of (i) above, and 1.4.(ii).

2.2. Definition.

The conditional expectation E_Q from $M \wedge Q'$ into M_Q defined in proposition 2.1. will be called conditional expectation associated with the stopping time $Q(.)$.

Notice that it is defined only up to an E_0-negligible projection in the center of M_Q. Its interpretation is that, by means of the stopping time, we have stopped a time evoluting family $\{E_t(x)\}_{t\geq 0}$ in the stochastic process. Next step is to apply this stopping machine to more general time evolutions.

As E_Q is obtained as a right sided Riemann integral, one is allowed to write:

$$E_Q(x) = \int_{\mathbb{R}_+} E_t^+(x)\ dQ(t)\ .$$

(understand the event $Q(t)$ as "the part where the stopping time is smaller than t" and the fiber $dQ(t)$ as the event "the stopping time is equal to t").

3.STOPPING PROCEDURES.

3.1.Definitions.

A family $\{X_t\}_{t\geq 0}$ of elements of M, indexed by times in $\mathbb{R}_+$, will be called

- bounded if $\sup_t ||X_t|| < +\infty$;
- measurable if, for any state ω on M, the map $\mathbb{R}_+ \ni t \to \omega(X_t) \in \mathbb{C}$ is Lebesgue measurable ;
- norm continuous (resp. uniformly norm continuous) if the map $\mathbb{R}_+ \ni t \to X_t \in M$ is continuous (resp.uniformly continuous) for the norm topology of M ;
- conditionally right norm continuous (resp. uniformly conditionally right norm continuous) if, for any t in $\mathbb{R}_+$, one has

$$\lim_{s \downarrow t} || \mathbb{E}_t(X_s - X_t) || = 0$$

(resp. and this convergence is uniform in t: $\lim_{s\downarrow 0}[\sup_t ||E_t(X_{t+s}-X_t)||] = 0$).

- commuting with the past of Q if $X_t.Q(s) = Q(s).X_t$, $\forall\ s\leq t$;
- adapted with the filtration if $X_t \in M_t$, $\forall\ t \geq 0$.

We shall be satisfied if we are able to stop a time indexed family in M which is bounded, conditionally right continuous, commuting with the past of Q and adapted with the filtration. One should be able to go further, but we need no more.

3.2.Notation.

Let $\{X_t\}_{t\geq 0}$ a bounded family in M commuting with the past of Q, and $\mathcal{S}$ a subdivision of $\mathbb{R}_+$, $\mathcal{S} = \{t_0=0<t_1<\dots<t_n<\dots\}$; then one puts

$$X_Q^{\mathcal{S}} = X_0 Q(0) + \sum_{n\geq 0} X_{t_{n+1}} [Q(t_{n+1})-Q(t_n)]\ .$$

Let us begin now with the most trivial case:

3.3. Lemma.

Let $\{X_t\}_{t\geq 0}$ be a bounded norm continuous family in M, commuting with the past of Q.

(i) The net of the Riemann sums $\{X_Q^S\}$ converges $*$-σ-strongly to a limit

$$X_Q = \int_0^\infty X_t^+ \, dQ(t)$$

along the filter of all subdivisions of $\mathbb{R}_+$.

(ii) For any fixed T>0, the net $\{X_Q^S \, Q(T)\}$ converges uniformly to its limit $X_Q Q(T)$ when the diameter of the subdivision S tends to zero.

(iii) If moreover the family $\{X_t\}_{t\geq 0}$ is uniformly norm continuous, the convergence in (i) above is norm convergence when the diameter of the subdivision tends to zero.

We omit the proof which is more or less trivial.

3.4. Lemma.

Let $\{X_t\}_{t\geq 0}$ be as above. Then

(i) for any subdivision $S = \{t_\circ=0<t_1<\ldots<t_n<\ldots\}$ or $\mathbb{R}_+$, $E_Q^S(X_Q^S)$ is equal to the Riemann sum

$$E_\circ(X_\circ)Q(0) + \sum_{n\geq 0} E_{t_{n+1}}(X_{t_{n+1}})[Q(t_{n+1})-Q(t_n)] \quad ;$$

(ii) The net $\{E_Q^S(X_Q^S).\Sigma_Q\}$ converges $*$-σ-strongly to $E_Q(X_Q)$ along the filter of all subdivisions of $\mathbb{R}_+$;

(iii) one has $\quad ||E_Q(X_Q)|| \leq \sup_t ||E_t(X_t)|| \quad .$

This lemma is to be understood as:

$$E_Q[\int_0^\infty X_t^+ \, dQ(t)] = \int_0^\infty E_t(X_t)^+ dQ(t) \quad .$$

Proof.

(i) is obvious. (ii) is a straightforward consequence of (i) above, proposition 2.1. and lemma 3.3. (iii) comes from (i) and (ii).

The last result is precisely the one we had in mind for our example. One could go much further assuming for instance that Q(.) is actually a spectral measure: the most regular is the stopping time, the less has to be the evoluting family one wants to stop. The rough definition we have chosen will avoid unuseful sophistication in the sequel, and allow us to work in non continuous quantum stochastic processes.

3.5. Proposition.

Let $\{x_t\}_{t\geq 0}$ be a family in M which is bounded, uniformly right continuous, commuting with the past of Q, measurable and adapted with the filtration.

Then the net of the Riemann sums $\{x_Q^S.\Sigma_Q\}$ converges $*$-σ-strongly, along the filter of all subdivisions S of $\mathbb{R}_+$, to a limit $x_Q = \int_0^\infty x_t^+ \, dQ(t)$ in M_Q.

Proof.

For $\lambda>0$, put
$$X_t(\lambda) = \int_0^\infty \lambda e^{-\lambda s} x_{t+s} ds \text{ and } x_t(\lambda) = E_t(X_t(\lambda)) = \int_0^\infty \lambda e^{-\lambda s} E_t(x_{t+s}) ds .$$

One has the obvious facts:

(a) the family $\{X_t(\lambda)\}_{t\geq 0}$ is bounded, uniformly norm continuous;

(b) $E_Q^S(X_Q^S(\lambda)) = x_Q^S(\lambda)$ for any subdivision S of $\mathbb{R}_+$;

(c) $\lim_{\lambda\to\infty}[\sup_t ||x_t(\lambda)-x_t||] = 0$.

By lemma 3.4.(iii), and (c), $\{E_Q(X_Q(\lambda))\}$ is a norm Cauchy net in M and has a norm limit when λ tends to infinity. Let x_Q be this limit, which lies in M_Q.

One has obviously $||x_Q^S(\lambda)-x_Q^S|| \leq \sup_t ||x_t(\lambda)-x_t||$; thus $x_Q^S(\lambda)-x_Q^S$ is norm convergent to zero when λ tends to $+\infty$, uniformly in S.

Moreover, by lemmas 3.3 and 3.4, and (b) above, for any fixed λ, the net $\{x_Q^S(\lambda).\Sigma_Q\}$ converges $*$-σ-strongly to $E_Q(X_Q(\lambda))$ along the net of all subdivisions of $\mathbb{R}_+$.

Combining those two facts, we get the $*$-σ-strong convergence of the net $\{x_Q^S.\Sigma_Q\}$ towards x_Q.

3.6. Remark.

With the notations of the proof above, one has:

(i) $\lim_{\lambda\to\infty} ||x_Q(\lambda)-x_Q|| = 0$;

(ii) $x_Q(\lambda) = E_Q(X_Q(\lambda))$, $\forall\ \lambda>0$, and, by 2.1.3
$$\lim_{\lambda\to\infty} E_\circ[X_Q(\lambda)-x_Q] = 0 .$$

4. THE CENTRAL DILATION THEOREM.

4.1. Notations.

From now on, we fix a separable C^*-algebra A and a pointwise norm continuous Markov semigroup on A, that is a semigroup $\{\phi_t\}_{t\geq 0}$ [ϕ_0=identity of A; $\phi_s \circ \phi_t = \phi_{s+t}$] of completely positive maps from A into itself such that:

- for any a in A, the map $t \to \phi_t(a)$ is continuous;
- for any state ω on A, any $t\geq 0$, $\omega \circ \phi_t$ is a state on A .

When A has a unit, this last condition is equivalent to $\phi_t(1_A)=1_A$; otherwise, it means that any ϕ_t extends to a unit preserving completely positive map from $\tilde{A}$ into itself.

Δ denotes the infinitesimal generator of the semigroup:
$\Delta(a) = \lim_{t\downarrow 0} \frac{1}{t}[\phi_t(a)-a]$ for a in its domain $\mathbb{D}(\Delta) = \{a\in A/\text{the limit exists}\}$, which is a dense subspace of A.

By double duality, the semigroup extends to a semigroup $\{\overline{\phi}_t\}_{t\geq 0}$ of normal, unit preserving, completely positive maps on the envelopping von Neumann algebra A^{**}.

In order to apply the stopping time theory, the first step is to build a covariant Markov process which dilates the semigroup, and preserves the center of A^{**} :

4.2. Proposition.

There exists a quantum process $\{M,(M_t)_{t\geq 0}, (E_t)_{t\geq 0}\}$ and a semigroup $\{\sigma_t\}_{t\geq 0}$ of normal $*$-endomorphisms of the algebra M [σ_0=identity of M; $\sigma_t(1_M)=1_M, \forall t\geq 0$; $\sigma_s \circ \sigma_t = \sigma_{s+t}, \forall s,t\geq 0$] with the following properties:

(i) [covariance] $\qquad \sigma_s \circ E_t = E_{t+s} \circ \sigma_s,\ \forall\ s,t \geq 0$.

(ii) [centrality] $\qquad$ center of $M_0 \subset$ center of M .

(iii) [dilation] $\qquad M_0 = A^{**}$ and, for any x in A^{**}, any $t\geq 0$,

$$\overline{\phi}_t(x) = E_0[\sigma_t(x)]$$

(thus $\phi_t(a) = E_0[\sigma_t(a)], \forall\ a \in A,\ \forall\ t\geq 0$).

An algebra M_t must be understood as the past of time t, i.e. the events depending only on the time interval $[0,t]$; σ_t is the time evolution towards the future: $\sigma_t(M)$ is the future of t, i.e. events depending on the time interval $[t,+\infty[$;

an element a of A evolves with the time in the algebra M: this time evolution $t \to \sigma_t(a)$ is adapted both to the past filtration $\{M_t\}_{t\geq 0}$ and to the future filtration $\{\sigma_t(M)\}_{t\geq 0}$, and is uniformly conditionally norm continuous:

$$\begin{aligned} E_t[\sigma_{t+s}(a)-\sigma_t(a)] &= E_t[\sigma_t(\sigma_s(a)-a)] \\ &= \sigma_t[E_\circ(\sigma_s(a)-a)] && \text{by 4.2.(i)} \\ &= \sigma_t(\phi_s(a)-a) && \text{by 4.2.(iii)} \end{aligned}$$

which, by pointwise norm continuity of the semigroup, is norm convergent to zero uniformly in t.

Sketch of a proof of 4.2.

If there were not the centrality condition (ii), this proposition would be just a corollary of theorem 3.1. in [6]. In view of condition (ii), the whole proof of [6]is to be written again with suited modifications, the main ones being as follows:

instead of an auxiliary state ω on A, fix an auxiliary conditional expectation ε from A^{**} onto its center Z ;

write again whole section 1 of [6], replacing the state ω by ε, the tensor products of Hilbert spaces by relative tensor products (with respect to Z: cf.[5] , [1]), and norm closures of algebras by σ-weak closures; then you get the following result:

4.2.1. To any normal, unit preserving, completely positive map ψ from a von Neumann algebra N into A^{**} is canonically associated a von Neumann algebra $N *_\psi^\varepsilon A^{**}$, generated by two faithful normal representations $\{N \ni y \to y*1_{A^{**}} \in N*_\psi^\varepsilon A^{**}\}$ and $\{A^{**} \ni x \to 1_N * x \in N*_\psi^\varepsilon A^{**}\}$, with the following properties:

(a) for any z in Z, $1_N * z$ belongs to the center of $N * A^{**}$.

(b) there exists a conditional expectation from $N *_\psi^\varepsilon A^{**}$ onto the range of A^{**} which dilates ψ ($E_\psi^\varepsilon(y*1_{A^{**}}) = 1_N * \psi(y)$, $\forall\, y \in N$), and characterized by the above property:

$\forall n\geq 1$, $\forall x_1,\dots,x_n \in A^{**}$ with $\varepsilon(x_2)=\dots=\varepsilon(x_n)=0$, $\forall y_1,\dots,y_n \in N$, then

$$E_\psi^\varepsilon[(y_n*1_{A^{**}}-1_N*\psi(y_n))(1_N*x_n)\dots(y_1*1_{A^{**}}-1_N*\psi(y_1))(1_N*x_1)] = 0 .$$

(c) if $N_\circ$ is a sub von Neumann algebra of N, and $\psi_\circ$ the restriction of ψ to $N_\circ$, then $N_\circ *_{\psi_\circ}^\varepsilon A^{**}$ is canonically isomorphic to the sub von Neumann algebra $N_\circ *_\psi^\varepsilon A^{**}$ of $N *_\psi^\varepsilon A^{**}$ generated by the ranges of $N_\circ$ and A^{**} ; $E_{\psi_\circ}^\varepsilon$ identifies with the restriction to it of E_ψ^ε .

(d) If $E: N \to N_\circ$ is a conditional expectation from N onto a sub von Neumann algebra, such that $\psi = \psi \circ E$, then there exists a canonical normal conditional expectation from $N *_\psi^\varepsilon A^{**}$ onto $N_\circ *_\psi^\varepsilon A^{**}$, denoted by E^*, satisfying

$$E_\psi^\varepsilon \circ E^* = E_\psi^\varepsilon$$

$$E^*(y * 1_{A^{**}}) = E(y) * 1_{A^{**}}, \ \forall\, y \in N$$

and characterized by the algorithmic property (cf.[6], 2.4.(iii)):
$\forall\, n\geq 1$, $\forall x_1,\dots,x_n \in A^{**}$ with $\varepsilon(x_2)=\dots=\varepsilon(x_n)=0$, $\forall\, y_1,\dots,y_n \in N$, then

$$E^*[Y_n(1_N * x_n)\dots Y_1(1_N * x_1)] = 0$$

where $Y_n = (y_n - E(y_n)) * 1_{A^{**}}$ and, for $k<n$, $Y_k = y_k * 1_{A^{**}} - 1_N * \psi(y_k)$.

From this point , you can repeat (mutatis mutandis) the proof of theorem 3.1. in [6] .

4.3. Basic properties of central covariant dilations.

4.3.1. Markov property:

For any X in M, one has $E_t[\sigma_t(X))] = \sigma_t(E_\circ(X)) \in \sigma_t(M_\circ)$: all the information contained in the future algebra $\sigma_t(M)$, read through E_t in the past algebra M_t, is contained in the present algebra $\sigma_t(M_\circ)$ of the events depending on the only time t.

The same calculation provides:

$$\sigma_t(x) \in M_t \cap \sigma_t(M), \ \forall\, x \in A^{**}, \ \forall\, t\geq 0 .$$

4.3.2. Basic computation:

$\forall\, n\geq 1$, $\forall\, x_1,\dots,x_n \in A^{**}$, $\forall\, t\leq t_1\leq\dots\leq t_n$ in $\mathbb{R}_+$,

$$E_t[\sigma_{t_1}(x_1)\dots\sigma_{t_n}(x_n)] =$$

$$\sigma_t[\overline{\Phi}_{t_1-t}(x_1\overline{\Phi}_{t_2-t_1}(x_2\dots\overline{\Phi}_{t_{n-1}-t_{n-2}}(x_{n-1}\overline{\Phi}_{t_n-t_{n-1}}(x_n))\dots))] .$$

Proof for n=1:

$$E_t[\sigma_{t_1}(x)] = E_t[\sigma_t(\sigma_{t_1-t}(x))] = \sigma_t[E_\circ(\sigma_{t_1-t}(x))] = \sigma_t[\overline{\Phi}_{t_1-t}(x)] .$$

For greater n, iterate this calculation:

$$\begin{aligned} E_t[\sigma_{t_1}(x_1)\dots\sigma_{t_n}(x_n)] &= E_tE_{t_{n-1}}[\sigma_{t_1}(x_1)\dots\sigma_{t_n}(x_n)] \\ &= E_t[\sigma_{t_1}(x_1)\dots\sigma_{t_{n-1}}(x_{n-1})\, E_{t_{n-1}}(\sigma_{t_n}(x_n))] \\ &= E_t[\sigma_{t_1}(x_1)\dots\sigma_{t_{n-2}}(x_{n-2})\sigma_{t_{n-1}}(x_{n-1}\overline{\Phi}_{t_n-t_{n-1}}(x_n))] \ \text{etc.} \end{aligned}$$

In other words, when you deal with ordered times, computations of expected values are intrinsic, i.e. do not depend on the choice of a central dilation.

4.3.3. Commutation properties.

Let p be a projection in the center of A^{**}: as an element of M, p is central (4.2.(ii)); for any $s \geq 0$, $\sigma_s(p)$ lies in the center of $\sigma_s(M)$, and thus commutes with $\sigma_t(M)$ for any $t \geq s$. This implies:

a/ for any s and t, $\sigma_s(p)$ commutes with $\sigma_t(p)$;

b/ for any $t \geq 0$, the projection $Q_p(t) = \bigvee_{s\in[0,t]} \sigma_s(p)$ lies in $M_t \cap \sigma_t(M)'$: the increasing family of projection $\{Q_p(t)\}_{t \geq 0}$ is a stopping time of which we can say that it commutes with the future.

4.4. A remark about measurability.

Let us come back to the proof of proposition 4.2: the construction given is purely algebraic and the process we get is certainly not continuous. However, all computations in it are explicit (though difficult to write) and , for any a in A, the time evolution $t \to \sigma_t(a)$ is σ-weakly Borel.

5. FIRST EXIT TIME.

5.1. Notations.

In addition to the C^*-algebra A and the Markov semigroup $\{\phi_t\}_{t \geq 0}$ considered in 4.1., let us fix:

- a central dilation $\{M, (M_t)_{t \geq 0}, (E_t)_{t \geq 0}, (\sigma_t)_{t \geq 0}\}$ of the semigroup, with $M_\circ = A^{**}$, provided by proposition 4.2.
- a closed bilateral ideal I in A:

$e(I)$ will be the support of I in A^{**}, i.e. the smaller projection in A^{**} such that $b.e(I) = b$, $\forall\, b \in I$;

$f(I) = 1_{A^{**}} - e(I)$ is the co-support of I, i.e. the bigger projection in A^{**} such that $b.f(I) = 0$, $\forall\, b \in I$.

The crucial fact is that e(I) and f(I) belong to the center of A^{**}.

By obvious identifications, I^{**} identifies with the reduced algebra $(A^{**})_{e(I)}$

$(A/I)^{**}$ identifies with the reduced algebra $(A^{**})_{f(I)}$

and we shall consider $f(I)$ as the support of the quotient algebra A/I in A^{**} .

By 4.3.3. above, the family of projections $\{\sigma_t[f(I)]\}_{t\geq 0}$ is commutative, and, for $t\geq 0$, we can define $Q_I(t) = \bigvee_{s\in[0,t]}\sigma_s[f(I)]$, the supremum of all projections $\sigma_s[f(I)]$ when s varies between 0 and t. The interpretation of the "event" $Q_I(t)$ is

"the support of the quotient algebra A/I has been reached at least once during the time interval [0,t] " , or

"the 'open subset' I of A has been left at least once during the time interval [0,t]" .

5.2.Lemma and definition.

The family of projections $\{Q_I(t)\}_{t>0}$ in M is a stopping time which will be called first exit time from the ideal I. It has the following properties:

(i) $Q_I(t) \in M_t \cap \sigma_t(M)'$, $\forall$ $t\geq 0$: i.e. the stopping time commutes with the future;

(ii) $Q_I(0) = f(I)$

(iii) $Q_I(t) - Q_I(s) = (1 - Q_I(s))\,\sigma_s[Q_I(t-s)]$, $\forall$ $s\leq t$;

(iv) $E_\circ(1-Q_I(t)) = \inf\,\{e\bar{\phi}_{t_1}(e\bar{\phi}_{t_2-t_1}(e\ldots\bar{\phi}_{t_n-t_{n-1}}(e)\ldots))$, $\forall$ $t>0$,

where $e = e(I)$, and the infimum is taken on all subdivisions $\{t_\circ=0<t_1<\ldots<t_n=t\}$ of the interval $[0,t]$ (the expression in the right hand side being decreasing with the refinement of subdivisions).

Proof.

(i) is 4.3.3 and (ii) is obvious.

For (iii), write $Q_I(t) = Q_I(s)\vee\sigma_s[Q_I(t-s)] = Q_I(s) + (1-Q_I(s))\sigma_s[Q_I(t-s)]$.

(iv) $E_\circ(1-Q_I(t))$ is the decreasing limit, along the net of all finite subdivisions $\{t_\circ=0<t_1<\ldots<t_n=t\}$ of $[0,t]$, of the $E_\circ[e\sigma_{t_1}(e)\ldots\sigma_{t_n}(e)]$: apply 4.3.2.

5.3.Proposition.

(i) For any a in A, the net of the Riemann sums

$$\sigma_{Q_I}(a).\Sigma_{Q_I} = [a.f + \sum_{n\geq 0}\sigma_{t_{n+1}}(a)\left(Q_I(t_{n+1})-Q_I(t)\right)]\Sigma_{Q_I}$$

(where Σ_{Q_I} is the final support of the conditional expectation E_{Q_I} defined in 2.1) converge $*$-σ-strongly, along the filter of all subdivisions of $\mathbb{R}_+$, to a limit

$$\sigma_{Q_I}(a) = \int_0^\infty \sigma_t^+(a)\, dQ_I(t)$$

which belongs to the von Neumann algebra M_{Q_I} .

(ii) The map $a \to \sigma_{Q_I}(a)$ is a $*$-representation of A into M_{Q_I}, with final support $(1-Q_I(+\infty))\Sigma_{Q_I}$.

(iii) For any s>0, one has

$$[\sigma_s(\sigma_{Q_I}(a)) - \sigma_{Q_I}(a)](1-Q_I(s)) = 0 .$$

Proof.

(i): all the conditions of proposition 2.5 are satisfied.

(ii) The Riemann sums $\sigma^S_{Q_I}$ are $*$-algebraic homomorphisms, and they commute asymptotically with Σ_{Q_I}. For the final support: one can extend Σ_{Q_I} to $\tilde{A}$, and $\sigma_{Q_I}(1_{\tilde{A}})$ is equal to $(1-Q_I(+\infty))\Sigma_{Q_I}$.

(iii) By 5.2.(iii), we have

$$\sigma_s[Q(t_{n+1})-Q(t_n)](1-Q(s)) = Q(t_{n+1}+s)-Q(t_n+s) ,$$

and thus $\sigma_s(\sigma^S_Q(a))(1-Q(s)) = \sigma^{S'}_Q(a)(1-Q(s))$ for any pair S and S' of subdivisions of $\mathbb{R}_+$ such that $S'\cap[s,+\infty[= S + s$. Go to the limit.

[From now on, we write in the proofs e instead of e(I), f instead of f(I), Q instead of Q_I, and so on.]

Remark.

Notice that, in the classical case, in order to define the "first exit time", you have to suppose some locality property of the semigroup for proving that almost all orbits of the process are right continuous. We delayed this difficulty, but all the computations of the classical case are contained in the following lemma:

5.4. Lemma.

Let b in $I_+\cap D(\Delta)$ be such that $\Delta(b)$ lies again in I. Then, for any $\varepsilon>0$, one can find $\delta>0$ such that

$$E_\circ[Q_I(t)\sigma_t(b)] \leqq t\varepsilon 1_{M_\circ} , \forall\, t \leqq \delta$$

Proof.

One has $f.b = f.\Delta(b) = 0$, with $f = f(I)$.

For $\varepsilon>0$, fix $\delta>0$ such that $||\frac{1}{t}(\phi_t(b)-b)-\Delta(b)||\leqq\varepsilon$ for any $t\leqq\delta$. Then, for such a t, one has $\phi_t(b) \leqq b + t\Delta(b) + t\varepsilon$, and $\phi_t(b).f \leqq t\varepsilon f$.

For $n\geqq 1$, and $s_1,\ldots,s_n\geqq 0$ such that $s_1+\ldots+s_n=t$, compute

$$e\phi_{s_n}(b) = \phi_{s_n}(b)-f.\phi_{s_n}(b) \geqq \phi_{s_n}(b) - s_n\varepsilon f$$

$$e\bar{\phi}_{s_{n-1}}(e\phi_{s_n}(b)) \geqq e\phi_{s_n+s_{n-1}}(b)-s_n\varepsilon e\bar{\phi}_{s_{n-1}}(f)\geqq \phi_{s_n+s_{n-1}}(b)-\varepsilon(s_n+s_{n-1})[f+e\bar{\phi}_{s_{n-1}}(f)]$$

$$\ldots$$

$$e\bar{\phi}_{s_1}(e\ldots\bar{\phi}_{s_{n-1}}(e\phi_{s_n}(b))\ldots) \geqq \phi_t(b) - t\varepsilon[f+e\bar{\phi}_{s_1}(f)+\ldots+e\bar{\phi}_{s_1}(e\ldots\bar{\phi}_{s_{n-2}}(e\bar{\phi}_{s_{n-1}}(f))\ldots)]$$

As $f+e\overline{\phi}_{s_1}(f)+..+e\overline{\phi}_{s_1}(e..\overline{\phi}_{s_{n-2}}(e\overline{\phi}_{s_{n-1}}(f))..) = 1-e\overline{\phi}_{s_1}(e..\overline{\phi}_{s_{n-2}}(e\overline{\phi}_{s_{n-1}}(e))..) \leqq 1_{M_\circ}$, we have shown (applying 4.3.2):

$$E_\circ[\sigma_{s_1}(e)\sigma_{s_1+s_2}(e)..\sigma_{s_1+..+s_{n-1}}(e)\ \sigma_t(b)] \geqq \phi_t(b) - \varepsilon t 1_{M_\circ} .$$

Going to the limit along the filter of the subdivisions of $[0,t]$, we get

$$E_\circ[(1-Q(t))\sigma_t(b)] = E_\circ[(\inf_{[0,t]}\sigma_s(e))\sigma_t(b)] \geqq \phi_t(b) - \varepsilon t 1_M$$

and the result.

We are now in order to prove that our stopping time has the properties one would expect from an exit time: under a rather weak locality assumption, the stopped process $\sigma_Q(a)$ depends only on the values of a "outside I", i.e. in A/I:

5.5.Proposition.

Suppose that there exists an approximate unit $\{b_n\}_{n\in\mathbb{N}}$ in the unit ball of I_+, such that $b_n \in \mathcal{D}(\Delta)$ and $\Delta(b_n) \in I$, $\forall\ n\geq 0$. Then

(i) For any b in I , $E_\circ[\sigma_{Q_I}(b)] = 0$.

(ii) Going to the quotient, $E_\circ o\ \sigma_{Q_I}$ defines a completely positive map Λ from the quotient algebra A/I into the envelopping von Neumann algebra A^{**} such that [lifting property] $\Lambda(\alpha).f(I) = \alpha$ in $(A/I)^{**}=(A^{**})_{f(I)}$, $\forall\ \alpha \in A/I$;

[compact convergence property] for any b in I, one has

$$\lim_{t\downarrow 0}||[\overline{\Psi}_t(\Lambda(\alpha))\ \Lambda(\alpha)]b|| = 0.$$

Proof.

(i) By 5.3.(ii), $\sigma_Q=\sigma_{Q_I}$ is a $*$-representation of A, and then norm continuous. It is enough to prove $E_\circ(\sigma_Q(b))= 0$ for b one of the elements of the approximate unit . For such a b, fix $\varepsilon>0$ and $\delta>0$ provided by lemma 5.4.

For any s and t such that $0\leqq t-s\leqq\delta$, one has

$$\begin{aligned} E_s[(Q(t)-Q(s))\sigma_t(b)] &\leqq E_s[\sigma_s(Q(t-s))\sigma_t(b)] && \text{by 5.2.(iii)} \\ &= \sigma_s[E_\circ(Q(t-s)\sigma_{t-s}(b))] \\ &\leqq (t-s)\varepsilon 1_M && \text{by lemma 5.4.} \end{aligned}$$

Fix T>0 : for any subdivision $\mathcal{S} = \{t_n\}_{n\geqq 0}$ of $\mathbb{R}_+$ whose diameter is less than δ, containing $T=t_N$, one will have

$$E_\circ[\sigma_Q^{\mathcal{S}}(b)Q(T)] \leqq \sum_{n=0}^{N-1}(t_{n+1}-t_n)\varepsilon 1_M = T\varepsilon 1_M$$

At the limit, $E_\circ[\sigma_Q(b)\ Q(T)] = 0$ for any T, and one gets the result.

(ii) Complete positivity comes from 5.3.(ii), and the lifting property from 5.2.(ii).

For the last property (which is compact convergence in I^{**}), it is enough to prove $\lim_{t\downarrow 0}||[\overline{\Phi}_t(\Lambda(\alpha))-\Lambda(\alpha)]b|| = 0$ for $b=b_n$ being one of the elements of the approximate unit provided by the statement of the proposition.

Let a be any lifting of α in A, and write , applying 5.3.(iii):

$$[\overline{\Phi}_t(\Lambda(\alpha))-\Lambda(\alpha)]b = E_\circ[(\sigma_t(\sigma_Q(a))-\sigma_Q(a))]b = E_\circ[(\sigma_t(\sigma_Q(a))-\sigma_Q(a))Q(t)]b \quad .$$

By lemma 5.4, $E_\circ[Q(t)\sigma_t(b)]$ is norm convergent to zero, thus

$E_\circ[(\sigma_t(\sigma_Q(a))-\sigma_Q(a))Q(t)\sigma_t(b)]$ is norm convergent to zero when $t\downarrow 0$.

By norm continuity of the semigroup, $E_\circ[(\sigma_t(b)-b)^2] = \phi_t(b^2)-b\phi_t(b)-\phi_t(b)b+b^2$ is norm convergent to zero, so $E_\circ[(\sigma_t(\sigma_Q(a))-\sigma_Q(a))Q(t)(\sigma_t(b)-b)]$ is norm convergent to zero when $t\downarrow 0$; hence the result.

5.6. Two last remarks.

5.6.1. Computation of $\Lambda(\alpha)$ involves only expected values of events depending on ordered times, and their limits: By 4.3.2, the lifting Λ is canonical, i.e. does not depend on the arbitrary choice of a central dilation made in 5.1.

5.6.2. Compact convergence property in proposition 5.4 has an important corollary: if the semigroup has the strong Feller property (i.e. $\overline{\phi}_t(A^{**}) \subset A,\ \forall\ t>0$), then for any α in A/I, $\Lambda(\alpha)$ restricted to I^{**} is a multiplier for I.

BIBLIOGRAPHY.

[1]. A.Connes , unpublished notes on correspondances.

[2]. E.B.Dynkin , Markov processes, vol.II, Springer-Verlag, 1965.

[3]. R.L.Hudson and K.R.Parthasarathy , Construction of quantum diffusions , Quantum Probability and Applications..., Lect.Notes in Maths. N°1055, p.173-198.

[4]. K.R.Parthasarathy et Sinha, in those Proceedings.

[5]. J.-L.Sauvageot, Produits tensoriels de Z-modules et applications, Operator Algebras and their connections with topology and Ergodic Theory, Lect. notes in Maths. n°1132, 1985, p.468-485.

[6]. J.-L.Sauvageot, Markov quantum semigroups admit covariant Markov C^*-dilations, Commun.Math.Phys. 106 , 1986, p.91-103.

[7]. W.F.Stinespring , Positive functions on C^*-algebras, Proc.Amer.Math.Soc. 6, 1975, p.211-216.

A Central Limit Theorem on the Free Lie Group

Michael Schürmann and Wilhelm von Waldenfels

Institut für Angewandte Mathematik
Universität Heidelberg
Im Neuenheimer Feld 294
D-6900 Heidelberg

Abstract

We introduce a "Fourier transform" for certain linear functionals on a non-commutative free algebra. The transform of a linear functional is a function on a group G of formal exponential series which can be regarded as a free group for Lie groups. A class of hermitian, conditionally positive linear functionals on the free algebra is introduced which includes the generators of non-commutative Wiener and Poisson processes, but provides also new examples. We compute the transforms of non-commutative Poisson and Gauss functionals and show that the transform of a gaussian functional vanishes on the ideal of the Weyl relations given by its covariance matrix.We prove a central limit theorem for twice continuously differentiable functions on G, avoiding the assumption of existence of all moments.

1. Introduction

In 1971, C.D. Cushen and R.L. Hudson proved a central limit theorem for canonical pairs of observables, that is for pairs of position and momentum operators on the space $L^2(R_+)$ of square integrable functions on the positive real line [6]. This theorem is a quantum mechanical version of the classical central limit theorem for sums of independent, identically distributed random variables. The limiting functional is a non-commutative gaussian (also called quasi-free) state with covariance matrix given by the covariances of the underlying state. The main part of the proof consists of the extension of convergence on the space of Weyl operators to convergence on the space of all bounded linear operators on $L^2(R_+)$. Not concentrating on these *analytic* difficulties but on the *algebraic* aspects of non-commutative central limit theorems, it turned out that it is not necessary to require the approximating quantities to fulfill some canonical commutation relations but that these commutation relations arise automatically in the limit. An algebraic non-commutative central limit theorem for arbitrary linear functionals on a free non-commutative algebra (i. e. a tensor algebra) was proved in [7] (see [17] for an alternative proof of this theorem using moment generating functions).The limiting functional vanishes on the ideal of the canonical commutation relations given by the covariance matrix of the underlying linear functional, and thus again can be regarded as a quasi-free state. An anti-commuting (Fermion) version of [6] was given in [9], and the anti-commuting case was treated algebraically in [16]. The results of [7] can be generalized to triangular arrays of linear functionals on the tensor algebra [3]. For a central limit theorem on coalgebras which unifies the limit theorems of [3,7,16] see [15]. A generalization into another direction could be established by allowing the approximating quantities to take values in a not necessarily commutative algebra; see [11] for the analytic aspects and [1,18] for an algebraic treatment of the subject.

The main topic of this paper is the "exponential" version of [7]; cf. [12]. Let $x_1, \ldots, x_d$ be non-commuting indeterminates. For two linear functionals μ and σ on the algebra T of all polynomials in $x_1, \ldots, x_d$ define the convolution product $\mu * \sigma$ by

$$(\mu*\sigma)(x_{k_1} \ldots x_{k_r}) = (\mu\otimes\sigma)((x_{k_1}\otimes 1 + 1\otimes x_{k_1}) \ldots (x_{k_r}\otimes 1 + 1\otimes x_{k_r})).$$

Denote by μ^{*N} the N-fold convolution of μ with itself. A result of [7] states that for a normalized, centralized linear functional μ

$$\mu^{*N}\left(\frac{x_{k_1}}{\sqrt{N}} \cdots \frac{x_{k_r}}{\sqrt{N}}\right) \xrightarrow[N\to\infty]{} \gamma_Q(x_{k_1} \ldots x_{k_r})$$

where γ_Q is the non-commmutative gaussian functional with covariance matrix Q given by $Q_{km} = \mu(x_k x_m)$. If V is the real vector space spanned by $x_1, \ldots, x_d$ the products of formal exponential series

$$EXP(iv) = \sum_{n=0}^{\infty} \frac{(iv)^n}{n!}$$

with $v \in V$ form a group G which we call the *free Lie group*; cf. [4]. (Notice that in general G is *not* a Lie group!) We show in Section 3 that in some cases the "Fourier transform" $\hat{\mu}$ of a linear functional μ on T can be defined as a function on G. We prove that the convolution product of linear functionals corresponds to the pointwise product of their transforms (see also [8]) and that positive linear functionals on T correspond to positive definite functions on G. We compute the transforms of non-commutative Gauss and Poisson functionals which arise from annihilation, creation, and number operators on the Bose Fock space over $L^2(R_+)$; cf. [10]. We give a formula for a new class of hermitian, conditionally positive linear functionals on T which includes as special cases the convolution logarithms of non-commutative Gauss and Poisson functionals on T.

In Section 4 we prove that if f is a function on G with continuous derivatives up to order 2 (in a sense made precise in this paper) and such that $f(1) = 0$ and $f'(1) = 0$ then

$$f^N\left(EXP\frac{iv_1}{\sqrt{N}} \ldots EXP\frac{iv_r}{\sqrt{N}}\right) \xrightarrow[N\to\infty]{} \hat{\gamma}_Q(EXP(iv_1) \ldots EXP(iv_r))$$

where the covariance matrix Q is given by the second order derivatives of f at 1. Moreover, regarded as a linear functional on the group ring of G, the limiting function $\hat{\gamma}_Q$ vanishes on the ideal of the Weyl relations given by the skew symmetric bilinear form $Q(v,w) - Q(w,v)$. As an application of our general theorem we get a central limit theorem for twice continuously differentiable functions on a Lie group.

In the classical case (probability theory on a euclidean space) the Fourier transform of a probability measure is twice continuously differentiable in 0 if the probability measure has finite moments up to order 2. This shows that our differentiability condition on the function f corresponds to the existence of *second* moments. So in our formulation of a central limit theorem we do not assume the existence of *all* moments as was done in [7].

2. Preliminaries

Let V be a real vector space. Denote by $V_{\underline{C}}$ its complexification consisting of elements $v_1 + iv_2$, $v_1, v_2 \in V$, $i = \sqrt{-1}$. The conjugation map on $V_{\underline{C}}$ is given by

$$v = v_1 + iv_2 \mapsto v^* = v_1 - iv_2 .$$

We denote by $T(V)$ the complex tensor algebra ($\underline{C}$ = "field of complex numbers")

$$T(V) = \underline{C} \oplus V_{\underline{C}} \oplus (V_{\underline{C}} \otimes V_{\underline{C}}) \oplus \ldots$$

over $V_{\underline{C}}$. Extending the conjugation map on $V_{\underline{C}}$ to an involution on $T(V)$, the algebra $T(V)$ becomes a *-algebra. The vector space tensor product $T(V) \otimes T(V)$ is turned into a *-algebra with multiplication given by

$$(a \otimes b)(a' \otimes b') = aa' \otimes bb'$$

and involution given by

$$(a \otimes b)^* = a^* \otimes b^*;\ a,b,a',b' \in T(V).$$

We define the *-algebra homomorphisms

$$\Delta: T(V) \to T(V) \otimes T(V)$$

and

$$\delta: T(V) \to \underline{C}$$

by the equations

$$\Delta v = v \otimes \mathbf{1} + \mathbf{1} \otimes v$$

and

$$\delta v = 0;\ v \in V.$$

With this additional structure the *-algebra $T(V)$ is called a **-bialgebra* [13]. For a vector space W over a field $\underline{K}$ we denote by W^* the space of all $\underline{K}$-valued linear functionals on W. For $\mu, \sigma \in T(V)^*$ the convolution $\mu * \sigma \in T(V)$ is defined by

$$\mu * \sigma = (\mu \otimes \sigma) \circ \Delta.$$

The convolution product turns $T(V)^*$ into an associative algebra with unit δ. For all $\mu \in T(V)^*$ and $a \in T(V)$ the series

$$\sum_{n=0}^{\infty} \frac{\mu^{*n}(a)}{n!}$$

converges to a limit denoted by $(\exp_* \mu)(a)$ and defines an element $\exp_* \mu$ in $T(V)^*$; see [13]. The space $T(V)$ is an $\underline{N}$-graded *-bialgebra ($\underline{N}$ = " set of non-negative integers") with the graduation given by

$$T(V) = \bigoplus_{n \in \underline{N}} T(V)^{\otimes n}$$

where $T(V)^{\otimes 0} = \underline{C}$, and $T(V)^{\otimes n}$, $n \geq 1$, is the n-fold tensor product of $T(V)$ with itself. We denote by $\overline{T(V)}$ the complete direct sum of the vector spaces $T(V)^{\otimes n}$, i. e. $\overline{T(V)}$ consists of all (not necessarily finite) sums

$$\sum_{n \in \underline{N}} d^{(n)};\ d^{(n)} \in T(V)^{\otimes n}.$$

For

$$a_k = \sum a_k^{(n)}$$

in $\overline{T(V)}$, $k = 1,2$, define the product $a_1 a_2$ to be the element $\sum b^{(n)}$ in $\overline{T(V)}$ with

$$b^{(n)} = \sum_{n_1 + n_2 = n} a_1^{(n_1)} a_2^{(n_2)}.$$

With this multiplication $\overline{T(V)}$ is an associative algebra. If $\{x_k \mid k \in I\}$ is a vector space basis of V the algebra $T(V)$ can be regarded as the algebra of polynomials (with complex coefficients) in the non-commuting indeterminates x_k, $k \in I$, and $\overline{T(V)}$ can be regarded as the algebra of formal power series in x_k, $k \in I$. For $v \in V_{\underline{C}}$ we denote by $EXPv$ the element

$$EXPv = \sum_{n \in \underline{N}} \frac{v^{\otimes n}}{n!}$$

in $\overline{T(V)}$. We consider the subset $G(V)$ of $\overline{T(V)}$ consisting of all elements of the form

$$EXP(iv_1)\ \dots\ EXP(iv_r)$$

where r runs through all positive integers and $v_k \in V$, $k = 1, \dots, r$. With the multiplication induced by $\overline{T(V)}$ the set $G(V)$ is a group with unit element $1 = EXP(0)$ and inverse

$$(EXP(iv_1)\ \dots\ EXP(iv_r))^{-1} = EXP(-iv_r)\ \dots\ EXP(-iv_1).$$

It was proved by K.-T. Chen in 1958 that $G(V)$ is isomorphic to the following group $E(V)$. As a set $E(V)$ consists of tuples $(v_1, \dots, v_r)$ where r runs through all positive integers and $v_k \in V$, $k = 1, .., r$, with for $r \geq 2$ the additional property that v_k and v_{k+1} are linearly independent for all $k = 1, .., r - 1$. The multiplication in $E(V)$ is defined inductively by the following rules

- $(v_1, \dots, v_r)(w_1, \dots, w_m) = (v_1, \dots, v_r, w_1, \dots, w_m)$

  for $(v_1, \dots, v_r),(w_1, \dots, w_m) \in E(V)$ such that v and w are linearly independent
- $(v)(w) = (v + w)$

  for $v,w \in V$ such that v and w are not linearly independent.

A homomorphism from $E(V)$ to $G(V)$ is readily seen to be given by the mapping

$$(v_1, \dots, v_r) \mapsto EXP(iv_1)\ \dots\ EXP(iv_r).$$

It is also injective; for a proof of this non-trivial fact see [4].

3. Transforms of Gauss and Poisson functionals

a) General properties of the transformation

First we want to define the transform of a linear functional on $T(V)$ as a function on $G(V)$, which will be possible only in special cases. Assume for a moment that V is equal to the field $\underline{R}$ of real numbers. In this case $T(V)$ is equal to the (commutative) polynomial algebra $\underline{C}[x]$ in one variable x. If μ is a probability measure on $\underline{R}$ such that $t \to t^n$ is integrable for all $n \in \underline{N}$, we define a linear functional on $\underline{C}[x]$, which we again denote by μ, by

$$\mu(x^n) = \int_{\underline{R}} t^n \, \mu(dt).$$

The Fourier transform $\hat{\mu}: \underline{R} \to \underline{C}$ is defined by

$$\hat{\mu}(s) = \int_{\underline{R}} exp(ist) \, \mu(dt)$$

and is a C^∞-function. Moreover, we have the well-known relations

$$\mu(x^n) = (-i)^n \frac{d^n}{ds^n} \hat{\mu}(s) \Big|_{s=0} .$$

If $\hat{\mu}$ is a power series

$$\hat{\mu}(s) = \sum_{n=0}^{\infty} \frac{i^n \mu(x^n)}{n!} s^n$$

which we can write formally as

$$\hat{\mu}(s) = \mu(exp(isx)).$$

In the general case we proceed as follows. For $\mu \in T(V)^*$ and $a = \sum a^{(n)}$ in $\overline{T(V)}$ we write $\mu(a)$ for the limit of the series

$$\sum_{n=0}^{\infty} \mu(a^{(n)})$$

if it exists. Denote by $G(V)_\mu$ the subset of $G(V)$ consisting of all elements $(v_1, \ldots, v_r)$ in $G(V)$ such that the series

$$\sum_{n_1, \ldots, n_r \in \underline{N}} \frac{|\mu(v_1^{\otimes n_1} \otimes \ldots \otimes v_r^{\otimes n_r})|}{n_1! \ldots n_r!}$$

converges. Define the function

$$\hat{\mu}: G(V)_\mu \to \underline{C}$$

by

$$\hat{\mu}((v_1, \ldots, v_r)) = \mu(EXP(iv_1) \ldots EXP(iv_r)).$$

Lemma 3.1. Let μ, μ_1, μ_2 be elements of $T(V)^*$. Then

(i) $G(V)_{\mu_1} \cap G(V)_{\mu_2} \subset G(V)_{\mu_1 * \mu_2}$ and

$$\widehat{\mu_1 * \mu_2}(g) = \hat{\mu}_1(g)\, \hat{\mu}_2(g) \tag{1}$$

for all $g \in G(V)_{\mu_1} \cap G(V)_{\mu_2}$.

(ii) $G(V)_{\mu} \subset G(V)_{exp_* \mu}$ and

$$\widehat{exp_* \mu}(g) = exp(\hat{\mu}(g))$$

for all $g \in G(V)_{\mu}$.

Proof: (i): We have

$$(\mu_1 * \mu_2)(v_1^{\otimes n_1} \otimes \dots \otimes v_r^{\otimes n_r}) = \sum_{s_1 = 0}^{n_1} \dots \sum_{s_r = 0}^{n_r} \binom{n_1}{s_1} \dots \binom{n_r}{s_r} \mu_1(v_1^{\otimes s_1} \otimes \dots \otimes v_r^{\otimes n_r})$$
$$\mu_2(v_1^{\otimes (n_1 - s_1)} \otimes \dots \otimes v_r^{\otimes (n_r - s_r)})$$

which gives

$$\sum_{n_1, \dots, n_r \in \underline{N}} \frac{|(\mu_1 * \mu_2)(v_1^{\otimes n_1} \otimes \dots \otimes v_r^{\otimes n_r})|}{n_1! \dots n_r!} < \infty$$

for $(v_1, \dots, v_r) \in G(V)_{\mu_1} \cap G(V)_{\mu_2}$ and the validity of (1). The statement (ii) follows from (i). □

Next we need the notion of differentiability for functions on $G(V)$. Let f be a complex-valued function on $G(V)$. For $q \in \underline{N}$ and $g = (v_1, \dots, v_r) \in G(V)$ we say that f is *q-times differentiable (analytic) in the direction of g* if the function

$$f_g : \underline{R}^r \to \underline{C}$$

given by

$$f_g(t_1, \dots, t_r) = f((t_1 v_1) \dots (t_r v_r))$$

is q-times continuously differentiable (analytic) in a neighbourhood of $0 \in \underline{R}^r$. For f q-times differentiable in *every* direction $g \in G(V)$ we write

$$D^{n_1, \dots, n_r}_{v_1, \dots, v_r} f = \frac{\partial^{n_1 + \dots + n_r}}{\partial t_1^{n_1} \dots \partial t_r^{n_r}} f((t_1 v_1) \dots (t_r v_r)) \Big|_{t_1 = \dots = t_r = 0}$$

if $r, n_k \in \underline{N}$, $n_1 + \dots + n_r \leq q$, $v_k \in V$, and

$$D_{v_1, \dots, v_r} = D^{1, \dots, 1}_{v_1, \dots, v_r} f$$

if $r \leq q$, $v_k \in V$. Notice that

$$D^{n_1, \dots, n_r}_{v_1, \dots, v_r} f = D_{\underbrace{v_1, \dots, v_1}_{n_1\text{-times}}, \dots, \underbrace{v_r, \dots, v_r}_{n_r\text{-times}}} f.$$

A complex-valued function f on $G(V)$ is said to be *q-times differentiable* if f is q-times differentiable in every direction *and* if the mapping

$$D^q f: V^q \to \underline{C}$$

given by

$$D^q f(v_1, \dots, v_q) = D_{v_1, \dots, v_q} f$$

is q-linear. We say that f is *analytic* if f is analytic in every direction *and* if f is q-times differentiable for all $q \in \underline{N}$. A function f which is q-times differentiable in every direction is certainly q-times differentiable if for all finite-dimensional subspaces W of V the functions

$$f_W: W^q \to \underline{C}$$

given by

$$f_W(w_1, \dots, w_q) = f((w_1) \dots (w_q))$$

are differentiable as functions on $\underline{R}^{nq}$ where we identify W with $\underline{R}^n$, $n = \dim W$.

Now we describe how certain functions on $G(V)$ can be used as *moment generating functions* for linear functionals on $T(V)$. For the following we choose a fixed basis $\boldsymbol{B} = \{x_k \mid k \in I\}$ of V. Denote by $M \subset T(V)$ the set of all monomials in x_k, $k \in I$, also containing the empty monomial $\boldsymbol{1}$ of length 0. The set $\boldsymbol{M}$ is a vector space basis of $T(V)$. Let f be q-times differentiable in every direction for all $q \in \underline{N}$. We define the mapping

$$T(f): \boldsymbol{M} \to \underline{C}$$

by

$$T(f)(\boldsymbol{1}) = f(\boldsymbol{1})$$

and

$$T(f)(y_1 \otimes \dots \otimes y_r) = (-i)^r D_{y_1, \dots, y_r} f;\ y_1, \dots, y_r \in \boldsymbol{B}.$$

We extend $T(f)$ to a linear functional on $T(V)$ which we again denote by $T(f)$. Of course

$$T(f)(v_1 \otimes \dots \otimes v_r) = (-i)^r D_{v_1, \dots, v_r} f$$

for all $r \geq 1$, $v_k \in V$, $k = 1, \dots, r$, holds if and only if f is also q-times differentiable for all $q \in N$. The following is immediate.

Proposition 3.1.

(i) Let μ be a linear functional on $T(V)$ such that $G(V)_\mu = G(V)$. Then $\hat{\mu}$ is analytic and $T(\hat{\mu}) = \mu$.

(ii) Let f be analytic. Then $G(V)_{T(f)} = G(V)$ and $\widehat{T(f)} = f$. □

As $T(V)$ is a *-bialgebra we have the notions of positive, hermitian and conditionally positive linear functionals on $T(V)$; see [13]. According to the general definition of [13], an element $\mu \in T(V)^*$ is *positive* if $\mu(a^*a) \geq 0\ \forall a \in T(V)$, *hermitian* if $\mu(a^*) = \overline{\mu(a)}\ \forall a \in T(V)$, and *conditionally positive* if $\mu(a^*a) \geq 0\ \forall a \in T(V)$ with $\delta a = 0$.

Theorem 3.1.

(i) Let μ be a positive (hermitian, conditionally positive) linear functional on $T(V)$ and assume that $G(V)_\mu = G(V)$. Then $\hat{\mu}$ is a positive definite (hermitian, conditionally positive definite) function on $G(V)$.

(ii) Let f be a positive definite (hermitian, conditionally positive definite) function on $G(V)$ and let f be q-times differentiable in every direction for all $q \in \underline{N}$. Then $T(f)$ is a positive (hermitian, conditionally positive) linear functional on $T(V)$.

Proof: (i): Let $\mu \in T(V)^*$ be positive and let $G(V)_\mu = G(V)$. Then we have for all $r \in \underline{N}$, $v_1, \ldots, v_r \in V$,

$$\sum_{n_1, \ldots, n_r \in \underline{N}} \frac{|\mu(v_1^{\otimes n_1} \otimes \ldots \otimes v_r^{\otimes n_r})|}{n_1! \ldots n_r!} < \infty. \tag{2}$$

To see this, we can assume without loss of generality that as an element of V^r the tuple $(v_1, \ldots, v_r)$ is of the form

$$(t_1^{(1)} w_1, \ldots, t_1^{(n)} w_1, \ldots, t_s^{(1)} w_s, \ldots, t_s^{(n)} w_s)$$

for $s,n \geq 1$, $t_k^{(l)} \in \underline{R}$, where w_k and w_{k+1} are linearly independent for $k = 1, .., s-1$. We have

$$\sum_{m_1^{(1)}, \ldots, m_s^{(n)} \in \underline{N}} \frac{|\mu((t_1^{(1)} w_1)^{\otimes m_1^{(1)}} \otimes \ldots \otimes (t_1^{(n)} w_1)^{\otimes m_1^{(n)}} \otimes \ldots \otimes (t_s^{(1)} w_s)^{\otimes m_s^{(1)}} \otimes \ldots \otimes (t_s^{(n)} w_s)^{\otimes m_s^{(n)}})|}{m_1^{(1)}! \ldots m_1^{(n)}! \ldots m_s^{(1)}! \ldots m_s^{(n)}!}$$

$$= \sum_{m_1, \ldots, m_s \in \underline{N}} \frac{|\mu(w_1^{\otimes m_1} \otimes \ldots \otimes w_s^{\otimes m_s})|}{m_1! \ldots m_s!} (\,|t_1^{(1)}| + \ldots + |t_1^{(n)}|\,)^{m_1} \ldots \ldots (\,|t_s^{(1)}| + \ldots + |t_s^{(n)}|\,)^{m_s}$$

$$= \sum_{m_1, \ldots, m_s \in \underline{N}} \frac{|\mu((t_1 w_1)^{\otimes m_1} \otimes \ldots \otimes (t_s w_s)^{\otimes m_s})|}{m_1! \ldots m_s!}$$

$$< \infty.$$

Let $n \geq 1$, $z_1, \ldots, z_n \in \underline{C}$ and $g_1, \ldots, g_n \in G(V)$ be given. We can assume that there exist an $r \geq 1$ and $v_{1,k}, \ldots, v_{r,k}$, $k = 1, \ldots, n$, such that

$$g_k = (v_{1,k}) \ldots (v_{r,k}).$$

Using the rearrangement theorem for absolutely convergent series, we have

$$\sum_{k,m} \bar{z}_k z_m \, \mu(g_k^{-1} g_m) = \lim_{N \to \infty} \mu(a_N^* \, a_N) \geq 0$$

where

$$a_N = \sum_{k=1}^{n} \sum_{n_1, \ldots, n_r = 0}^{N} z_j \frac{i^{n_1 + \ldots + n_r}}{n_1! \ldots n_r!} v_{1,k}^{\otimes n_1} \otimes \ldots \otimes v_{r,k}^{\otimes n_r}.$$

The cases when μ is hermitian or conditionally positive are treated similarly.

(ii): For $q \geq 1$ and a function

$$h: \underline{R}^q \to \underline{C}$$

which is q-times continuously differentiable in a neighbourhood of $0 \in \underline{R}^q$ the following holds

$$\frac{\partial^q h}{\partial t_1 \ldots \partial t_q}(0) = \lim_{t \to 0} t^{-q} \Big(\sum_{A \subset \{1, \ldots, q\}} (-1)^{q - \#A} \, h_A(t) \Big) \tag{3}$$

where $\#A$ denotes the number of elements of the set A and

$$h_A(t) = h(\chi_A(1)\, t, \ldots, \chi_A(q)\, t)$$

(here χ_A denotes the characteristic function of the set A). For $t \in \underline{R}$ let ρ_t be the mapping from the set $\boldsymbol{M}$ of monomials in x_k, $k \in I$, to the group $G(V)$ given by

- $\rho_t(\boldsymbol{1}) = 0$
- $\rho_t(x_{k_1} \otimes \ldots \otimes x_{k_r}) = (t x_{k_1}) \ldots (t x_{k_r})$

  for $r \geq 1$, $k_1, \ldots, k_r \in I$.

We set

$$|x_{k_1} \otimes \ldots \otimes x_{k_r}| = r,$$

$|\boldsymbol{1}| = 0$, and introduce an order relation $\prec$ on $\boldsymbol{M}$. For $M, N \in \boldsymbol{M}$ the relation $N \prec M$ is said to hold if $N = \boldsymbol{1}$ or if there exist $m, n \geq 1$, $n \leq m$, $k_1, \ldots, k_m \in I$ and a subset $\{u(1), \ldots, u(n)\}$ of $\{1, \ldots, m\}$, $u(1) < \ldots < u(n)$, such that

$$N = x_{k_{u(1)}} \otimes \ldots \otimes x_{k_{u(n)}}$$

$$M = x_{k_1} \otimes \ldots \otimes x_{k_m}.$$

For f q-times differentiable in every direction for all $q \in \underline{N}$ we have by (3)

$$T(f)(M) = (-i)^{|M|} \lim_{t \to 0} t^{-|M|} \sum_{N \prec M} (-1)^{|M| - |N|} f(\rho_t(N)).$$

Let $a \in T(V)$ be given. Then

$$a = \sum_{k=1} c_k M_k$$

for some $M_k \in \boldsymbol{M}$, $c_k \in \underline{C} \setminus \{0\}$, $k = 1, \ldots, r$. Using (4) we get

$T(f)(a^*a)$

$$= \sum_{k,l=1}^{r} \bar{c}_k c_l \, T(f)(M_k^* M_l)$$

$$= \sum_{k,l=1}^{r} \bar{c}_k c_l \, (-i)^{|M_k| + |M_l|}$$

$$\Big(\lim_{t \to 0} t^{-(|M_k| + |M_l|)} \sum_{N < M_k,\, L < M_l} (-1)^{|M_k| + |M_l| - |N| - |L|} f(\rho_t(N)^{-1} \rho_t(L)) \Big)$$

$$= \lim_{t \to 0} \sum_{(k,N),(l,L)} \bar{d}_{k,N} d_{l,L} \, f(\rho_t(N)^{-1} \rho_t(L))$$

where (k,N) and (l,L) run through all elements of the finite subset $\{(k,N) \mid 1 \le k \le r, N \prec M_k\}$ of $N \times M$ and

$$d_{k,N} = c_k \, (i t^{-1})^{|M_k|} (-1)^{|N|}.$$

If f is positive definite the above equations show that $T(f)(a^*a) \ge 0 \;\; \forall a \in T(V)$. If δ vanishes on a then $M_k \neq 1$ for $k = 1, \ldots, r$. But then

$$\sum_{(k,N)} d_{k,N} = \sum_{k=1}^{r} c_k (i t^{-1})^{|M_k|} \Big(\sum_{N \prec M_k} (-1)^{|N|} \Big) = 0$$

because

$$\sum_{N \prec M} (-1)^{|N|} = 0$$

if $M \neq 1$. It follows that $T(f)(a^*a) \ge 0 \;\; \forall a \in T(V)$ with $\delta a = 0$ if f is conditionally positive definite. It is clear that $T(f)$ is hermitian if f is.

b) Dirac, Gauss, and Poisson functionals

If $\mu \in T(V)^*$ is hermitian and conditionally positive then the (1-parameter) convolution semi-group $\phi_t = exp_*(t\mu)$ consists of positive linear functionals; see [13,14]. Moreover, to any convolution semi-group of states on $T(V)$ one can associate a non-commutative stochastic process with independent, stationary increments in the sense of [2], consisting of operators on a Hilbert space. We compute the transforms of some special hermitian and conditionally positive linear functionals on $T(V)$.

Let α be a linear functional on $V_{\underline{C}}$, i. e. $\alpha \in (V_{\underline{C}})^*$. We define $d_\alpha \in T(V)^*$ by

- $d_\alpha(v) = \alpha(v)$ for $v \in V$

- $d_\alpha / V^{\otimes n} = 0$ for $n \in \underline{N} \setminus \{1\}$.

If α is real-valued on V, i. e. $\alpha \in V^*$, then d_α is hermitian and conditionally positive. The unique extension of $\alpha \in (V_{\underline{C}})^*$ to a homomorphism from $T(V)$ to $\underline{C}$ (mapping 1 into 1) will be denoted by δ_α. We have $exp_* d_\alpha = \delta_\alpha$ and $\delta_0 = \delta$. Clearly

$$G(V)_{d_\alpha} = G(V)$$

and for $(v_1, \ldots, v_r) \in G(V)$

$$\widehat{d}_\alpha((v_1, \ldots, v_r)) = i\,\alpha\Big(\sum_{1 \leq k \leq r} v_k\Big)$$

and

$$\widehat{\delta}_\alpha((v_1, \ldots, v_r)) = exp\Big(i\,\alpha\Big(\sum_{1 \leq k \leq r} v_k\Big)\Big).$$

We call the linear functionals δ_α, $\alpha \in (V_C)^*$, the *Dirac functionals* on $T(V)$.

We now introduce a new class of hermitian, conditionally positive linear functionals on $T(V)$. For a sesquilinear form Q on $V_{\underline{C}}$ and a linear functional σ on $T(V)$ we define the linear functional $h_{Q,\sigma}$ as follows:

- $h_{Q,\sigma}(1) = 0$
- $h_{Q,\sigma}(v) = 0$ for $v \in V$
- $h_{Q,\sigma}(v_1 \otimes v_2) = Q(v_1, v_2)\, \sigma(1)$ for $v_1, v_2 \in V$
- $h_{Q,\sigma}(v_1 \otimes \ldots \otimes v_r) = Q(v_1, v_r)\, \sigma(v_2 \otimes \ldots \otimes v_{r-1})$ for $r \geq 3$, $v_1, \ldots, v_r \in V$.

If Q and σ are positive $h_{Q,\sigma}$ is hermitian. We show that in this case $h_{Q,\sigma}$ is also conditionally positive. Let

$$a = \sum_{n=1}^{d} c_n\, v_1^{(n)} \otimes \ldots \otimes v_{r(n)}^{(n)}$$

($r(n) \geq 1$, $c_n \in \underline{C}$, $v_k^{(n)} \in V$) be in $T(V)$ such that $\delta a = 0$. Then

$$h_{Q,\sigma}(a^* a) = \sum_{n,m=1}^{d} \bar{c}_n c_m A_{n,m} B_{n,m}$$

where A, B are the $d{\times}d$-matrices given by

$$A_{n,m} = Q\big(v_{r(n)}^{(n)}, v_{r(m)}^{(m)}\big)$$
$$B_{n,m} = \sigma\big(v_{r(n)-1}^{(n)} \otimes \ldots \otimes v_1^{(n)} \otimes v_1^{(m)} \otimes \ldots \otimes v_{r(m)-1}^{(m)}\big).$$

By assumption A and B are positive definite, so it follows that $(A_{n,m} B_{n,m})_{n,m}$ is positive definite, and we have $h_{Q,\sigma}(a^* a) \geq 0$.

From now on we consider the special case when σ is a homomorphism.

Theorem 3.2. Let α be a linear functional and let Q be a sesquilinear form on $V_{\underline{C}}$. We put $h_{Q,\alpha} = h_{Q,\delta_\alpha}$. Then $G(V)_{h_{Q,\alpha}} = G(V)$ and for $r \geq 1$, $v_1, \ldots, v_r \in V$

$$\hat{h}_{Q,\alpha}((v_1) \ldots (v_r))$$

$$= \sum_{1 \leq k \leq r} Q(v_k, v_k) \frac{exp(i\,\alpha(v_k)) - i\,\alpha(v_k) - 1}{\alpha(v_k)^2}$$

$$+ \sum_{1 \leq k < m \leq r} Q(v_k, v_m) \frac{exp(i\,\alpha(\sum_{k \leq l \leq m} v_l)) - exp(i\,\alpha(\sum_{k < l < m} v_l))}{\alpha(v_k)\,\alpha(v_m)}$$

(where we extend by continuity if the denominator of one of the fractions is equal to 0).

Proof: We have

$$EXPv_1 \ldots EXPv_r$$

$$= \sum_{n_1, \ldots, n_r \in \underline{N}} \frac{v_1^{\otimes n_1} \otimes \ldots \otimes v_r^{\otimes n_r}}{n_1! \ldots n_r!}$$

$$= 1 + \sum_{1 \leq k \leq r} \sum_{n \geq 1} \frac{v_k^{\otimes n}}{n!}$$

$$+ \sum_{1 \leq k < m \leq r} \sum_{n_k, \ldots, n_m \in \underline{N},\, n_k, n_m \geq 1} \frac{v_k^{\otimes n_k} \otimes \ldots \otimes v_m^{\otimes n_m}}{n_k! \ldots n_m!}$$

and

$$\sum_{n_k, \ldots, n_m \in \underline{N},\, n_k, n_m \geq 1} i^{n_k + \ldots + n_m} \frac{\alpha\left(v_k^{\otimes(n_k - 1)} \otimes v_{k+1}^{\otimes n_{k+1}} \otimes \ldots \otimes v_{m-1}^{\otimes n_{m-1}} \otimes v_m^{\otimes(n_m - 1)}\right)}{n_k! \ldots n_m!}$$

$$= \sum_{n_k, \ldots, n_m \in \underline{N},\, n_k, n_m \geq 1} \frac{(i\,\alpha(v_k))^{n_k} \ldots (i\,\alpha(v_m))^{n_m}}{\alpha(v_k)\alpha(v_m)\, n_k! \ldots n_m!}$$

$$= \frac{exp(i\,\alpha(\sum_{k \leq l \leq m} v_l)) - exp(i\,\alpha(\sum_{k < l < m} v_l))}{\alpha(v_k)\,\alpha(v_m)}$$

which yields the formula given in the theorem. □

Let $Q \neq 0$ be positive and let α be in V^*. We give a realization of the non-commutative stochastic process with independent, stationary additive increments whose generator is $h_{Q,\alpha}$. Denote by Λ_t, A_t, and A_t^+ the gauge, annihilation, and creation process on the Bose Fock space; see [10]. First we restrict ourselves to the case $Q_{km} = Q(x_k,x_m) = \bar{z}_k z_m$ for $z_k \in \underline{C}$, $k,m \in I$. Consider the operators

$$\Pi_t^{(k)} = \alpha_k \Lambda_t + \bar{z}_k A_t + z_k A_t^+ + \beta_k t$$

where $\alpha_k = \alpha(x_k)$ and $\beta_k \in \underline{R}$, $k \in I$, and denote by $\mathcal{A}$ the algebra of all polynomials in the $\Pi_t^{(k)}$ where $\Pi_t^{(k)}$ is regarded as an operator on $\mathcal{A}\phi^{(0)}$ with $\phi^{(0)}$ the vacuum state. For $s,t \in \underline{R}_+$, $s \leq t$, define *-algebra homomorphisms

$$j_{st}: T(V) \rightarrow \mathcal{A}$$

by

$$j_{st}(x_k) = \Pi_t^{(k)}.$$

Then $(\mathcal{A}, j_{st}, \phi^{(0)})$ is a continuous independent, stationary increment process over $T(V)$ in the sense of [2], and the generator of this process is $h_{Q,\alpha} + d_\beta$, i. e.

$$(h_{Q,\alpha} + d_\beta)(x_{k_1} \otimes \dots \otimes x_{k_r}) = \frac{d}{dt} \langle \phi^{(0)}, \Pi_t^{(k_1)} \dots \Pi_t^{(k_r)} \phi^{(0)} \rangle \Big|_{t=0}$$

(where $\beta \in V^*$ is given by $\beta(x_k) = \beta_k$). Assume that $dim V < \infty$. If Q is a positive sesquilinear form on $V_{\underline{C}}$ it can be written in the form

$$Q_{km} = \sum_p \overline{z_k^{(p)}} z_m^{(p)}$$

for some complex numbers $z_k^{(p)}$. Using many dimensional gauge, annihilation and creation processes, we find a realization of the process with generator $h_{Q,\alpha}$.

We consider two special cases of $h_{Q,\alpha}$. First assume $\alpha(x_k) = 1$, $Q_{km} = z_k z_m$, $z_k = |z_k|$, $\beta_k = z_k^2$. The process $\Pi_t^{(k)}$ is a Poisson process of intensity z_k and $(\Pi_t^{(k)} \mid k \in I)$ may be called a *non-commutative Poisson process*; see [10]. Indeed, by Theorem 3.2

$$(h_{Q,\alpha} + d_\beta)((tx_k)) = z_k^2(exp(it) - 1)$$

in this case, which as a function in $t \in \underline{R}$ is the logarithm of the Fourier transform of a Poisson distribution with intensity z_k, and $h_{Q,\alpha} + d_\beta$ restricted to the (commutative) polynomial algebra $\underline{C}[x_k]$ is the generator of a Poisson process. Next assume $\alpha(x_k) = 0$, and let Q be an arbitrary sesquilinear form on $V_{\underline{C}}$. The functional $h_{Q,\alpha}$ which we denote by g_Q vanishes on all monomials not of length 2 and its covariance matrix is given by Q. The convolution exponential $\gamma_Q = exp_* g_Q$ is a *gaussian functional* in the sense of [7]. By Lemma 3.1 (ii) and Theorem 3.2

$$\hat{\gamma}_Q((v_1, \dots, v_r)) = exp(g_Q((v_1, \dots, v_r)))$$

is equal to

$$exp\Big(-\Big(\frac{1}{2}\sum_{1\le k\le r} Q(v_k,v_k) + \sum_{1\le k<m\le r} Q(v_k,v_m)\Big)\Big)$$

$$= exp\Big(-\frac{1}{2}\Big(F\Big(\sum_{1\le k\le r} v_k\Big) + i\sum_{1\le k<m\le r} B(v_k,v_m)\Big)$$

where $F(v,w) = \frac{1}{2}(Q(v,w) + Q(w,v))$, $F(v) = F(v,v)$, and $B(v,w) = -i(Q(v,w) - Q(w,v))$.

(If Q is hermitian, $F= Re\, Q$ and $B = Im\, Q$.) For $dim V = 2$ and

$$Q = \begin{pmatrix} 1 & i \\ -i & 1 \end{pmatrix}$$

the process associated to g_Q is

$$(A_t + A_t^+, i(A_t - A_t^+))$$

which is a *quantum Wiener process* in the sense of [5]. Denote by $\underline{C}G(V)$ the group ring of $G(V)$, i. e. the *-algebra consisting of formal finite sums

$$\sum_{g\in G(V)} \alpha(g)g,$$

$\alpha(g)\in\underline{C}$, with multiplication given by the group multiplication and involution given by $g^* = g^{-1}$. We identify complex-valued functions on $G(V)$ and elements of $(\underline{C}G(V))^*$ in the obvious way. $\underline{C}G(V)$ is a *-bialgebra with comultiplication Δ given by $\Delta g = g\otimes g$ and counit δ given by $\delta g = 1$ for $g\in G(V)$. A function $f\colon G(V) \to \underline{C}$ is positive definite (hermitian, conditionally positive definite) if and only if the corresponding linear functional on $\underline{C}G(V)$ is positive (hermitian, conditionally positive).

Theorem 3.3. Let S be a complex-valued skew symmetric bilinear form on V. Then the function

$$f_S\colon G(V) \to \underline{C}$$

given by

$$f_S((v_1) \dots (v_r)) = exp\Big(\sum_{1\le k<m\le r} S(v_k,v_m)\Big) \tag{5}$$

$v_1, \dots, v_r\in V$, vanishes on the ideal in $\underline{C}G(V)$ generated by the elements

$$(v)(w) - exp(2S(v,w))(w)(v)$$

where v and w run through all vectors in V.

<u>*Proof*</u>: As $f_S((v_1) \dots (v_r)) = (exp{*}g_{-S})(EXP(iv_1) \dots EXP(iv_r))$ the definition (5) of f_S makes sense. For $v,w,v_1,\dots,v_r,w_1,\dots w_n\in V$ we have

$$f_S((v_1) \dots (v_r)(v)(w)(w_1) \dots (w_n))$$

$$= exp\Big(\sum_{1 \le k < m \le r} S(v_k,v_m) + S\Big(\sum_{1 \le k \le r} v_k ,v + w \Big) + S(v,w)$$

$$+ S\Big(v + w, \sum_{1 \le l \le n} w_l \Big) + \sum_{1 \le l < s \le n} S(w_l, w_s) \Big).$$

But this is equal to $exp(2S(v,w))$ times the same expression with v and w exchanged, so it is equal to

$$exp(2S(v,w))\ f_S((v_1) \dots (v_r)(w)(v)(w_1) \dots (w_n)). \square$$

We have the following result which is the analogue of Theorem 2 in [7].

Corollary 3.1. For a sesquilinear form Q on $V_{\underline{C}}$ the transform $\hat{\gamma}_Q$ of the gaussian functional γ_Q vanishes on the ideal in $\underline{C}G(V)$ generated by the elements

$$(v)(w) - exp(-iB(v,w))(w)(v);\ v,w \in V. \square$$

Let Q be a positive sesquilinear form on $V_{\underline{C}}$. Denote by

$$\pi_Q: \underline{C}G(V) \to U(H)$$

the GNS-representation of $\hat{\gamma}_Q$ (here H is the Hilbert space of the representation and $U(H)$ is the group of unitary operators on H). For $v \in V$ we *set* $W(v) = \pi_Q(v)$. Then by Corollary 3.1

$$W: V \to U(H)$$

is a representation of the canonical commutation relations with

$$W(v)W(v') = exp(-iB(v,v')W(v')W(v)$$

for $v,v' \in V$.

4. Central limit theorem

For a complex number z denote by κ_z the homomorphism on $T(V)$ given by

$$\kappa_z(v) = zv;\ v \in V.$$

The mapping κ_z can be extended to a homomorphism on $\overline{T(V)}$ in the obvious way. A linear functional μ on $T(V)$ is called normalized if $\mu(1) = 1$. It was proved in [7] that for any $a \in T(V)$ and any normalized $\mu \in T(V)^*$

$$\lim_{N \to \infty} \mu^{*N}(\kappa_{1/N}(a)) = \delta_{\alpha_\mu}(a)$$

(Law of large numbers)

and

$$\lim_{N \to \infty} \mu_0^{*N}(\kappa_{1/\sqrt{N}}(a)) = \gamma_{Q_\mu}(a)$$

(Central limit theorem)

where $\alpha_\mu(v) = \mu(v)$ and $Q_\mu(v,w) = \mu(v \otimes w) - \mu(v)\mu(w)$, and μ_0 denotes the centralized functional $\mu_0 = \delta_{-\alpha_\mu} * \mu$. We prove the corresponding result for functions on $G(V)$, using Taylor series expansions for f_g, $g \in G(V)$. A function $f: G(V) \to \underline{C}$ is called normalized if $f(1) = 1$.

Theorem 4.1. Let $g = (v_1, ..., v_r)$ be in $G(V)$ and let f be a normalized complex-valued function on $G(V)$.

(i) If f is differentiable in the direction of g

$$\lim_{N \to \infty} f\left(\left(\frac{v_1}{N}, ..., \frac{v_r}{N}\right)\right)^N = \exp\left(\sum_{1 \le k \le r} D_{v_k} f\right).$$

(Law of large numbers)

(ii) If f is twice differentiable in the direction of g

$$\lim_{N \to \infty} \exp\left(-\sqrt{N} \sum_{1 \le k \le r} D_{v_k} f\right) f\left(\left(\frac{v_1}{\sqrt{N}}, ..., \frac{v_r}{\sqrt{N}}\right)\right)^N$$

$$= \exp\left(\frac{1}{2}\left(\sum_{1 \le k < r} D^2_{v_k} f - \left(\sum_{1 \le k \le r} D_{v_k} f\right)^2\right) + \sum_{1 \le k < m \le r} D_{v_k, v_m} f\right).$$

(Central limit theorem)

Proof: (i): By assumption $f_g: \underline{R}^r \to \underline{C}$, $g = (v_1, ..., v_r)$, is continuously differentiable in a neighbourhood of 0. Thus for $\sum_{1 \le k \le r} t_k^2$ sufficiently small

$$f_g(t_1, ..., t_r) = f(1) + \sum_{1 \le k \le r} (D_{v_k} f)\, t_k + R(t_1, ..., t_r)$$

where

$$\lim_{(t_1, ..., t_r) \to 0} \left(\sum_{1 \le k \le r} t_k^2\right)^{-1/2} R(t_1, ..., t_r) = 0.$$

This gives for N sufficiently large

$$f\left(\left(\frac{v_1}{N}, ..., \frac{v_r}{N}\right)\right) = 1 + \frac{1}{N} \sum_{1 \le k \le r} D_{v_k} f + R_N$$

where $\lim_{N \to \infty} N R_N = 0$. But then

$$f\left(\left(\frac{v_1}{N}, ..., \frac{v_r}{N}\right)\right)^N = \left(1 + \frac{1}{N}\left(\sum_{1 \le k \le r} D_{v_k} f + NR_N\right)\right)^N$$

$$\xrightarrow[N \to \infty]{} \exp\left(\sum_{1 \le k \le r} D_{v_k} f\right).$$

(ii): By assumption f_g is twice continuously differentiable in a neighbourhood of 0. First assume

$Df = 0$. Then for small $\sum t_k^2$

$$f_g(t_1, \ldots, t_r)$$

$$= 1 + \frac{1}{2} \sum_{1 \le k \le r} (D^2_{v_k} f)\, t_k^2 + \sum_{1 \le k < m \le r} (D_{v_k, v_m} f)\, t_k t_m + R(t_1, \ldots, t_r)$$

where

$$\lim_{(t_1, \ldots, t_r) \to 0} \left(\sum t_k^2\right)^{-1} R(t_1, \ldots, t_r) = 0.$$

Thus

$$f\left(\left(\frac{v_1}{\sqrt{N}}, \ldots, \frac{v_r}{\sqrt{N}}\right)\right)$$

$$= 1 + \frac{1}{N}\left(\frac{1}{2} \sum_{1 \le k \le r} D^2_{v_k} f + \sum_{1 \le k < m \le r} D_{v_k, v_m} f\right) + R_N$$

where $\lim_{N \to \infty} N R_N = 0$. This gives

$$f\left(\left(\frac{v_1}{\sqrt{N}}, \ldots, \frac{v_r}{\sqrt{N}}\right)\right)^N$$

$$\xrightarrow[N \to \infty]{} \exp\left(\frac{1}{2} \sum_{1 \le k \le r} D^2_{v_k} f + \sum_{1 \le k < m \le r} D_{v_k, v_m} f\right).$$

In the general case we apply this result to the function $f_0 = gf$ where

$$g((v_1, \ldots, v_r)) = \exp\left(- \sum_{1 \le k \le r} D_{v_k} f\right).$$

The identity

$$D_{v,w} f_0 = D_{v,w} f - (D_v f)(D_w f)$$

for $v, w \in V$, yields

$$\frac{1}{2} \sum_{1 \le k \le r} D^2_{v_k} f_0 + \sum_{1 \le k < m \le r} D_{v_k, v_m} f_0$$

$$= \frac{1}{2}\left(\sum_{1 \le k \le r} D^2_{v_k} f - \left(\sum_{1 \le k \le r} D_{v_k} f\right)^2\right) + \sum_{1 \le k < m \le r} D_{v_k, v_m} f$$

which proves (ii). □

For a differentiable function $f\colon G(V) \to \underline{C}$ we denote by f_0 the centralized function $\delta_{iDf} f$.

Corollary 4.1. Let f be a normalized, complex-valued function on $G(V)$.

(i) If f is differentiable, for $N \to \infty$ the functions $(f \circ \kappa_{1/N})^N$ converge pointwise to the transform $\hat{\delta}_\alpha$ of the Dirac functional δ_α with $\alpha = -i\,Df$.

(ii) If f is twice differentiable, for $N \to \infty$ the functions $(f_0 \circ \kappa_{1/\sqrt{N}})^N$ converge pointwise to the transform $\hat{\gamma}_Q$ of the gaussian functional γ_Q with $Q(v,w) = -D^2 f_0(v,w) = Df(v)\,Df(w) - D^2 f(v,w)$.□

If in (ii) of the previous corollary f is positive definite, then also the limiting functional $\hat{\gamma}_Q$ is positive definite, which by Theorem 3.1 and [13] holds if and only if Q is positive definite.

As an application of Corollary 4.1 we treat the case of functions on a Lie group. Let $\mathfrak{G}$ be a real Lie group of dimension $d \in \underline{N}$. Denote by $\mathfrak{g}$ its Lie algebra and by

$$exp_{\mathfrak{G}} : \mathfrak{g} \to \mathfrak{G}$$

its exponential mapping. Regarding $\mathfrak{g}$ as a d-dimensional real vector space, the mapping

$$\eta_{\mathfrak{G}} : G(\mathfrak{g}) \to \mathfrak{G}$$

defined by

$$\eta_{\mathfrak{G}}((v_1, \ldots, v_r)) = (exp_{\mathfrak{G}}\, v_1) \ldots (exp_{\mathfrak{G}}\, v_r)$$

is a group homomorphism which is surjective if $\mathfrak{G}$ is connected. For a function

$$\tilde{f} : \mathfrak{G} \to \underline{C}$$

we define the function

$$f : G(\mathfrak{g}) \to \underline{C}$$

by

$$f = \tilde{f} \circ \eta_{\mathfrak{G}}.$$

It is clear that f is q-times differentiable (analytic) if $\tilde{f}$ is q-times continuously differentiable (analytic) in a neighbourhood of $1 \in \mathfrak{G}$, and that f is positive definite (hermitian, conditionally positive definite) if $\tilde{f}$ is. We have

Corollary 4.2. Let $\tilde{f}$ be a complex-valued function on $\mathfrak{G}$. If $\tilde{f}$ is twice continuously differentiable in a neighbourhood of 1 and $f(1) = 1$ then we have for $(v_1, \ldots, v_r) \in G(\mathfrak{g})$

$$\lim_{N \to \infty} (f)_0 \Big(\Big(\frac{v_1}{\sqrt{N}}, \ldots, \frac{v_r}{\sqrt{N}} \Big) \Big)^N$$

$$= exp\Big(\frac{1}{2} \Big(D^2 f\big(\sum_{1 \le k \le r} v_k \big) - Df\big(\sum_{1 \le k \le r} v_k \big)^2 + \sum_{1 \le k < m \le r} Df([v_k, v_m]) \Big) \Big).$$ □

References

[1] Accardi, L. and Bach, A., Quantum limit theorems for strongly mixing random variables, Z. Wahrscheinlichkeitstheorie verw. Gebiete 68, 393-402 (1985)

[2] Accardi, L., Schürmann, M. and von Waldenfels, W., Quantum independent increment processes on superalgebras, submitted for publication

[3] Canisius, J., Algebraische Grenzwertsätze und unbegrenzt teilbare Funktionale, Diplomarbeit, Heidelberg 1979

[4] Chen, K.-T., Exponential isomorphism for vector spaces and its connection with Lie groups, J. London Math. Soc. 33, 170-177 (1958)

[5] Cockcroft, A.M. and Hudson, R.L., Quantum mechanical Wiener processes, J. Multivariate Anal. 7, 107-124 (1977)

[6] Cushen, C.D. and Hudson, R.L., A quantum-mechanical central limit theorem, J. Appl. Prob. 8, 454-469 (1971)

[7] Giri, N. and von Waldenfels, W., An algebraic version of the central limit theorem, Z. Wahrscheinlichkeitstheorie verw. Gebiete 42, 129-134 (1978)

[8] Hegerfeldt, G.C., Noncommutative analogs of probabilistic notions and results, J. Funct. Anal. 64,436-456 (1985)

[9] Hudson, R.L., A quantum-mechanical central limit theorem for anti-commuting observables, J. Appl. Prob. 10, 502-509 (1973)

[10] Hudson, R.L. and Parthasarathy, K.R., Quantum Ito's formula and stochastic evolutions, Comm. Math. Phys. 93, 301-323 (1984)

[11] Quaegebeur, J., A non-commutative central limit theorem for CCR-algebras, J. Funct. Anal. 57, 1-20 (1984)

[12] Schürmann, M., Positiv definite Funktionen auf der freien Lie-Gruppe und stabile Grenzverteilungen, Diplomarbeit, Heidelberg 1979

[13] Schürmann, M., Positive and conditionally positive linear functionals on coalgebras, in Lect. Notes in Math. 1136, Springer, Berlin 1985

[14] Schürmann, M., Über *-Bialgebren und quantenstochastische Zuwachsprozesse, Dissertation, Heidelberg 1985

[15] Schürmann, M., A central limit theorem for coalgebras, in Lect. Notes in Math. 1210, Springer, Berlin 1986

[16] von Waldenfels, W., An algebraic central limit theorem in the anti-commuting case, Z. Wahrscheinlichkeitstheorie verw. Gebiete 42, 135-140 (1978)

[17] von Waldenfels, W., Proof of an algebraic central limit theorem by moment generating functions, in Lect. Notes in Math. 1250, Springer, Berlin 1985

[18] von Waldenfels, W., Non-commutative algebraic central limit theorems, in Lect. Notes in Math. 1210, Springer, Berlin 1986

ENTROPY, OBSERVABILITY AND THE GENERALISED SECOND LAW OF THERMODYNAMICS

by Geoffrey L. Sewell

Department of Physics, Queen Mary College, London E1 4NS

ABSTRACT

We derive the Generalised Second Law (GSL) of Thermodynamics, pertaining to processes in which observable systems exchange matter with unobservable regions of space-time, i.e. Black Holes. Our derivation of this Law is based exclusively on a treatment of *observable* quanties, and thus avoids Bekenstein's [1] assumption that a Black Hole has an entropy, proportional to its surface area. In fact, we demonstrate that it follows from simple statistical thermodynamic and general relativistic properties of the *open* system comprising the matter and fields in the region outside a Black Hole that, in suitable units, the sum of the entropy of that region and the surface area of the Hole never decreases. This is formally equivalent to the GSL of Ref.[1], though in our scheme, the area term represents mechanical work done by the exterior system, not entropy. Since our derivation of the GSL depends on the Hawking thermalisation effect [2], we devote the Appendix to a sketch of our earlier model-independent treatment [3] of quantum fields on manifolds, which covers this effect.

1. INTRODUCTION

It has been appreciated for some time that a serious thermodynamical problem arises in connection with processes, demanded by General Relativity, in which observable systems exchange matter with unobservable space-time regions, i.e. Black Holes (cf.[1]). The heart of the problem is that, according to General Relativity, the only observables of a Black Hole that can be perceived from outside are its mass, electric charge and angular momentum [4,P.876]. From the standpoint of an external observer, this implies that its entropy is indeterminate, and that, consequently, the standard form of the Second Law of Thermodynamics has no predictive value for processes in which observable systems discharge matter into the Hole [5].

On the other hand, the Relativistic Mechanics of Black Holes [6] exibits remarkable analogies with Classical Thermodynamics, in which the area of a Hole plays the role of an entropy. Bekenstein [1] argued that these analogies carry physical content by showing that, in certain Gedankenexperiments in which matter was discharged into a Black Hole, the sum of the entropy, S, of the exterior region and the area, A, of the Hole, multipled by a universal constant, λ, never decreased. This led him to propose that the Black Hole had an entropy λA, and that processes involving the Hole obeyed a Generalised Second Law, of the form

$$\Delta(S + \lambda A) \geq 0. \tag{1}$$

This proposal was supported by Hawking's [2] subsequent demonstration that a Black Hole acted as a thermal source, which raised an ambient quantum field to a temperature that corresponded precisely to the assumed form, λA, of the Black Hole entropy. At the microscopic level, Bekenstein's [1] interpretation of this entropy, now widely accepted [7,8], is that it is an information-theoretic quantity, representing the external observer's ignorance of the microstate of the Hole.

We shall now argue, however, that the assumption of a Black Hole entropy cannot be accommodated within the framework of Statistical Thermodynamics. The essential point is that, according to quantum statistics, the entropy of a system is a function of its microstate. To be specific, the state is given by a density matrix, ρ, whose form may be obtained from measurements, and its entropy is $-kTr(\rho \ln \rho)$, where k is Boltzmann's constant [9,Ch.5]. This entropy is, of course, an information-theoretic quantity; but most importantly, it represents a *structural property* of the microstate, and provides a measure of its degree of impurity, or disorder [10,P.57]. Indeed, it is the structure of the state, not the observer's ignorance, that governs the thermodynamic behaviour of the system. Now, in the case of a Black Hole, the concept of a microstate cannot have any operational meaning, since no microstructure is observable from outside. Hence, the statistical mechanical concept of entropy, depending as it does on that of microstate, cannot be applicable to Black Holes.

In view of this situation, we propose a different approach, based exclusively on observable quantities, to the whole question of thermodynamical processes involving Black Holes. This is centred on a test-body, Σ, which is placed in the Hawking radiation of a Black Hole, B, and is able to exchange energy, charge and angular momentum with both B and the radiation. Thus Σ is an *open* system. This point is crucial, since the key thermodynamical potential of an open system is its Gibbs free energy, rather than it entropy: for the Second Law of Classical Thermodynamics tells us that the Gibbs potential of such a system decreases in irreversible processes and remains unchanged in reversible ones [11,Ch.2]. Accordingly, we shall formulate the thermodynamics of processes, in which Σ exchanges matter with B, on the basis of

(a) Classical Thermodynamics, which tells us that the Gibbs free energy of Σ cannot increase spontaneously;

(b) the Hawking effect [2], whereby a Black Hole emits radiation at a temperature given by equation (3) below - an effect which we have shown [3] to ensue from the basic axioms of Quantum Theory, Relativity and Statistical Thermodynamics; and

(c) the Laws of Black Hole Mechanics [6], stemming from General Relativity, for the surface area, energy, charge and angular momentum of B, *without thermodynamical interpretation*.

On this basis, we derive the GSL (1), though now the area term in that formula represents work done on B by the exterior system, not entropy. Thus, we conclude that the GSL (1) is a consequence of Quantum Theory, General Relativity and Statistical Thermodynamics, which does not involve any notion of Black Hole entropy.

We shall present our argument as follows. In section 2, we shall formulate the above principles (b) and (c). In Sections 3, we shall derive the GSL from (a)-(c), and we shall comment further on the interpretation of this Law.The argument here is carried out at the thermodynamic level. In order to exhibit its connection with the underlying quantum mechanisms, we shall provide an Appendix, in which we sketch our earlier *model-independent* treatment of quantum fields on manifolds [3], which covers processes, such as the Hawking effect, in which these fields are thermalised by gravitational forces.

2. PRELIMINARIES ON BLACK HOLES

We start by recalling that a Black Hole, B, is a bounded region of a curved, stationary, asymptotically flat space-time, Γ, which corresponds to a collapsed state of matter, that is invisible from outside; i.e. no light-ray can emerge from B in a time that is finite for a fixed distant observer ([4],Ch.33). The Laws of Black Hole Mechanics, [6], described from the standpoint of such an observer, are that B has the following general properties.

(B1) In any process, the total energy, electric charge and angular momentum, comprising the contributions from B and the exterior region, are conserved.

(B2) The surface gravity, α, of B, which is the acceleration of a freely infalling body at the surface of B, is constant over that surface.

(B3) The surface area, A, of B is a function of the energy, E_B, charge Q_B and angular momentum, J_B, of the Hole, which satisfies the differential formula

$$\alpha c^2 dA/8\pi\gamma = dE_B - \Phi dQ_B - \omega . dJ_B \tag{2}$$

where c is the speed of light, γ is the gravitational constant, and Φ and ω are the electric potential and angular velocity, respectively, of B.

(B4) The area A increases in irreversible processes and remains unchanged in reversible ones.

Although these Laws have striking analogies with Classical Thermodynamics, with A and α playing the roles of entropy and temperature, respectively, we shall not assume any thermodynamical interpretation of them.

Suppose now that we have a quantum field, F, on the space-time Γ. Then the classical gravitational field of B, corresponding to the geometry of Γ, will thermalise F, by Hawking effect [2,3], to a temperature which, for an observer in the asymptotically flat region, takes the value

$$T = \hbar\alpha/\pi kc, \tag{3}$$

where $\hbar$ is Planck's constant. Furthermore, the symmetries of Γ ensure that both the angular momentum and electric charge of F are conserved quantities, and the conditions on the boundary of B fix the values of the thermodynamically conjugate variables of this field, namely its angular velocity and electric potential, to be the same as those of B, i.e. ω and Φ, respectively. Thus, the quantum field behaves as thermal radiation, providing a heat bath of temperature T, electric potential Φ and angular velocity ω for bodies far from B.

3. DERIVATION OF THE GSL.

We consider the thermodynamics of a system Σ, placed in the asymptotically flat region. Such a system will be immersed in the Hawking radiation, which, as we have just noted, has temperature T, electric potential Φ and angular velocity ω. Σ is therefore an open system, and its Gibbs potential is

$$G = E - TS - \Phi Q - \omega . J, \tag{4}$$

where E, S, Q and J are its energy, entropy, electric charge and angular momentum, respectively. Thus, in any process, in which work ΔW is done on Σ, the resultant changes in G, E, S, Q and J satisfy the inequality [11,Ch.2]

$$\Delta W \geqslant \Delta G = \Delta E - T\Delta S - \Phi\Delta Q - \omega . \Delta J. \tag{5}$$

Suppose now that the process results in the exchange of energy, charge and angular momentum between Σ and B. Then it follows from the above conservation law (B1) that

$$\Delta E + \Delta E_B = \Delta W; \ \Delta Q + \Delta Q_B = 0; \text{ and } \Delta J + \Delta J_B = 0 \tag{6}$$

and therefore, by (5) that

$$T\Delta S + \Delta E_B - \Phi\Delta Q_B - \omega . \Delta J_B \geqslant 0. \tag{7}$$

By equations (2) and (3), this reduces to the form

$$\Delta S + kc^3\Delta A/4\gamma\hbar \geqslant 0. \tag{8}$$

which is precisely Bekenstein's GSL.

Thus we see that the GSL ensues from the Classical Thermodynamics of open systems, together with the Laws of Black Hole Mechanics and the Hawking radiation theory. We conclude with the following comments about our derivation and interpretation of the GSL.

(1) Since the area term in the GSL (8) is T^{-1} times the R.H.S. of (2), we see that this term represents the work done by Σ on B. We thus interpret this term not as an entropy, but as a contribution to the mechanical work done by Σ, with the coefficient of ΔA playing the role of a surface tension.

(2) In comparing our derivation of the GSL with standard treatments of open systems, we note that, in the usual situation, one has a system Σ in contact with a much larger one, R, which acts as a thermal reservoir. In that case, one may treat Σ *either* as part of the closed composite sytem (Σ+R) *or* as an open system subjected to forces (pressures etc.) imposed by R. Now,

in the situation we have considered here, where R is replaced by the Black Hole, B, and its Hawking radiation, the second kind of treatment is applicable, and that is the one we have employed. By contrast, the previous works on the GSL were based on the first kind of treatment, and so required the assumption of a Black Hole entropy, which we have criticised in Section 1.

(3) Our derivation of the GSL may easily be generalised to situations where some of the energy, etc., of Σ is transferred to the Hawking radiation field. For, in that case, equation (6) and the GSL formula (8) remain valid if ΔS, ΔE, ΔQ and ΔJ are defined to be the total changes in entropy energy, charge and angular momentum of the exterior region, comprising the contributions of Σ and the radiation field.

APPENDIX: THERMALISATION OF QUANTUM FIELDS BY GRAVITATIONAL FORCES.

We shall sketch here our general treatment [3] of the thermalisation of quantum fields by the classical gravitational forces carried by a certain class of space-times. For simplicity, we shall deal mainly with the case where these latter forces correspond, via the Principle of Equivalence, to a uniform acceleration in flat space-time, and will just indicate how the treatment can be extended to fields on a Black Hole background. In fact, we shall see that the formula (3) for the temperature of the quantum field is equally applicable whether the gravitational field corresponds to a uniform acceleration, α, in Minkowski space or to the geometry of a Black Hole, with surface gravity α. In the former case, this result amounts to a generalised Unruh effect (cf.[12].)

Our treatment of the thermalisation of quantum fields by gravitation falls into three parts, the first being quantum statistical, the second geometrical and the third quantum field theoretical.

Quantum Statistics. In a standard way [13], we take a quantum dynamical system to consist of a quadruple ($\mathbf{H}$, $\mathbf{A}$, U, Ω), where $\mathbf{H}$ is a Hilbert space, $\mathbf{A}$ is a $*$-algebra of operators in $\mathbf{H}$, U is a continuous unitary representation of $\mathbf{R}$ in H, such that the transformations $A \to U(t)AU(-t) \equiv A(t)$ are automorphisms of A, and Ω is a unit vector in H, that is cyclic w.r.t. $\mathbf{A}$ and invariant under $U(\mathbf{R})$. Thus, $\mathbf{A}$ and U represent the observables and dynamics, respectively, of the system, and Ω corresponds to a stationary state. We denote the infinitesimal generator of $U(\mathbf{R})$ by $iH/\hbar$, H being the Hamiltonian operator. Evidently, $H\Omega = 0$, since Ω is invariant under $U(\mathbf{R})$.

The vector Ω corresponds to a <u>ground state</u> if H is positive, in which case the time correlation functions $t \to (\Omega, AB(t)\Omega)$ admit analytic continuations to the upper half of the complex t-plane, for all A,B in **A**. On the other hand, Ω corresponds to a thermal state if it satisfies the Kubo-Martin-Schwinger (KMS) condition, which may be formulated as (cf.[14]).

$$(U(\tfrac{1}{2}i\beta)A\Omega, U(\tfrac{1}{2}i\beta)B\Omega) = (B^*\Omega, A^*\Omega), \text{ with } U(\tfrac{1}{2}i\beta) \equiv \exp(-\tfrac{1}{2}\beta H), \qquad (A1)$$

<u>The Geometry</u>. We consider here the case when the space-time, M, is Minkowskian. Thus, denoting the points of M by m = (x,y,z,ct), where x,y,z are the spatial Cartesian coordinates and t is the temporal one, the metric is given by the following standard formula for the "arc-length" ds:

$$ds^2 = c^2dt^2 - dx^2 - dy^2 - dz^2. \qquad (A2)$$

Lorentz boosts are given by the one-parameter group of isometries $m \to m(\lambda)$ of M, as defined by

$$m(\lambda) = (x\cosh\lambda + ct\sinh\lambda, y, z, ct\cosh\lambda + x\cosh\lambda) \qquad (A3)$$

To describe uniform accelerations along Ox, we introduce the Rindler wedge W= $\{m \mid x > c|t| \}$. The boundaries, $x = \pm ct$, of this region, are event horizons, in the sense that light signals emitted from W cannot cross the latter surface, while signals cannot be received by W from across the former one. The geometry of W is conveniently described in Rindler coordinates ρ (> 0) and τ, which are related to x,t in W by the formula

$$x = \rho\cosh\tau \; ; \; ct = \rho\sinh\tau . \qquad (A4)$$

The metric in W is therefore given by

$$ds^2 = \rho^2 \, d\tau^2 - d\rho^2 - dy^2 - dz^2, \qquad (A5)$$

which signifies that τ is a temporal coordinate for W. in fact [4,P.166], the curves in W for which ρ,y and z are constant are trajectories of uniform acceleration.

$$\alpha = c^2 /\rho , \qquad (A6)$$

Along Ox, and proper time

$$\tau_p = \tau\rho/c$$

We note that, by (A3) and (A4), *time translations $\tau \rightarrow \tau + \lambda$, for the accelerated observer in W, correspond to Lorentz boosts for an inertial one.*

Quantum Field Theory. We formulate the theory of a quantum field in **M** within the terms of Wightman's [15] axiomatic scheme. For notational simplicity, we shall confine ourselves to real scalar fields. Thus, we take the ingredients of the field theory to be a quadruple ($\mathbf{H}$, U, Φ, Ω), where **H** is a separable Hilbert space, U is a continuous unitary representation in **H** of the proper Poincaré group G, of isometries of M, $m \rightarrow \Phi(m)$ is a Hermitian operator-valued tempered distribution in **H**, and Ω is a normalised vector in **H**, that is cyclic w.r.t. the polynomials of the operators, $\Phi(f)$, obtained by integrating Φ against **S**-class test functions, f. In this setting Φ is a real scalar field, and Ω is a state vector, which will be characterised as the vacuum by axiom's (II) and (III) below. The Hamiltonian, H, is defined as $-i\hbar$ times the generator of the one-parameter subgroup of U(G) that represents time translations. Assuming that the "smeared field" operators $\Phi(f)$ are self-adjoint (not merely symmetric) for real f, we take the algebras of observables, **A** and **A**(W), for the entire system and for the Rindler wedge, W, to be the Von Neumann algebras generated by the Weyl operators $\exp i\Phi(f)$, with f having support in M and W, respectively. We assume that the system ($\mathbf{H}$, U, Φ, Ω) satisfies Wightman's axioms, which may be stated as follows.

(I) $\Phi(m)$ and $\Phi(m')$ intercommute if m and m$'$ have spacelike separation. This is a requirement of relativistic causality.

(II) $$U(g)\Phi(m)U(g)^* = \Phi(gm) \quad \text{and} \quad U(g)\Omega = \Omega \text{ for } g \text{ in } G. \tag{A8}$$

These are the requirements for a relativistically covariant field and invariant vacuum. Evidently, the second condition implies that $H\Omega = 0$.

(III) H is a positive operator, i.e. the vacuum is a ground state.

The field theory is centred on the Wightman distributions

$$D(m_1, \ldots, m_n) = (\Omega, \Phi(m_1) \ldots \Phi(m_n)\Omega), \tag{A9}$$

since the cyclicity of Ω w.r.t. the field algebra ensures that these carry all the properties of the states corresponding to the vectors and density matrices in **H**. The key to the theory is that axioms (I)-(III) imply that the D's have excellent analyticity properties [15]. These first arise from the fact that, by

(III), Ω is a ground state (cf. discussion just before (A1)), and are enhanced by (I) and (II). Most importantly, they permit an analytic continuation of $D(m_1(\lambda_1), \ldots m_n(\lambda n))$ By (3), under Lorentz boosts, given by (A3), to imaginary values of the λ's. By (3), such a continuation converts boosts into rotations in the x-t plane; and, in the particular case where the λ's take the value $i\pi$, to space-time inversions $\to -x$, $t \to -t$.

PCT and Thermalisation. The implementation, described above, of space-time inversion, by analytic continuation of the Wightman functions, has led to a general derivation of the celebrated PCT theorem [15] from axioms (I)-(III). It has also led to a closely related theorem, due to Bisognano and Wichmann [16], which tells us that, if $V(\lambda)$ is the element of $U(g)$ that implements the Lorentz boost $m \to m(\lambda)$, given by (A3), then

$$(V(i\pi)A\Omega,\ V(i\pi)B\Omega) = (B^*\Omega,\ A^*\Omega), \text{ for all } A, B \text{ in } \mathbf{A}(W) \qquad \text{(A10)}$$

On comparing this formula with (A1), we see that it signifies that the restriction of the vacuum state to W satisfies the KMS condition w.r.t. Lorentz boosts, for $\beta = 2\pi/\hbar$. Furthermore, as noted after (A7), these boosts correspond to time-translations for a uniformly accelerated observer in W. Hence, on converting to the proper time scale, as given by (A7), for such an observer, we see that he perceives the inertial observer's vacuum as a thermal state of temperature

$$T = \alpha\hbar/2\pi kc. \qquad \text{(A11)}$$

This is the generalised Unruh effect, whereby the gravitational field, corresponding to the acceleration α, thermalises the quantum field to this temperature T.

The Hawking Effect. The axiomatic theory [3] of quantum fields on a Black Hole background has some parallels with that of fields in flat space-time, the exterior manifold corresponding to the Rindler wedge and its Kruskal extension [17] to the Minkowski space. It also has some important differences from the flat space-time theory, most notably because the lack of global time translational isometries of its extended manifold precludes a definition of a global vacuum along the lines of the Wightman prescription. However, a boundary condition [12] at the surface of the Black Hole, representing its origination from the collapse of a Star, leads to a Wightman theory, for the field on the past horizon (corresponding to the surface x+ct = 0 of Minkowski space) and hence to an Unruh-type thermalisation effect there [3]. Furthermore, this extends, by continuity and stability, to the whole of the exterior region, with the result that the field there has the Hawking temperature, as given by (3). The essential reason why this temperature has the same form as that of Unruh is that the Hawking effect arises from the action of the surface gravity, which corresponds to a constant acceleration (cf. (B2), Sec.2), on the quantum field.

REFERENCES

1. J. Bekenstein: Phys. Rev. D7, 2333 (1973); Phys. Rev. D12, 3292, (1974)
2. S.W. Hawking: Commun.Math.Phys. 43, 199, (1975)
3. G.L. Sewell: Ann.Phys. 41, 201, (1982)
4. C.W. Misner, K.S. Thorne and J.A. Wheeler: "Gravitation", Freeman, San Francisco, 1973
5. This observation was credited to J.A.Wheeler by J.Bekenstein: Physics Today, P.24, Jan.1980
6. J.M. Bardeen, B. Carter and S.W. Hawking: Commun.Math.Phys.31, 161 (1973)
7. S.W. Hawking: Phys.Rev. D13, 2188 (1976)
8. P.C.W. Davies: Rep.Prog.Phys. 41, 1313, (1978)
9. J. Von Neumann: "Mathematical Foundations of Quantum Mechanics", Princeton University Press, 1955
10. W. Thirring: "Quantum Mechanics of Large Systems", Springer, N.Y., Vienna, 1983
11. L.D. Landau and E.M. Lifschitz: "Statistical Physics", Pergamon Press, London, Paris, 1959
12. W. Unruh: Phys.Rev. D14, 870, (1976)
13. G.G. Emch: "Algebraic Methods in Statistical Mechanics and Quantum Field Theory", Wiley-Interscience,London, N.Y., 1971
14. R. Haag, N.M. Hugenholtz and M. Winnink: Commun.Math.Phys. 5, 215, (1967)
15. R.F. Streater and A. Wightman: "PCT, Spin and Statistics and All That". Benjamin, N.Y., Amsterdam, 1964
16. J.J. Bisognano and E.H. Wichmann: J.Math.Phys. 16, 985, (1975)
17. M.D. Kruskal: Phys.Rev. 119, 1743, (1960)

REMARKS ON ASYMPTOTIC PROPERTIES OF GROUPS OF THE BOGOLIUBOV TRANSFORMATIONS OF CAR C*-ALGEBRAS

A.G.Shuhov Yu.M.Suhov
Institute for Problems of Information Transmission
USSR Academy of Sciences
101447 Moscow GSP - 4

A convergence theorem is proven for states of CAR C*-algebras subjected to a "non-degenerate" group of the Bogoliubov transformations. Also the case of "degenerate" spherically symmetric groups is considered for which the analogous result is established.

1. Introduction

Groups of the linear canonical, or Bogoliubov, transformations (LCT) of quasilocal C*-algebras form the simplest class of *-automorphisms which admit a detailed investigation. From the point of view of Statistical Mechanics the question about the asymptotic behavior of the time-shifted state is of a major interest. A natural example of an LCT group arises when one deals with the free motion of quantum systems. It is well-known (see, e.g., [1], [2]) that the state Q_t obtained in the course of the free evolution from an initial state Q which has a property of decay of space correlations converges, as $t \to \pm\infty$, to a stationary quasi-free state P. It is interesting to investigate how general is such a situation (note that the group of identity LCT's leaves invariant any state).

General LCT groups commuting with space translations on CAR C*-algebras were studied in[3](see also [4], [5]). Under certain conditions of non-degeneracy of the generator of the group it was proven that, as for the free motion, the state Q_t obtained in the course of time from an initial state having properties of space mixing converges to a quasifree state P.

This paper contains generalizations of the results of [3]. The most important remark is that, in the case of CAR C*-algebras over $\ell_2(\mathbb{Z}^\nu)$ and $L_2(\mathbb{R}^\nu)$ with $\nu \geqslant 2$, one does not need to use the singularity theory. Also the class of initial states for which the convergence theorem holds is extended. This is the subject of Section 2.

On the other hand, there exist natural examples of degenerate groups for which the convergence theorem is still valid. As a rule, such examples are related to additional "symmetries" of the LCT group. Having a "good" symmetry one can weaken non-degeneracy conditions. Such a possibility is realized in Section 3 where the case of the spherical symmetry is considered. The spherical symmetry arises naturally when one deals with the wave and Klein - Gordon equations. Asymptotic properties of the corresponding groups were studied (in the classical situation) in [6] - [9].

2. Convergence to the quasifree state for non-degenerate LCT groups

Let us remind basic facts which are used below. By $\mathfrak{A}\ (=\mathfrak{A}_-)$ we denote the CAR C*-algebra with the generators $a^+(f)$, $a(g)$, $f, g \in \mathfrak{V}$, where $\mathfrak{V}$ is a complex Hilbert space. We suppose that a unitary group $\mathbb{U} = \{U_x\}$ is given acting in $\mathfrak{V}$ where $x \in \mathbb{Z}^\nu$ (lattice case) or $\mathbb{R}^\nu$ (continuous case), and that $\mathbb{U}$ has the simple Lebesgue spectrum. The element $a^+(f)$ is linear in f and $a(g)$ antilinear in g.

Let $\{\mathbb{T}_t,\ t \in \mathbb{R}^1\}$ be a one parameter group of (2 X 2)-matrices acting in $\mathfrak{V} \oplus \mathfrak{V}$:

$$\mathbb{T}_t = \begin{pmatrix} T_t^{(1)} & T_t^{(2)} \\ T_t^{(2)} & T_t^{(1)} \end{pmatrix} \qquad (1)$$

where $T_t^{(1)}$ is a linear and $T_t^{(2)}$ antilinear bounded operator $\vartheta \longrightarrow \vartheta$ commuting with U_x, continuous in t in the strong topology, and the following relations hold: $T_t^{(2)*}T_t^{(1)} + T_t^{(1)*}T_t^{(2)} = T_t^{(1)}T_t^{(2)*} + T_t^{(2)}T_t^{(1)*} = 0$, $T_t^{(2)*}T_t^{(2)} + T_t^{(1)*}T_t^{(1)} = T_t^{(2)}T_t^{(2)*} + T_t^{(1)}T_t^{(1)*} = I$. The formula

$$\mathcal{T}_t\, a^+(f) = a^+(T_t^{(1)}f) + a(T_t^{(2)}f) \qquad (2)$$

defines the group $\{\mathcal{T}_t,\ t \in \mathbb{R}^1\}$ of *-automorphisms of the C*-algebra $\mathfrak{U}$ (the group of the Bogoliubov transformations (LCT)).

The group $\{\mathbb{T}_t\}$ is determined by its infinitesimal matrix $\mathbb{D}$: $\mathbb{T}_t = \exp(it\mathbb{D})$, $t \in \mathbb{R}^1$, which reads as

$$\mathbb{D} = \begin{pmatrix} B & C \\ C & B \end{pmatrix}$$

where B is a linear and C an antilinear operator in ϑ.

In the spectral (Fourier) representation when ϑ is realized as $L_2([-\pi,\pi)^\nu)$ (lattice case) or $L_2(\mathbb{R}^\nu)$ (continuous case) and U_x as the multiplication by $\exp(i(x\cdot\theta))$, $\theta \in [-\pi,\pi)^\nu$ or $\mathbb{R}^\nu$, the operator B is the multiplication by a real function b and C is the product of the operator $J : f \longrightarrow \bar{f}(-\cdot)$ and the multiplication by an odd function c. The operator $T_t^{(1)}$ is the multiplication by

$$e^{itb_2}\left(\cos(tw) + \frac{ib_1}{w}\sin(tw)\right) \qquad (3)$$

and $T_t^{(2)}$ is the product of J and the multiplication by

$$i e^{it b_2} \frac{c}{w} \sin(tw) \tag{4}$$

where

$$b_1(\theta) = \tfrac{1}{2}\,(b(\theta) + b(-\theta)),\quad b_2(\theta) = \tfrac{1}{2}\,(b(\theta) - b(-\theta)), \tag{5}$$

$$w = (b_1^2 + |c|^2)^{1/2}. \tag{6}$$

The non-degeneracy conditions are formulated in the sequel in terms of the functions

$$\omega_\pm = b_2 \pm w. \tag{7}$$

Given a state Q of $\mathfrak{A}$, the formula

$$Q_t(A) = Q(\mathcal{T}_{-t} A), \quad A \in \mathfrak{A},$$

defines the family of states $\{Q_t,\ t \in \mathbb{R}^1\}$, the time evolution of $Q\ (= Q_0)$. It is not hard to check that a quasifree state P is invariant ($P_t = P$) iff its lower moment functionals $K_P^{(m,n)}$, $m+n \leq 2$, are invariant. The moment functionals, $K_Q^{(m,n)}$, of a state Q are given by

$$K_Q^{(m,n)}(f_1,\dots,f_m; g_1,\dots,g_n) = Q\Big(\prod_{1\leq j\leq m} a^+(f_j) \prod_{1\leq k\leq n} a(g_k)\Big),$$

$$f_j, g_k \in \mathcal{V},\ 1 \leq j \leq m,\ 1 \leq k \leq n,\quad m, n = 0, 1, \dots \tag{8}$$

We study the problem of convergence of the state Q_t, as $t \to \pm\infty$, to an invariant quasifree state P. The main restriction imposed on the initial state Q is related to the decay of the space (spectral) correlations in Q. Let us pass to the dual realisation of $\mathcal{V}$, i.e., assume that $\mathcal{V}$ is realised as $\ell_2(\mathbb{Z}^\nu)$ (lattice) or $L_2(\mathbb{R}^\nu)$ (continuum) and $\mathbb{U}$ as the group of the space translations. Given a state Q and $m, n = 0, 1, \dots,$ we set

$$\rho_Q^{(m,n)}(r,s) = \sup_{y \in \mathbb{R}^\nu} \sup |Q(A_1 A_2) - (-1)^{d(A_1,A_2)} Q(A_1)\, Q(A_2)|.$$

Here the internal $\sup$ is taken over the pairs of the monomials A_1, A_2 of the form

$$A_i = \prod_{0 \le j \le m(i)} a^+(f_j^{(i)}) \prod_{0 \le k \le n(i)} a(g_k^{(i)}), \quad 0 \le m(i) \le m,$$

$$0 \le n(i) \le n, \quad i = 1,2, \quad m(1) + m(2) = m, \quad n(1) + n(2) = n,$$

with $\|f_j^{(i)}\| \le 1, \|g_k^{(i)}\| \le 1$, $\operatorname{supp} f_j^{(1)} \subseteq I(y,r)$, $\operatorname{supp} g_k^{(1)} \subseteq I(y,r)$,

$\operatorname{supp} f_j^{(2)} \subseteq I(y, r+s)^c$, $\operatorname{supp} g_k^{(2)} \subseteq I(y, r+s)^c$, where $I(y,v)$ denotes the ν-dimensional cube $\times_{i=1}^{\nu} [y_i - v/2, y_i + v/2)$ (or its intersection with $\mathbb{Z}^\nu$). The exponent $d(A_1, A_2) = n(1)(m(2) + n(2))$.

In this section we assume that the coefficients $\rho_Q^{(m,n)}(r,s)$ satisfy the following condition. For $\nu \geqslant 2$

$$\lim_{s \to \infty} \rho_Q^{(m,n)}(s, \alpha s) = 0, \quad \forall \alpha > 0, \; m, n = 0, 1, \ldots, \tag{9}$$

and for $\nu = 1$

$$\lim_{s \to \infty} \limsup_{r \to \infty} \rho_Q^{(m,n)}(r,s) = 0, \quad \forall\, m, n = 0, 1, \ldots . \tag{10}$$

Denote

$$\beta_2(\omega_\varepsilon) = \left\{ \theta : \det \frac{\partial^2 \omega_\varepsilon}{\partial \theta_j \partial \theta_k} = 0 \right\}, \quad \varepsilon = \pm . \tag{11}$$

The result of this section is

Theorem 1. Let $b, c \in C^\infty$, $\beta_2(\omega_+) \cup \beta_2(\omega_-)$ be a set of zero Lebesgue measure and an initial state Q obey (9), (10). Then the state Q_t converges (in the w^*-topology), as $t \to \pm\infty$, to a quasifree state P iff the lower moment functionals $K_{Q_t}^{(m,n)}$, $m + n \le 2$, (weakly) converge to $K_P^{(m,n)}$.

Sufficient conditions for convergence of the lower moment functionals are formulated in [3] (they are related to a "periodicity" property w.r.t. $\mathbb{U}$).

The proof of Theorem 1 is reduced to checking the condition of Theorem 1 from [10] (see also [3], Proposition 1.1). This is a non-commutative version of the Bernstein method (see [11], Ch.XVIII). It is convenient to realize $\mathfrak{V}$ as $l_2(\mathbb{Z}^\nu)$ or $L_2(\mathbb{R}^\nu)$. We introduce the partition of $\mathbb{Z}^\nu$ or $\mathbb{R}^\nu$ into cubic "rooms" and complementary "corridor" which depend on t (see Fig. 1).

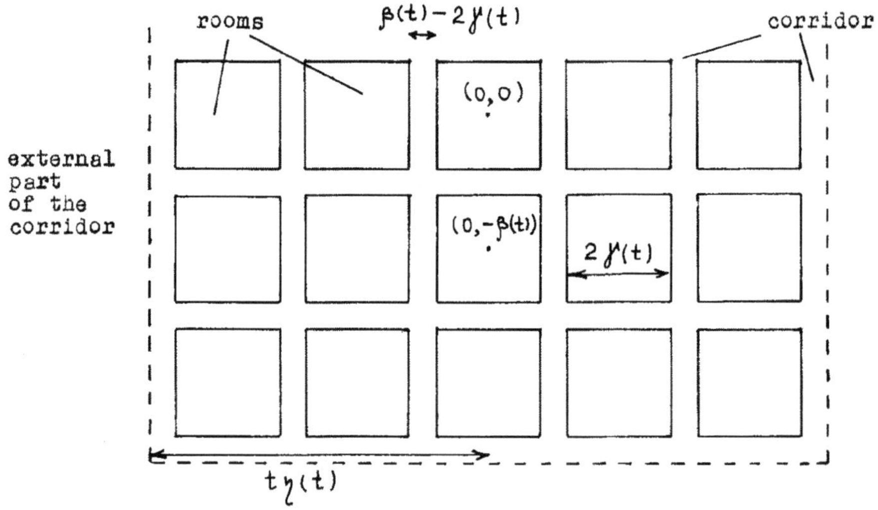

Figure 1

Our aim is to prove that the value $K_{Q_t}^{(m,n)}(f_1,\dots,f_m;g_1,\dots,g_n)$ converges, as $t \longrightarrow \pm\infty$, to $K_P^{(m,n)}(f_1,\dots,f_m;g_1,\dots,g_n)$ $(m+n \geq 3)$. Clearly, it is sufficient to check this fact for a dense set $\tilde{\mathfrak{V}} \subseteq \mathfrak{V}$. It is convenient to take

$$\tilde{\mathfrak{V}} = \left\{ f \in \mathfrak{V} : \hat{f} \in C_0^\infty,\ (\operatorname{supp}\hat{f}) \cap (\beta_2(\omega_-) \cup \beta_2(\omega_+)) = \emptyset \right\}.$$

For definiteness, assume that $t > 0$. According to the partition, the value $K_{Q_t}^{(m,n)}(f_1,\dots,f_m;g_1,\dots,g_n)$ is written as an $(m+n)$-tuple sum. The parameters $\beta(t)$, $\gamma(t)$, $\eta(t)$ are chosen in such a way that the contribution of the corridor as well as that of multiple products of the creations and annihilations related to a single room vanishes as $t \longrightarrow \infty$. The remainder is investigated by means of relations (9), (10) .

Formally, the first step of the proof, the estimation of the contribution of the external part of the corridor $(I(0,2t\eta(t))^c)$ proceeds by repeating the arguments used on the first stage of the proof of Theorems 1 - 6 from [3] . Next, for any $f \in \tilde{\mathcal{V}}$ uniformly in $x \in \mathbb{R}^\nu$ the following bound holds for sufficiently large t :

$$\left|\int d\theta\, e^{-i\theta\cdot x + it\omega_\pm(\theta)} \hat{f}(\theta)\right| < const \cdot t^{-\nu/2}, \quad f \in \tilde{\mathcal{V}}, \tag{12}$$

(see, e.g., [12] , Ch. III, Theorem 3.1). Then the rest of the proof is performed in the same way as for Theorems 1 - 6 of [3].

Remark. The conditions of smoothness of functions b and c may be weakened: as one can see from the proof, any singularities are allowed on a closed set of zero Lebesgue measure.

3. The spherically symmetric case

In this section we consider the multidimensional continuous case ($\mathcal{V} = L_2(\mathbb{R}^\nu)$, $\nu \geqslant 2$) under the additional assumption: the functions $b, c : \mathbb{R}^\nu \longrightarrow \mathbb{C}^1$ which determine the infinitesimal generator $\mathbb{D}$ of $\{\mathbb{T}_t\}$ depend only on the norm $\|\theta\|$ of $\theta \in \mathbb{R}^\nu$: $b(\theta) = b^0(\|\theta\|)$, $c(\theta) = c^0(\|\theta\|)$ and hence, $\omega_\pm(\theta) = \omega_\pm^0(\|\theta\|)$. We assume that the initial state Q satisfies the following condition: there exists $\delta_0 > 0$ such that

$$\lim_{s\to\infty} s^{\delta_0} \rho_Q^{(m,n)}(s^{1-\delta_0}, \alpha s^{1-\nu\delta_0}) = 0, \ \forall \alpha > 0, \quad m, n = 0, 1, \dots \tag{13}$$

Theorem 2. *Let $\ell^0, c^0 \in C^\infty$, the derivatives $(\omega^0_\pm)'$ vanish on a set of zero Lebesgue measure only and Q obey (13). Then the assertion of Theorem 1 holds: the state Q_t converges to a quasifree state P iff the lower moment functionals $K^{(m,n)}_{Q_t}$, $m+n \leq 2$, converge to $K^{(m,n)}_P$.*

Remark. Conditions of Theorem 2 allow us to consider, e.g., the LCT group with $c \equiv 0$ and $\ell(\theta) = \|\theta\|$. This was not possible to do by means of Theorem 1.

We start with an auxiliary assertion which is perhaps of interest itself. Again assume that $t > 0$.

Lemma 3.1. *Let $\nu \geq 2$, $\varphi \in C_0^\infty(\mathbb{R}^\nu)$ and $(\operatorname{supp} \varphi) \cap \{0\} = \emptyset$. Given $y \in \mathbb{R}^\nu$, we set*

$$\mathcal{J}(y) = \int_{\mathbb{R}^\nu} d\theta \, e^{-i\theta\cdot y + it\|\theta\|} \varphi(\theta) . \tag{14}$$

Then, for t large enough, the following bounds are fulfilled:

a) $|\mathcal{J}(y)| < c_1 t^{-(\nu-1)/2}$ *uniformly in* $y \in \mathbb{R}^\nu$,

b) *for any* $N > 0$, *any* $\delta \in (0,1)$ *and any* $y \in \mathbb{R}^\nu$ *with* $|\,\|y\| - t\,| > t^\delta$, $|\mathcal{J}(y)| < c_2 t^{-N}$

where $c_1, c_2 = c_2(N,\delta) > 0$ *are constants.*

Remark. The assertion a) of Lemma 3.1 is well-known in the literature and proven, e.g., in [13] (see [13], Theorem XI.18)). Also the assertion b) for $\delta = 1$ (replacing t^δ by εt, $\varepsilon \in (0,1)$) was proven (see [13], Theorem XI.18a)). However, we need to take $\delta \in (0,1)$. The proof is based on standart arguments and given here for completeness.

Proof of Lemma 3.1. Passing to spherical coordinates ($r = \|\theta\|$, $a = \|y\|$, α = the angle between θ and y), we write $\mathcal{J}(y)$ in the form

$$\mathcal{J}(y) = \int d\psi \int d\alpha \int dr\, e^{-ira\cos\alpha + itr}\, r^{\nu-1} (\sin\alpha)^{\nu-2}\, \varphi_{(y)}(r,\alpha,\psi) \qquad (15)$$

where $d\psi$ is the surface element of the unit sphere in the $(\nu-1)$ - dimensional subspace orthogonal to y and $\varphi_{(y)}$ is a C_0^∞ -function which is bounded uniformly in y together with its derivatives and vanishes at 0 .

To estimate (15) it suffiches to estimate the integral

$$\mathcal{J}_1(y) = \int_0^\pi d\alpha \int_{r_0}^\infty dr\, e^{i(tr - ra\cos\alpha)}\, r^{\nu-1} (\sin\alpha)^{\nu-2}\, \varphi^{(y)}(r,\alpha) \qquad (16)$$

where $\varphi^{(y)} \in C_0^\infty$ is uniformly bounded together with its derivatives and $r_0 = \inf\{r : r \in \operatorname{supp} \varphi^{(y)}\}$. If $a < t/2$, then $|t - a\cos\alpha| > t/2$ and $|\mathcal{J}_1(y)|$ decreases faster than t^{-N} for any $N > 0$ (see [12], Ch. III, Lemma 1.1). Therefore, we can assume that $a > t/2$. Integrating by parts we rewrite (16) as

$$\mathcal{J}_1(y) = \int_{r_0}^\infty dr\, e^{itr} (ia)^{-1} \Big\{ e^{-ira\cos\alpha}\, r^{\nu-2} (\sin\alpha)^{\nu-3}\, \varphi^{(y)}(r,\alpha) \Big|_0^\pi - $$
$$- \int_0^\pi d\alpha\, e^{-ira\cos\alpha}\, r^{\nu-1} \frac{\partial}{\partial\alpha}\big((\sin\alpha)^{\nu-3} \varphi^{(y)}(r,\alpha)\big) \Big\} . \qquad (17)$$

For $\nu > 3$ the boundary term vanishes. Integrating by parts $[(\nu-1)/2]$ times we get that, for even number ν,

$$|\mathcal{J}_1(y)| \le \mathcal{J}_2(y) = a^{-(\nu-2)/2} \Big| \int_0^\pi d\alpha \int_{r_0}^\infty dr\, e^{itr - iar\cos\alpha}\, \varphi_1^{(y)}(r,\alpha) \Big| \qquad (18)$$

and for odd ν

$$|\mathcal{J}_1(y)| \le a^{-(\nu-2)/2} \Big| \int_{r_0}^\infty dr\, e^{irt} \big(e^{-ira\cos\alpha}\, \varphi_2^{(y)}(r,\alpha)\big) \Big|_0^\pi \Big| +$$
$$+ a^{-(\nu-1)/2} \Big| \int_{r_0}^\infty dr \int_0^\pi d\alpha\, e^{irt - ira\cos\alpha}\, \varphi_3^{(y)}(r,\alpha) \Big| \qquad (19)$$

where $\varphi_1^{(y)}, \varphi_2^{(y)}, \varphi_3^{(y)}$ are C^∞-functions having the same properties as $\varphi^{(y)}$ in (16).

Fix $\delta \in (0,1)$. If $|t-a| > t^\delta$, then the boundary term in (19) decreases faster than t^{-N} for any $N>0$ (see [12], Ch. III, Lemma 1.1). If, on the contrary, $|t-a| \leq t^\delta$, then it does not exceed $\mathrm{const}\cdot t^{-(\nu-1)/2}$. Hence, to finish the proof of Lemma 3.1 it is sufficient to estimate the RHS of (18) (the second term in the RHS of (19) does not essentially differ on it). As before, $J_2(y) < \mathrm{const}\cdot t^{-N}$ for any $N>0$ provided $a < t - t^\delta$. The uniform bound is obtained in the following way:

$$J_2(y) < a^{-(\nu-1)/2} \sup_{r \in \operatorname{supp} \varphi_1^{(y)}} \Big\{ \Big| \Big(\int_0^{\pi/4} + \int_{3/4\,\pi}^{\pi} \Big) d\alpha\, e^{-ira\cos\alpha} \times$$

$$\times \varphi_1^{(y)}(r,\alpha) \Big| + \Big| \int_{\pi/4}^{3\pi/4} d\alpha\, e^{-ira\cos\alpha} \varphi_1^{(y)}(r,\alpha) \Big| \Big\} <$$

$$< c_1 a^{-(\nu-1)/2} < c_2 t^{-(\nu-1)/2}$$

where c_1, c_2 are some constants. The second inequality follows from the bound (12) and from Lemma 1.1 from [12], Ch. III.

Let now $a > t + t^\delta$ with $0 < \delta < 1$. We set $\alpha_0 = \arccos(t/a)$ and write the integral in (18) as the sum

$$\int_{r_0}^{\infty} dr \int_0^{\alpha_0/2} d\alpha + \int_{r_0}^{\infty} dr \int_{\alpha_0/2}^{\pi/2} d\alpha + \int_{r_0}^{\infty} dr \int_{\pi/2}^{\pi} d\alpha . \tag{20}$$

We have

$$\inf_{\alpha \in [0,\alpha_0/2]} |t - a\cos\alpha| = a|\cos\alpha_0 - \cos(\alpha_0/2)| \geq a \frac{\alpha_0}{2} \sin(\alpha_0/2) >$$

$$> a \sin^2(\alpha_0/2) = (a/2)(1-\cos(\alpha_0/2)) = \frac{a}{2}\Big(\frac{a-t}{a}\Big) > \frac{1}{2} t^\delta .$$

Integrating subsequently by parts (in r) we get that the first term in (20) decreases faster than t^{-N} for every $N>0$. The third term is estimated via the bound $\inf_{\alpha\in[\pi/2,\pi]} |t-a\cos\alpha| \geq t$.

Finally, for estimating the second term in (20), we write it in the form

$$\int_{r_0}^{\infty} dr\, e^{irt} \left\{ e^{-ira\cos\alpha} \frac{\vartheta_1^{(\gamma)}(r,\alpha)}{ira\sin\alpha} \Big|_{\alpha_0/2}^{\pi/2} - \int_{\alpha_0/2}^{\pi/2} d\alpha\, e^{-ira\cos\alpha} \times \right. \tag{21}$$

$$\times \frac{1}{ira} \left(\frac{\vartheta_1^{(\gamma)}(r,\alpha)}{\sin\alpha} \right)'.$$

Since $|t-a\cos\alpha_0/2| > \frac{1}{2} t^{\delta}$, the first addend in (21) decreases faster than t^{-N} for any $N>0$. Using the second mean-value theorem and integrating subsequently m times by parts we obtain that the integral term does not exceed

$$\text{const}\cdot a^{-m}(\sin\alpha_0/2)^{-2m} < \text{const}\cdot t^{-\delta m}.$$

Lemma 3.1 is proven.

Lemma 3.2. <u>Let</u> $\omega_{\pm}(\theta) = \|\theta\|$ <u>and the state</u> Q <u>satisfy</u> (13). <u>Then the assertion of Theorem 1 holds true</u>.

<u>Proof of Lemma</u> 3.2. It is sufficient to verify the conditions of Theorem 1 from [10]. This proceeds essentially in the same way as the proof of Theorem 1 from Section 2. The main difference is connected with the construction of a new room-corridor partition. Fix $\delta\in(0,1)$ and monotonic functions $\beta(t)$, $\gamma(t)$ increasing to ∞ with $t\to\infty$ such that

(a) $\lim_{t\to\infty} \beta(t)\, t^{-1+\delta/\nu-1} = 0$, $\lim_{t\to\infty} (\beta(t)-2\gamma(t)) = 0$,

(b) $\lim_{t\to\infty} (t\beta(t)^{-1})^{\nu-1} \rho_Q^{(m,n)}(2\gamma(t)(1+t^{\delta-1}), \beta(t)(1-t^{\delta-1})) = 0$,

$$(c).\ \lim_{t\to\infty} t^{\delta}\left(1-(2\gamma(t)/\beta(t))^{\nu-1}\right) = 0$$

(such functions exist due to (13)). The role of the external part of the corridor is played now by the complement $(\mathcal{O}^{(t)})^c$ where $\mathcal{O}^{(t)} = \{x \in \mathbb{R}^{\nu} : |\|x\| - t| < t^{\delta}\}$. An appropriate partition of $\mathcal{O}^{(t)}$ may be constructed in many ways. For example, one can start with a partition of the sphere $S^{(t)} = \{x : \|x\| = t\}$ and take conic "extentions" of its elements (see Fig. 2a). Moreover, the partition of $S^{(t)}$ may be constructed by dividing it, at first, into 2 hemispheres, $S_{\pm}^{(t)} = \{x = (x_1, \dots, x_{\nu}) \in S^{(t)} : x_{\nu} \gtrless 0\}$ and by fixing homeomorphisms $S_{\pm}^{(t)} \longleftrightarrow D^{(2t)}$ where $D^{(2t)}$ is a $(\nu-1)$-dimensional disk $\{y = (y_1, \dots, y_{\nu-1}) : \|y\| \le 2t\}$ under which all distances increase at most twicely (one can take, e.g., the "stereographic" projections from the poles $(0, \dots, 0, \pm t)$ onto the hyperplanes $W_{\pm} = \{x \in \mathbb{R}^{\nu} : x_{\nu} = \pm t\}$). After this it remains to construct a suitable partition of the disk $D^{(2t)}$. For example, it is convenient to take $\beta(t)$ times diluted $(\nu-1)$-dimensional lattice $\mathbb{Z}^{\nu-1}$ and to extract from every cell the cube of the edge length $2\gamma(t)$ (cf. constructions of Section 2). See Fig. 2b. This completes the proof of Lemma 3.2.

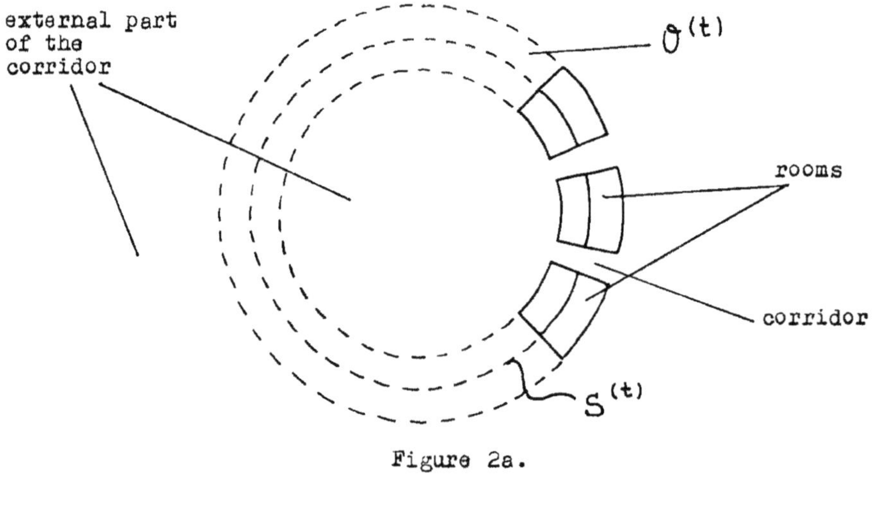

Figure 2a.

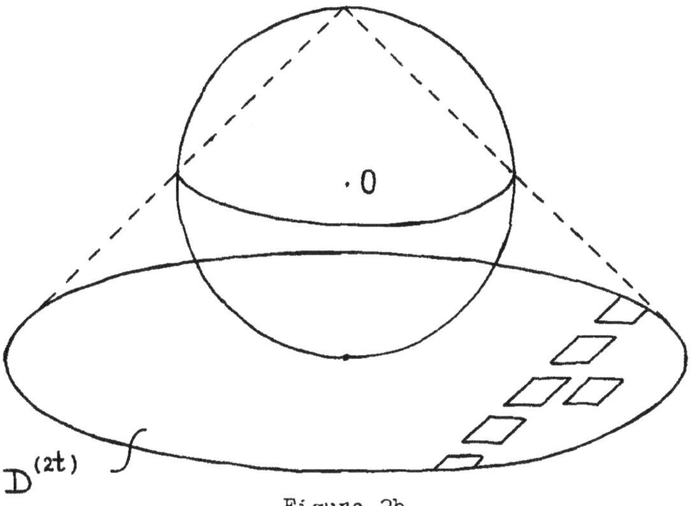

Figure 2b.

Proof of Theorem 2. Notice that the condition $(\omega_{\pm}^{0})'(\|\theta\|)\cdot(\omega_{\pm}^{0})''(\|\theta\|) = 0$ is equivalent to the condition $\det\left(\frac{\partial^2\omega_{\pm}}{\partial\theta_j\partial\theta_k}\right)(\theta) \neq 0$ which figures in Theorem 1. Thereby, the proof of Theorem 2 is reduced to a combination of the arguments used in the proof of Theorem 1 of Section 2 (i.e., of Theorems 1 - 6 from [3]) and in the proof of Lemma 3.2. Here we need a "combined" partition in which the two structures coexist: the cubic one figuring in the proof of Theorem 1 and the spherical one figuring in the proof of Lemma 3.2.

Remark. The methods of this paper allow us to prove the convergence theorem under the assumption that the functions $\omega_{\pm}$ satisfy the conditions of Theorems 1 and 2 only in a "local" sense. More precisely, for every θ, except a closed set of zero Lebesgue measure, one must guarantee the existence of a neighbourhood where the functions $\omega_{\pm}$ either have non-vanishing Hessian or are of the form $d_{\pm}\|\theta\| + c_{\pm}$ where $c_{\pm}$, $d_{\pm}$ are constants.

R E F E R E N C E S

1. Lanford O.E., Robinson D.W. - Commun. Math. Phys., 1982, 24, 193-210.
2. Haag R., Kadison R.V., Kastler D. - Commun. Math. Phys., 1973, 33, 1 - 22.
3. Shuhov A.G., Suhov Yu.M. - Ann. Phys., 1987, 175.
4. Suhov Yu.M., Shuhov A.G. - Doklady AN SSSR (Russian), 1985, 284, N° 5, 1076-1079.
5. Shuhov A.G., Suhov Yu.M. - In: Statistical Physics and Dynamical Systems. Rigorous Results. 1985. Boston - Basel - Stuttgart: Birkhäuser, 83-104.
6. Ratanov N.E. - Uspekhi Mat. Nauk (Russian), 1984, 39, N° 4, 112.
7. Ratanov N.E. - Uspekhi Mat. Nauk (Russian), 1984, 391 N° 1, 151-152.
8. Kopylova E.A. - Uspekhi Mat. Nauk (Russian), 1985, 40, N° 5, 240-241.
9. Kopylova E.A. - Vestnik MGU (Russian), ser. I (Mat., Mech.), 1986, N° 2, 92-95.
10. Suhov Yu.M. - Izvestia AN SSSR (Russian), ser. Mat., 1984, 48, N° 1, 155-191.
11. Ibragimov I.A., Linnik Yu.V. Independent and stationary sequences of random variables. 1971. Groningen: Wolters-Noordhoff Publishing.
12. Fedoriuk M.V. Method perevala (Russian). 1977. Moscow: Nauka.
13. Reed M., Simon B. Methods of modern mathematical physics. III: Scattering theory. 1979. New York-San Francisco-London: Academic Press.

Linear and non-linear stochastic processes

R.F. STREATER

Dept. of Mathematics, King's College, London

There are two main problems in non-equilibrium statistical mechanics. In the first, a system S having few degrees of freedom is placed in thermal contact with a large system B in a thermal state, the heat bath. S warms up or cools down to approach the thermal state at the same beta as B. In the second problem we ask how did the heat-bath get to its thermal state in the first place ? This is the question of the approach to equilibrium of an autonomous system.

There are two ways to tackle these problems, which might be described as fundamentalist and stochastic, respectively. In the first we describe B by a large but finite number of degrees of freedom, with Hamiltonian H_B and state ω_β S is described by its Hamiltonian H_S in some initial state ω_o; the combined system initially is in the state $\omega_\beta \otimes \omega_o$, so that B and S are initially independent. Let H_1 be an operator (or function, in classical mechanics) involving some dynamical variables of both B and S; these are then coupled by taking for the combined system the Hamiltonian

$$H = H_B + H_S + \lambda H_1$$

Commonly, H_B is modelled by Bosonic harmonic oscillators and H_1 is bilinear, describing the exchange of quanta between B and S.

One can study the time-development of dynamical variables of S for small λ as the number of degrees of freedom of B becomes large. One aims to prove that for longish times the dynamical variables of S appear to be nearly thermal most of the time.

In the stochastic method, one introduces randomness by hand.

Thus for the first problem one directly describes the dynamics of S by a contraction semi-group of completely positive maps on the algebra of observance of S[1]; these are linear maps, whose dual action describes dissipative motion on the state space. Models including noise are also constructed, by setting up a suitable Langevin equation; or in quantum mechanics, a stochastic equation driven by quantum noise [2]. These two approaches are neatly related by the dilation programme [3]: given a continuous one-parameter semigroup of completely positive maps, its generator determines some operators giving the Lindblad form. These operators can be used to construct a quantum stochastic differential equation in the sense of Hudson and Parthasarathy, and the solution is a unitary dilation of the semigroup [3]. One might interpret this as a Hamiltonian motion, and the noise operators as modelling B.

In spite of the success of this programme the concept of quantum Brownian motion and its Fermion analogue has aroused some hostility. Why is this ? The origin is R. Kubo's telling intervention at the All Japan Conference on Statistical Mechanics [4]. Kubo's remark is the quantum version of an old debate in the theory of noise in classical sytems. Mathematicians like to use Brownian motion, B_t, which is very Markovian; indeed, it is a martingale with independent increments. Its derivative $\dot{B}_t$ is white noise, that is, the generalised Gaussian process with

$$E(\dot{B}_t) = 0, \quad E(\dot{B}_s\dot{B}_t) = \delta(s-t)$$

The covariance of B is not the boundary value of an analytic function in the upper half-plane: this is because the spectrum (the support of its Fourier transform in s-t) contains the whole line, and is not restricted to positive energy. Since any real system has positive energy, Brownian motion is not a possible choice for noise.

An attempt to construct an "analytic noise" in quantum mechanics was made by Senitzsky [5]. Stevens [6] studied a classical model using an L-C circuit coupled to a large number of transmission lines. These appeared as a damping force and noise when the number became infinite. A satisfactory treatment was formulated by Ford, Kac and Mazur [6], and it was successfully quantized by Lewis and Thomas [7]. Maassen [8] has been able to add a non-linear force to the Lewis-Thomas process. In this work the energy in the noise is positive at zero temperature, so that one can formulate well-defined temperature states $\omega_\beta = Tr(e^{-\beta H}.)/Tr(e^{-\beta H})$. This obeys the KMS condition with respect to time-evolution $a \to a_t$; namely for any two operators a,b, $F_{ab}(t) = \omega_\beta(a_t b)$ is an analytic function of complex time in the strip $0 < Im\ t < \beta$; it is continuous on the boundary and obeys a periodicity condition coming from the cyclicity of the trace. The heat-bath of the Lewis-Thomas process is a good physical system, an infinite string attached to the oscillator. The system S does indeed approach the temperature of the string, and the dynamics of S is embedded in a Hamiltonian system, but one with an infinite number of degrees of freedom.

This is an example of a useful stability property of KMS states of infinite systems: a local disturbance dissipates to infinity and the state returns to equilibrium as $t \to \infty$. This is due to disappearance at infinity and not local dissipation; in this way an actually infinite model of a heat bath is better than a finite one. The dissipation of a finite Hamiltonian system is only apparent: it returns to coherence in due course, whereas for a realistic system, small random extraneous forces confirm the randomness and prevent order returning, by destroying correlations.

The incompatibility of Brownian noise, and indeed martingales of any sort, with the KMS condition is brought home by a theorem of Takesaki: if a conditional expectation from the whole algebra of observables A, onto the observabless up to time t, A_t, is compatible with a KMS state ω_0, then the KMS automorphism group

σ_s(= the time-evolution) leaves A_t invariant. This is contrary to the interpretation of A_t as the observables up to the time t: the automorphism of time-evolution should leap forward into new chance: $\sigma_s A_t = A_{t+s}$ is wanted.

This phenomenon was encountered in the theory of quantum K-systems [9], there, the KMS automorphism of each A_t was interpreted as 'microscopic' time, and the time-development of the process as "macroscopic" time. However, this is difficult to justify since in models with positive energy [7], [10], no analogues of these two times appears - the KMS automorphism moves the filtration $\{A_t\}$ forward, as it should. The interpretation of the quasi-free process [11], [12] as a heat-bath is also untenable, for the same reason.

There is no practical numerical difference between white noise and experimental results [13]; Frigerio (this volume) argues that martingales arise as certain limits of realistic models. No doubt Brownian motion and its quantum versions will continue to be studied and used as models of noisy systems.

Someone sincere in his belief in the use of probability theory in physics might say that a Hamiltonian theory of approach to equilibrium needs justifying. He would not believe that the heat bath is exactly harmonic; or that the coupling is bilinear, etc. Only if the reduced dynamics of S is independent of these details does he believe the model - the boot is on the other foot.

In quantum physics, impure states are the order of the day. A pure state of the larger system will usually be impure on restriction to the smaller system. Thus the view that every system is actually, out there in Nature, in a pure state (but we might not know which) is untenable in quantum theory. Indeed, this is the lesson of the EPR paradox. Quantum theory is already a probabilistic theory, and it is natural to use quantum noise to model unknowables, such as initial conditions, impurities, boundary conditions, unspecified couplings, cosmic rays etc. The stochastic equations one uses are non-linear evolution equations for the operators; but they can be embedded in automorphisms of the algebra of observables, and by duality, lead to linear maps on state space. We call these linear processes, and they are natural tools for the study of the first problem.

For the second problem, it is fair to say that only the dilute gas has any body of theory, and that is based on the Boltzmann equation; see Spohn [14], and for certain Fermion systems, Hugenhotz [15].

The Boltzmann equation is non-linear in the state; it is irreversible and gives an increase in entropy. None of these things is true of Hamiltonian dynamics. This has led some people to argue that the Boltzmann equation is merely an approximation, and that the true dynamics "out there" is reversible.

However, theoretical physics gives results, not directly about Nature, but about our description of it; they concern the model we use. No meaning can be given, for example, to the entropy of an actual bit of gas, outside the context of a model. In a detailed Hamiltonian theory in which the system initially is in a pure state, the entropy $s = \mathrm{Tr}\rho\log\rho$ is zero and remains zero for all time. In a stochastic model of the same gas, entropy will increase with time. Both might describe different aspects of experiment quite well, and other aspects rather poorly. The Hamiltonian description is very poor for long times.

I have studied [16] three Boltzmann equations referring to classical, quantum and second quantized systems respectively. Time is a discrete parameter, $t = 0,1,2,\ldots$

(a) A dilute gas of very many classical atoms each capable of being in a state j with energy $E_j, j = 0,1,\ldots N$. For definiteness we take $E_j = j$. The phase space of the atom is thus the sample space $\Omega = \{0,1,\ldots N\}$. We assume that a fraction p_j of the atoms lie in the state j, $j \in \Omega$, and interpret p_j as the probability that an atom, picked at random from the gas, has energy j. The Stosszahlansatz, or hypothesis of molecular chaos, then says that the probability that two atoms, picked at random from the gas, have energies j and k is $p_j p_k$. This is a probability on $\Omega \times \Omega = \Omega^2$. Suppose that the probability that two particles of energies j and k collide and produce two particles of energies ℓ and m is $T_{jk\ell m}$. T can create more entropy. We take T to be a symmetric and hence bistochastic matrix acting on Ω^2, so that $T_{jk,\ell m} = T_{\ell m,jk} = T_{kj,\ell m}$ and $T_{jk,\ell m} \geq 0$, $\sum_{j,k} T_{jk,\ell m} = 1$. We express energy conservation by having $T_{jk\ell m} = 0$ unless $j + k = \ell + m$. After scattering, the joint distribution of the two particles is $p'_{jk} = T_{jk\ell m} p_\ell p_m$. The new one-particle distribution is the marginal distribution of this:

$$p'_j = \sum_{\substack{k \\ j+k=\ell+m}} \sum_{\ell m} T_{jk\ell m} p_\ell p_m$$

Suppose time is in discrete units, a unit being large compared to the time needed to complete an atomic collision, and short compared to the time between collisions (possible if the gas is dilute). Then if $p_k(t)$ is the one-particle distribution at time t, then $p'_j = p_j(t + 1)$ is the distribution at time $t + 1$. It can be shown that the average energy is conserved: $\sum j p_j(t) = \sum j p_j(t + 1)$.

The map $p \to p'$ is a non-linear map on the state-space of Ω, called the Boltzmann map, τ. It is the combination of three stages.

1. the stosszahlansatz: $p \to p \otimes p$ on Ω^2
2. a linear bistochastic map $p \otimes p \to Tp \otimes p$
3. reduction to the marginal distribution of the first particle.

For example, if we have three energy-levels 0,1,2, then the state-space is the two-dimensional set

$$\{(p_0,p_1,p_2):p_j \geq 0, p_0+p_1+p_2 = 1\}.$$

The typical energy-conserving scattering matrix is

$T_{02,11} = \lambda = T_{11,02}$ etc, the rest of the off-diagonal elements being zero.

The Boltzmann map then becomes

$$p_0' = p_0 - \lambda(p_0p_2 - p_1^2)$$
$$p_1' = p_0 + 2\lambda(p_0p_2 - p_1^2)$$
$$p_2' = p_2 - \lambda(p_0p_2 - p_1^2)$$

from which

$p_0' + p_1' + p_2' = 1$ and $0p_0' + 1p_1' + 2p_2' = 0p_0 + 1p_1 + 2p_2$ is evident. The latter expresses energy conservation. The parameter $x = p_0p_2 - p_1^2$ is called a "disequilibrium parameter". Clearly if $x = 0$, the distribution (p_0,p_1,p_2) is in equilibrium. In that case, $p_0/p_1 = p_1/p_2 = e^{\beta}$ say, for some β. So this equilibrium state is a Gibbs state.

Return now to the general model. One can prove that τ increases entropy and, if T is ergodic on each energy subspace of Ω^2, then the iterated map τ^n takes any initial distribution to the Gibbs state with the same energy.

(b) In a quantum theory with Hilbert space $\mathcal{H}$ we describe a state by a density matrix ρ. Then the state of 2 independent particles is $\rho\otimes\rho$, a density matrix on $\mathcal{H}\otimes\mathcal{H}$. The collision is then given by a map $\rho\otimes\rho \to T^*(\rho\otimes\rho)$, where T is a completely positive (linear) bistochastic map on $B(\mathcal{H}\otimes\mathcal{H})$, the algebra of all bounded operators on $\mathcal{H}\otimes\mathcal{H}$, and the adjoint is relative to the Hilbert-Schmidt scalar product, $\langle a,b\rangle = Tr(a^*b)$.

Let H be the Hamiltonian on $\mathcal{H}$. Then the two-particle energy is $H_0 = H\otimes 1 + 1\otimes H$ on $\mathcal{H}\otimes\mathcal{H}$. If the collision is caused by a local potential then the S-operator conserves energy, and so commutes with H_0. The scattering automorphism $A \to SAS^*$ then leaves fixed the spectral resolutions of H_0. A wide class of completely positive bistochastic maps is obtained by averaging such maps:

$$TA = \sum_{\alpha} \mu(\alpha)\, S(\alpha)\, AS(\alpha)^{-1}, \text{ where } \mu \geqslant 0,\ \sum_{\alpha} \mu(\alpha) = 1.$$

We can imagine $S(\alpha)$ is the S-matrix of a random Hamiltonian $H_0+V(\alpha)$. As such it leaves the spectral projections $P(E)$, of H_0 invariant, and so does its adjoint map $A \to S(\alpha)^{-1}A\, S(\alpha)$. This leads us to define an energy-conserving bistochastic map T as one such that
$TP(E) = P(E) = T^*P(E)$, where $H_0 = \sum_E EP(E)$.

We obtain a non-linear map τ on density matrices by combining the maps

1. $\rho \to \rho\otimes\rho$
2. $\rho\otimes\rho \to T^*(\rho\otimes\rho)$, T an energy-conserving bistochastic (completely positive map on $B(\mathcal{H}\otimes\mathcal{H})$)
3. $T^*(\rho\otimes\rho) \to Tr_2\, T^*(\rho\otimes\rho)$
   where Tr_2 is the trace over the second factor.
   So $\tau\rho = Tr_2(T^*\rho\otimes\rho)$.

Again, under ergodic assumptions one shows that $\tau^n\rho$ converges to the Gibbs state, as $n \to \infty$, with the same energy as ρ, namely $\rho(H)$.

(c) In the second quantized version, we have n oscillators (boson or fermion) with energies $\omega_1,\ldots\omega_n$, and the Hilbert space is $K = L^2(\mathbb{R}^n)$. The Hamiltonian is $H_0 = \sum \omega_j a_j^* a_j = \sum EP(E)$. A doubly stochastic map $T\colon B(K) \to B(K)$ will conserve energy if $TP(E) = P(E) = T^*P(E)$ for each spectral resolution $P(E)$ of H_0. We claim that the Stosszahlansatz should remove the interparticle correlations after each collision. Let ρ be a state on $B(K)$ with finite two-point functions. Let $Q\rho$ be the quasifree state with the same two-point functions as ρ. We prove that the entropy of $Q\rho$ is $\geqslant$ the entropy of ρ. Since any bistochastic map T is entropy non-decreasing, the map $\tau\rho = QT^*\rho$ is non-decreasing in entropy, and in fact increases it unless ρ is quasifree. If T has some ergodic properties, we prove [16] that $\tau^n\rho$ converges to a Gibbs state if all the ratios ω_j/ω_k are mutually rational. (This is the condition for instability of the corresponding classical theory).

In each of these three models we obtain a non-linear map τ on the state space. This is what is wanted - no linear map has the property that for any initial state ρ, $\tau^n\rho$ converges to the Gibbs state (of energy $\varepsilon = \rho(H)$).

But some people find non-linear maps disturbing: Some mistake here, surely ? The answer lies in the remark that in each case the map τ involves only the one-particle distribution. The full state of the gas cannot be described by the initial one-body distribution. Different physical situations with the same initial one-body distribution p will evolve differently. Thus consider the two situations with the same p:

1. All molecules are in the same container in thermal contact with each other; p_j is the probability of finding a particle of energy E_j.
2. The container is divided into two cells, thermally isolated from each other; in the first cell, chosen with probability λ, the distribution is q_j. In the second cell, chosen with probability $(1-\lambda)$, the distribution is q'_j. Suppose $\lambda q_k + (1-\lambda)q'_k = p_k$.

Then the statistics of the one-body energy distribution is the same for both situations. Yet 1. would converge to a single Gibbs state, and 2. would converge to two, selected with probabilities λ and $(1-\lambda)$. Clearly, we would apply the Boltzmann map only to the first situation.

In fact, in situations where the Boltzmann equation has been justified in some well-defined sense [Spohn, Hugenholtz] the need for all parts to be in thermal contact is evident from the proof.

One might ponder how one can go about 'justifying' the Boltzmann map along the lines of dilation theory: is there a non-linear process X(t), whose distributions in the initial state obey this non-linear equation ? Such a question was posed first by McKean [17], in the classical case (with continuous time).

References

[1] E.B.Davies, Quantum Theory of Open Systems. Academic Press (1976).

[2] G.Sewell, Quantum Theory of Collective Phenomena. Oxford University Press (1986).

[3] B. Kümmerer (this volume)

D.E.Evans and J.T.Lewis. Dilations of dynamical semigroups; Commun. Math. Phys. 50,219-227 (1976)

R.L.Hudson and K.R.Parthasarathy Acta Appl. Math, 2, 353-378 (1984).

[4] R.Kubo; J. Phys. Soc. Japan 26, Suppl 1-5 (1969).

[5] I.R.Senitzky. Phys. Rev. 119,670 (1960).

[6] K.W.H.Stevens, Proc. Phys. Soc. 77, 515 (1961)
G.Ford, M.Kac and P.Mazur, J. Math. Phys. 6,504 (1965).

[7] J.T.Lewis and L.C.Thomas. Ann. Inst. H Poincaré, A22, 241-8 (1975).

J.T.Lewis, How to construct a heat bath, in: Proc. Conf. on Functional Integration (Ed. A.M.Arthurs), Academic, 1973.

[8] H.Massen; Ph.D. thesis, Groningen, 1982.

[9] G.G.Emch, S.Albeverio and J.-P.Eckman Quasi-free generalized K-Flows. Rep. Math. Phys. 13,73-85 (1978).

[10] R.F.Streater; Damped oscillator with quantum noise, J.Phys. A15, 1477-85 (1982).

[11] R.L.Hudson and J.M.Lindsay, J.Functl. Anal. 61-202 (1985).

[12] C.Barnett, R.F.Streater and I.F.Wilde J.Functl. Analysis 52, 19-47 (1983).

[13] John Smythe, Frank Moss and P.V.E.McClintock. Phys. Rev. Lett. 51, 1062 (1983).

[14] H.Spohn; Reviews of Modern Physics 52, 569-615 (1980).

[15] N.M.Hugenholtz. J.Stat. Phys. 32, 231-254 (1983).

[16] R.F.Streater, Publ. RIMS (Kyoto) 20, 913 (1985).

R.Alicki and J.Messer. J.Stat. Phys. 32, 299 (1983).

R.F. Streater, Commun. Math. Phys. 98, 177-185 (1985).

R.F.Streater. Entropy and the central limit theorem in quantum mechanics; to appear in J.Phys.A. See also E.H. Wichmann, J. Math. Phys. 4, 884-896 (1963).

R.F.Streater. A Boltzmann map for quantum oscillators; to appear in J.Stat. Phys.

[17] H.McKean. Proc. Natl. Acad. Sci. U.S.A. 56, 1907 (1966).

DETAILED BALANCE AND CRITICAL SLOWING DOWN

A. Verbeure
Institute for Theoretical Physics
University of Leuven, B-3030 Leuven, Belgium

I INTRODUCTION

In this lecture we are interested in the study of the following function :

$$t \in \mathbb{R} \to \omega(A^{*}A(t)) \in C$$

where t stands for the time parameter, A is an observable, as being an element of a C^{*}- algebra or von Neumann algebra; A(t) is the time evolved observable under a time evolution which we will specify later; ω is a state of the system under consideration. In the physics literature $\omega(A^{*}A(t))$ is called the auto correlation function and from an experimental point of view this is an interesting object because it can be measured rather directly. What one finds is that if $\omega = \omega_\beta$, an equilibrium state at inverse temperature $\beta = 1/kT$ for large t :

$$\omega_\beta(A^{*}A(t)) - |\omega_\beta(A)|^2 \simeq e^{-t\,C(\beta,A)} \tag{1}$$

The quantity $\tau(\beta,A) = \frac{1}{C(\beta,A)}$ is called the relaxation time of the observable A at β. It is interesting to note that if the systems show a phase transition, at some critical value β_c, of a certain type, then there might exist an observable A such that

$$\lim_{\beta\to\beta_c} C(\beta,A) = 0$$

or

$$\lim_{\beta\to\beta_c} \tau(\beta,A) = \infty$$

This is the phenomenon of critical slowing down. One has developed many models in order to compute numerically or to obtain by heuristic arguments the behavior of $\tau(\beta,A)$ around T_c (see e.g. [1,2]). Suppose that $\tau(\beta,A) \simeq (T - T_c)^{-\gamma}$ then γ is called a dynamical critical exponent. As far as we know there are not so many rigorous results known about this phenomenon. It is our aim to discuss some results we obtained in collaboration with our coworkers.

We are discussing classical lattice systems as well as quantum systems. As everything is commutative for classical lattice systems, we have a Hamiltonian describing the interactions but we have no natural time evolution. Hence we have to invent one. For quantum systems a Hamiltonian H does describe a natural time evolution by the unitary operator: $\exp itH$. But except for maybe very special cases this evolution does not yield the behavior as described in (1). Again one should invent the right model to investigate some dynamical aspects of the equilibrium states near the critical point. It will turn out from this analysis that any evolution satisfying the detailed balance condition will have the right property to obtain the above described phenomenon.

II CLASSICAL LATTICE SYSTEMS

As usual consider a countable lattice S. At each site $j \in S$ we describe the particles by means of their phase space K, which we suppose to be a compact space. Given Λ a finite subset of S, denote by K_Λ the product $\underset{j\in\Lambda}{\times} K_j$ where K_j is a copy of K attached to the site $j \in \Lambda$. Denote by $C(K_\Lambda)$ the set of real-valued continuous functions on K_Λ and $C(K_S) = \bigcup_\Lambda C(K_\Lambda)$. The set of states of the systems is given by the probability measures ω on K_S and we use the notations

$$\omega(f) = \int_{K_S} f \, d\omega = \langle f \rangle$$

First we define the dynamics. Let Γ be a family $\{\Lambda\}$ of finite subsets of S such that for each Λ there exists a map L_Λ of $C(K_S)$ and

$$L = \sum_\Lambda L_\Lambda \tag{2}$$

satisfying the following conditions :

(i) $L_\Lambda f = 0$ for all $f \in C(K_{S\setminus\Lambda})$; <u>locality</u>

(ii) $\forall\Lambda \in \Gamma$ and $\forall f \in C(K_S)$

$L_\Lambda(f^2) \geqslant 2 f L_\Lambda(f)$; <u>dissipativity</u>

$L_\Lambda(1) = 0$; <u>unity preserving</u>

(iii) $\forall i \in S$: $\sum_{\substack{\Lambda\in\Gamma \\ i\in\Lambda}} \|L_\Lambda\| < \infty$

where $\|L_\Lambda\|$ is the sup-norm on $C(K_S)$

(iv) the <u>condition of detailed balance</u> with respect to a measure ω of K_S :

$\forall f,g \in C(K_S)$, $\forall \Lambda \in \Gamma$:

$$\omega(fL_\Lambda g) = \omega(L_\Lambda(f)g)$$

<u>Example</u> : The generalized Glauber dynamics.

Consider $S = Z^\nu$ and $K = \{0,1,\ldots,n\}$. Take for Q the set of invertible transformations of the configuration space K_S leaving it invariant except in a finite number of points of the lattice S. One could restrict ourself to the set of transformations $\{\tau_i \mid i \in Z^\nu\}$ changing only at the individual lattice points.

Suppose one is given a relative Hamiltonian i.e. a map h from Q into $C(K_S)$ such that

$$h(\tau_1\tau_2) = \tau_1 h(\tau_2) + h(\tau_1)$$

Remark that it is equivalent to give a relative Hamiltonian h or to give a set of local Hamiltonian H_Λ and that their relation is expressed by

$$h(\tau) = \lim_\Lambda(\tau H_\Lambda - H_\Lambda)$$

It is easily checked that ω_β is an equilibrium state at inverse temperature β or a DLR-state at β for $\{H_\Lambda\}_\Lambda$ or h iff :

$$\omega_\beta(\tau^{-1}f) = \omega_\beta(f\, e^{-\beta h(\tau)})$$

for all $\tau \in Q$ and for all $f \in C(K_S)$.

Now a solution for the maps L_Λ (2) satisfying the conditions (i)-(iii) and (iv) with respect to a DLR-state ω_β is given by :

$$L_\tau^\beta = e^{-\beta h(\tau)/2}(\tau - 1) + e^{-\beta h(\tau^{-1})/2}(\tau^{-1} - 1)$$

where the finite volume Λ is equal to the support of $\tau - 1$.

A solution for L (2) constructed by means of these L_τ^β can be called a generalized Glauber dynamics [3].

Remark also that if ω is any measure satisfying

$$\omega(L_\tau^\beta(f)) = 0$$

for all $f \in C(K_S)$ and $\tau \in Q$ then ω is a DLR-state ω_β [4]. In this sense the solutions of the detailed balance equation characterize the DLR-states.

Let us now come back to the general problem of the map L as defined in (2).
Let

$$\mathcal{H}_\omega = L^2(K_S, \omega)$$

with the scalar product

$$<f,g> = \omega(\overline{f}\, g) \ .$$

The operator L induces on $\mathcal{H}_\omega$ a symmetric, negative definite operator, hence there exists a unique self-adjoint, negative definite operator which we denote by the same symbol L and which is expontiable yielding

$$\{\gamma_t = \exp tL \mid t \in \mathbb{R}^+\}$$

a well defined strongly continuous one parameter semigroup of self-adjoint contractions on $\mathcal{H}_\omega$ leaving invariant the unit function [5].
In this scheme we are interested in the following auto correlation function :

$$f_\psi(t) = \frac{<\psi, \gamma_t \psi>}{<\psi,\psi>} \ , \ \psi \in \mathcal{H}_\omega \ , \ <\psi,1> = 0$$

Consider the spectral resolution $\{E(\lambda) \mid \lambda t \ \mathbb{R}^+\}$ of $-L$ then

$$f_\psi(t) = \int_0^\infty e^{-\lambda t}\, d\, m_\psi(\lambda)$$

with

$$dm_\psi(\lambda) = \frac{\langle \psi, dE(\lambda)\psi \rangle}{\langle \psi,\psi \rangle}$$

Suppose there exists a finite gap $\ell_0(\psi,\omega)$ in the spectrum of $-L$, then

$$f_\psi(t) = \int_{\ell_0}^\infty e^{-\lambda t}\, dm_\psi(\lambda) \leqslant e^{-\ell_0(\psi,\omega)t} \qquad (3)$$

and one has exponential decay.
Our main result is the derivation of an upper bound for the energy gap $\ell_0(\psi,\omega)$ in terms of the fluctuation of ψ for the particular observables ψ.
Take the observable ϕ defined by the function :

$$\phi : X \in S \to \phi(X) \in C(K_X)$$

Denote for any finite set Δ :

$$\phi_\Delta = \sum_{X \subset \Delta} \phi(X)$$

and suppose that ϕ satisfies the condition :

$$Q = \sup_j \sum_{k \in S} \sum_{\substack{X \subset S \\ j,k \in X}} \| \phi(X) \| < \infty \tag{4}$$

<u>Lemma II.1</u>

Let ω be a state of $C(K_S)$ and L a detailed balance generator as in (2) satisfying also the condition

$$M = \sup_{j \in S} \sum_{k \in S} \sum_{\substack{\Lambda \in \Gamma \\ j,k \in \Lambda}} \| L_\Lambda \| < \infty \tag{5}$$

then for each observable ϕ, one has

$$0 \leqslant - \langle \phi_\Delta , L \phi_\Delta \rangle \leqslant M \, Q^2 \, |\Delta|$$

where $|\Delta| = \# \Delta$.

Proof : The proof consists in a straightforward computation of a upper bound of

$$|\langle \phi_\Delta , L \, \phi_\Delta \rangle|$$

This expression does contain a triple summation over the volumes Λ of Γ. Essentially this should be brought back to a single one. We limit ourself here to a conceptual explanation of the proof, the technical one can be found in [6]. One summation is killed due to the locality property of L acting on ϕ_Δ, so that L ϕ_Δ reduces to a single sum. Then one applies the detailed balance condition

$$\langle \phi_\Delta , L \, \phi_\Delta \rangle = \langle L \, \phi_\Delta , \phi_\Delta \rangle$$

and uses again the locality of L to eliminate an other summation. This means that $\langle \phi_\Delta , L \, \phi_\Delta \rangle$ must be of the order $|\Delta|$, the volume of Δ. The rest of the proof consists of tedious but straightforward majoration of the remaining terms.

∎

This lemma yields the technical tool to derive the following theorem.

Theorem II.2

Let ω be a measure on K_S and γ_t a detailed balance evolution with respect to ω, then for any finite subset Δ of S and ϕ as before in the Lemma (conditions (4) and (5)) :

$$\frac{<\phi_\Delta,\gamma_t\phi_\Delta> - <\phi_\Delta>^2}{<(\phi_\Delta - <\phi_\Delta>)^2>} \geq \exp - \frac{MQ^2|\Delta|t}{<(\phi_\Delta - <\phi_\Delta>)^2>}$$

Proof : In formula (3) take $\psi = \phi_\Delta - <\phi_\Delta>$, using the convexity of the exponential function one gets :

$$f_{\phi_\Delta - <\phi_\Delta>}(t) \geq \exp \frac{<\phi_\Delta, L\ \phi_\Delta>t}{<(\phi_\Delta - <\phi_\Delta>)^2>}$$

Using the Lemma, one gets the proof of the theorem.

■

Corollary II.3

With the notations and conditions of above one gets

$$\underline{\lim}_{\Delta} \frac{<\phi_\Delta - <\phi_\Delta>\ ,\ \gamma_t(\phi_\Delta - <\phi_\Delta>)}{<(\phi_\Delta - <\phi_\Delta>)^2>} \geq \exp - \frac{MQ^2t}{\chi_{\phi,\omega}}$$

where $\chi_{\phi,\omega} = \overline{\lim}_{\Delta} \frac{<(\phi_\Delta - <\phi_\Delta>)^2>}{|\Delta|}$ is the fluctuation of ϕ in ω.

Proof : Immediate from the theorem.

Corollary II.4

If the state is such that the operator L has a finite energy gap then

$$\ell_0(\phi,\omega) \leq \frac{MQ^2}{\chi_{\phi,\omega}}$$

Proof : Follows from a combination of the inequality (3) and corollary II.3.

■

Application

Suppose that $\omega = \omega_\beta$ is an equilibrium state or DLR state at inverse temperature β for some relative Hamiltonian. Then clearly the fluctuation of the observable ϕ in the state ω_β is temperature dependent. Denote $\chi_{\phi,\omega} = \chi(\phi,T)$, suppose the system shows abnormal fluctuations at some critical temperature T_c i.e. $\lim_{T\to T_c} \chi(\phi,T) = \infty$, then it follows from Corollary II.4 that $\lim_{T\to T_c} \ell_0(\phi,T) = 0$ or the energy gap tends to zero. This is an exact result about the phenomenon on critical slowing down. An example of this situation is given by the three-dimensional Ising model. There one can choose :

$$L = \sum_{i\in\mathbb{Z}^3} e^{-\beta h(\tau_i)} (\tau_i - 1)$$

$$\phi_\Delta = \sum_{i\in\Delta} \sigma_i$$

where σ_i is the Ising spin, and τ_i is the spin-flip at the $i^{\underline{th}}$ site; $h(\tau_i)$ is the relative Hamiltonian of the Ising model

$$h(\tau_i) = \sum_{\substack{j\in\mathbb{Z}^3\\(i,j)}} 2J\, \sigma_i\, \sigma_j$$

$\sum_{\substack{j\in\mathbb{Z}^3\\(i,j)}}$ is the sum over all j neighbours of the site i.

III QUANTUM SYSTEMS

The basic theorem of the former section should also hold for quantum systems. In [7] we solved the detailed balance condition for quantum systems. However in the proof of the theorem the locality property of the generator was crucial. If we concentrate on the state ω being an equilibrium state which is time invariant for the natural time evolution, the generator must also be time-invariant. However this property destroys the locality property. We have not solved yet the locality problem, therefore we have not the quantum version of the theorem. What we do have are rigorous results about models.

We studied the free Boson gas [8] and found an observable such that $\ell_0 \simeq (T - T_c)^2$. We studied also the Dicke-Maser model, defined by the Hamiltonian :

$$H_N = \sum_{k=-N}^{N} (a_k^\star a_k + \varepsilon \sigma_k^+ \sigma_k^-) + \frac{\lambda}{2N+1} \sum_{k,\ell=-N}^{N} (a_k^\star \sigma_\ell^- + h.c.) \tag{6}$$

where $\sigma_i^\pm$ are the Pauli-matrices; $a_k^{(\star)}$ Boson creation and annihilation operts with $0 < \varepsilon < \lambda^2$. This model has a phase transition occuring at the critical temperature T_c :

$$\beta_c = \frac{1}{kT_c} = (\frac{2}{\varepsilon})\ ch^{-2} \frac{\varepsilon}{\lambda^2}$$

As generator of the semigroup we take the following operator defined in the representation of an equilibrium state ω_β with $T < T_c$:

$$L = \int_{-\infty}^{\infty} dt\ F(t)\ M_s\ \{x_t[\cdot, x_{t+S}] + [x_S, \cdot]x_{t+S}\}$$

where $F(t) = \exp\{-t^2 + \beta t/2\}$

$$x = \alpha\ a^\star(\psi) + \alpha^\star a(\psi)$$

$$\psi \in \ell^2(Z)$$

$$\alpha = \text{Weak}^\star - \lim_N \frac{1}{2N} \sum_{i=-N} a_k$$

$$M_s f(S) = \text{Weak} - \lim_R \frac{1}{2R} \int_{-R}^{R} dS\ f(S)$$

$$x_t = \text{Weak} - \lim_N\ e^{itH_N}\ x\ e^{-itH_N}$$

The operators $\alpha^{(\star)}$ are called the creation and annihilation operators of the condensed mode. One proves :

Theorem [9]

The spectrum of the map L is discrete and given by the set $\{-\varepsilon_\psi(\beta)p \mid p \in \mathbb{N}\}$. Moreover $\varepsilon_\psi(\beta) \simeq (T - T_c)^1$ if $T \to T_c$.

Proof : The proof is based on the fact that the natural time evolution $x \to x_t$ is explicitly known and given by :

$$a^{(\star)}(\psi)_t = a^{(\star)}(e^{it\psi})$$

Clearly we can now compute explicitly the action of L on monomials in the creation or annihilation operators. As these monomials are dense in the representation space one gets the full spectrum of L. Explicitly one finds that the spectrum is discrete with equal spacing $\varepsilon_\psi(\beta)$ given by :

$$\varepsilon_\psi(\beta) = \hat{F}(1)\ (1 - e^{-\beta})(\psi,\psi)\ |\ \omega_\beta(\alpha)\ |^2$$

An explicit computation in the equilibrium state ω_β learns that $|\omega_\beta(\alpha)|^2 \simeq T - T_c$ if T tends to T_c, proving the theorem.

■

REFERENCES

[1] L. Van Hove; Phys. Rev. 95, 1374 (1954)
[2] T. Schneider; Phys. Rev. B9, 3819 (1974)
[3] R.J. Glauber; J. Math. Phys. 4, 294 (1963)
[4] A. Verbeure; Commun. Math. Phys. 95, 301 (1984)
[5] M. Reed, B. Simon; Methods of Modern Mathematical Physics Vol. II (Academic Press, New York, 1975)
[6] R. Alicki, M. Fannes, A. Verbeure; J. Stat. Physics 41, 263 (1985)
[7] J. Quaegebeur, G. Stragier, A. Verbeure; Ann. Inst. H. Poincaré 41, 25 (1984)
[8] J. Quaegebeur, A. Verbeure; Lett. Math. Phys. 9, 93 (1985)
[9] D. Goderis, A. Verbeure, P. Vets; Relaxation of the Dicke Maser Model; KUL preprint; jan. '87

QUANTUM MARTINGALES AND STOCHASTIC INTEGRALS

Ivan F. Wilde
Department of Mathematics, King's College
Strand, London WC2R 2LS

Various theories of non-commutative stochastic integration have now been developed and it is of interest to pursue their mathematical structure. Indeed, many of the results valid for, say, the stochastic calculus of Brownian motion have analogues or generalizations within the various non-commutative theories [1,2,3,10]. One such result is the martingale representation theorem. The first non-commutative martingale representation theorem was obtained for the fermion Itô-Clifford theory [2]. The non-Fock (quasi-free) stochastic integration for both fermions and bosons was developed in [3] and martingale representation theorems were proved in [12] and [8,13].

We present here completely elementary proofs of the analogue of Itô's theorem on the representation of Brownian functionals for the Itô-Clifford theory and for the quasi-free fermion theory. As a corollary, we obtain elementary proofs of the martingale representation theorems in these cases. For the quasi-free case our result here is new in that we do not require that the one-particle operator determining the quasi-free representation be bounded away from 0 or 1 as in [12], nor indeed do we even require the state to be strictly non-Fock.

Our method of proof, inspired by [16], is to look at certain elements for which the result is manifestly obvious, and then extend the result by continuity using the isometry relations valid for these theories.

For the Fock theories [1,10], there are no isometry relations known, and, in view of "counterexamples" [11] it is not at all clear what the real situation is. Partial results have been obtained in [9,14].

§1. The CAR algebra

Let $\mathfrak{h}$ be a complex Hilbert space and let $\mathfrak{A}(\mathfrak{h})$ denote the CAR C^*-algebra over $\mathfrak{h}$: thus $\mathfrak{A}(\mathfrak{h})$ is generated by elements $b^*(u)$, $b(u)$, $u \in \mathfrak{h}$, satisfying $b^*(\alpha u+v) = \alpha b^*(u)+b^*(v)$, $b^*(u) = b(u)^*$ and the canonical anticommutation relations $\quad b(u)b(v) + b(v)b(u) = 0$

$$b^*(u)b(v) + b(v)b^*(u) = (v,u)\mathbb{1}$$

for $u,v \in \mathfrak{h}$, $\alpha \in \mathbb{C}$.

In particular, $b(u)^2 = b^*(u)^2 = 0$, $u \in \mathfrak{h}$. We also note that $\|b(u)\| = \|b^*(u)\| = \|u\|$, for all $u \in \mathfrak{h}$.

We shall be concerned with (the Clifford subalgebra of) $\mathfrak{A}(\mathfrak{h})$ in the Fock representation and certain quasi-free representations of $\mathfrak{A}(\mathfrak{h})$ for the case $\mathfrak{h} = L^2(\mathbb{R}^+)$.

§2. Itô-Clifford Theory [2]

Let $\mathfrak{h} = L^2(\mathbb{R}^+)$ and let $b_0^*(u)$, $b_0(u)$, $u \in L^2(\mathbb{R}^+)$, denote the creation and annihilation operators in the Fock representation. We define $\Psi(u) = b_0(\bar{u}) + b_0^*(u)$, for $u \in L^2(\mathbb{R}^+)$. Then one sees that $\Psi(u)^* = \Psi(\bar{u})$ and

$$\Psi(u)\Psi(v) + \Psi(v)\Psi(u) = 2(\bar{u},v)\mathbb{1}.$$

In particular, $\Psi(u)^2 = (\bar{u},u)\mathbb{1}$, $u \in L^2(\mathbb{R}^+)$.

For $t > 0$, set $\Psi_t = \Psi(\chi_{[0,t]})$, and define $\mathcal{C}_t = \{\Psi(u): \text{supp } u \subset [0,t]\}''$, and $\mathcal{C} = \{\Psi(u): u \in L^2(\mathbb{R}^+)\}''$. $\{\Psi_t: t > 0\}$ is the Clifford process. Using the linearity and continuity of the map $u \mapsto \Psi(u)$, one can verify that $\mathcal{C}_t = \{\Psi_s: 0 < s < t\}''$. $\{\mathcal{C}_t: t > 0\}$ is our filtration of von Neumann algebras, and is the non-commutative analogue of a filtration of σ-algebras. Indeed, if we were considering bosons rather than fermions, $\{\Psi_t: t > 0\}$ would be isomorphic to Brownian motion and the corresponding filtration of von Neumann algebras would be precisely $\{L^\infty(\Sigma_t): t > 0\}$ where Σ_t is the σ-algebra generated by the Brownian motion up to time t.

Let Ω_0 denote the Fock vacuum vector. Then the map $x \mapsto (\Omega_0, x\Omega_0)$ defines a faithful central state on $\mathcal{C}$.

Define $\|x\|_2^2 = \|x\Omega_0\|^2$ for $x \in \mathcal{C}$. Then $\|\cdot\|_2$ is a norm on $\mathcal{C}$ and $L^2(\mathcal{C})$ is defined to be the completion of $\mathcal{C}$ with respect to this norm. Similarly, one defines $L^2(\mathcal{C}_t)$ to be the corresponding completion of $\mathcal{C}_t$, $t > 0$. Thus $L^2(\mathcal{C}_t)$ is a closed subspace of the Hilbert space $L^2(\mathcal{C})$. (One can show that the map $x \mapsto x\Omega_0$ extends to a unitary map between $L^2(\mathcal{C})$ and the Fock space.)

The orthogonal projection: $L^2(\mathcal{C}) \to L^2(\mathcal{C}_t)$ is the conditional expectation $\mathbb{E}_t$. (Under the above isomorphism, $\mathbb{E}_t$ is equivalent to the quantization of the operator of multiplication by $\chi_{[0,t]}$.)

An $L^2(\mathcal{C})$-valued process $\{X_t: t > 0\}$ is said to be a martingale if $\mathbb{E}_s X_t = X_s$ for all $0 < s < t$. One can see that $\mathbb{E}_s \Psi_t = \Psi_s$,

$0 < s < t$, and so $\{\Psi_t : t > 0\}$ is a martingale.

Let I be a finite interval in $\mathbb{R}^+$ and let $x \in \mathcal{C}_{\inf I}$. Then $x\chi_I$ is called an elementary adapted ($\mathcal{C}$-valued) process. Given such a process, we define its Itô-Clifford stochastic integral as

$$\int_0^\infty x\chi_I(s)\,d\Psi_s = x\Psi(I)$$

where $\Psi(I)$ denotes $\Psi(\chi_I)$.

A simple ($\mathcal{C}$-valued) adapted process is a finite linear combination of elementary processes. The stochastic integral of a simple process is then defined by linearity.

Theorem. (Isometry property) [2]

Let f be a simple adapted process. Then

$$\|\int_0^\infty f(s)\,d\Psi_s\|_2^2 = \int_0^\infty \|f(s)\|_2^2\,ds.$$

The proof is a straightforward computation using the martingale property of Ψ_s.

This theorem allows us to extend the definition of the Itô-Clifford stochastic integral to any element belonging to the completion, $\mathcal{K}$, of the space of simple adapted processes with respect to the norm $f \mapsto (\int_0^\infty \|f(s)\|_2^2 ds)^{\frac{1}{2}}$; the isometry property still being valid (as in the usual theory of the Itô-integral).

This completion can be identified with the adapted elements of $L^2(\mathbb{R}^+, ds; L^2(\mathcal{C}))$. ($f: \mathbb{R}^+ \to L^2(\mathcal{C})$ is adapted if $f(s) \in L^2(\mathcal{C}_s)$ a.e.)

If $t > 0$, the integral $\int_0^t f(s)\,d\Psi_s$ is defined to be $\int_0^\infty \chi_{[0,t]}(s) f(s)\,d\Psi_s$.

We note that $(\mathbf{1}, \int_0^\infty f d\Psi) = 0$, i.e. $\mathbb{E}_0(\int_0^\infty f d\Psi) = 0$. For further details and proofs of the above results, we refer to [2].

Theorem [2]

Let $f \in L^2_{loc}(\mathbb{R}^+, ds; L^2(\mathcal{C}))$ be adapted. Then $\{\int_0^t f(s)\,d\Psi_s;\ t > 0\}$ is a centred martingale.

We wish to obtain the converse to this result.

Theorem. Let $X \in L^2(\mathcal{C})$. Then there exist unique $\alpha \in \mathbb{C}$ and $f \in \mathcal{K}$ such that

$$X = \alpha\mathbf{1} + \int_0^\infty f\,d\Psi.$$

Proof. The idea is to look at a suitable total set in $L^2(\mathcal{C})$ for which the result is obvious and then argue by continuity.

Now, $\mathcal{C}$ is generated by linear combinations of products $\Psi(u_1)\ldots\Psi(u_n)$, $u_i \in L^2(\mathbb{R}^+)$. But each u_i can be approximated by a step-function, so by the linearity of $\Psi(.)$, we see that $\mathcal{C}$ is generated by linear combinations of products of the form $\Psi(I_1)\ldots\Psi(I_n)$. Furthermore, by taking refinements and further linear combinations, if necessary, we may assume that the I_i's are pairwise distinct or equal; i.e. $I_i = I_j$ or $I_i \cap I_j = \emptyset$. But using the anticommutation relations for Ψ and $\Psi(I)^2 = \int_I ds$, we need only consider $\mathbb{1}$ or $\Psi(I_1)\ldots\Psi(I_k)$ with the I_j's disjoint. In other words, we have deduced that $\mathbb{1}$ and products of the form $\Psi(I_1)\ldots\Psi(I_k)$ with the I_j's disjoint form a total set $\mathcal{L}$, say, in $\mathcal{C}$ and hence in $L^2(\mathcal{C})$.

Again, using the anticommmutation relations, we may suppose that I_k is later than all the other intervals $I_1,\ldots,I_{k-1}$. But then

$$\Psi(I_1)\ldots\Psi(I_k) = \int_0^\infty z\chi_{I_k}(s)d\Psi_s$$

where $z = \Psi(I_1)\ldots\Psi(I_{k-1}) \in \mathcal{C}_{\inf I_k}$; i.e. $\Psi(I_1)\ldots\Psi(I_k)$ is a stochastic integral.

Now let $X \in L^2(\mathcal{C})$. Then there is a sequence $X_n \in \operatorname{span}\mathcal{L}$ such that $X_n \to X$ in $L^2(\mathcal{C})$ and each X_n can be written as

$$X_n = \alpha_n\mathbb{1} + \int_0^\infty f_n(s)d\Psi_s .$$

We have $(\Omega_0, X_n\Omega_0) = \alpha_n \to (\mathbb{1}, X) \equiv \alpha$ and so $X_n - \alpha_n\mathbb{1} \to X - \alpha\mathbb{1}$ in $L^2(\mathcal{C})$. But then $(\int_0^\infty f_n(s)d\Psi_s)$ is a Cauchy sequence in $L^2(\mathcal{C})$. By the isometry property, we deduce that (f_n) is a Cauchy sequence in $\mathcal{K}$ and so there is $f \in \mathcal{K}$ such that $X - \alpha\mathbb{1} = \int_0^\infty f(s)d\Psi_s$.

The uniqueness of α is clear, and that of f follows again by the isometry property.

QED.

<u>Corollary</u>. Let $\{X_t: t \geq 0\}$ be an $L^2(\mathcal{C})$-valued martingale. Then there is a unique adapted f in $L^2_{loc}(\mathbb{R}^+, ds; L^2(\mathcal{C}))$ such that

$$X_t = X_0 + \int_0^t f(s)d\Psi_s, \qquad t \geq 0.$$

<u>Proof</u>. Fix $t \geq 0$. Then, by the theorem,

$$X_t = \alpha\mathbb{1} + \int_0^\infty f_t(s)d\Psi_s,$$

and $\alpha\mathbb{1} = \mathbb{E}_0 X_t = X_0$.

Then $X_t = \mathbb{E}_t X_t = X_0 + \int_0^t f_t(\tau)d\Psi_\tau$ and, for $0 \leq s \leq t$,

$$X_s = \mathbb{E}_s X_t = X_0 + \int_0^s f_t(\tau)d\Psi_\tau .$$

But, again by the theorem,

$$X_s = X_0 + \int_0^s f_s(\tau)d\Psi_\tau.$$

By the isometry property, $f_s(\tau) = f_t(\tau)$ on $[0,s]$ a.e., i.e. there is a unique adapted f in $L^2_{loc}(\mathbb{R}^+, ds; L^2(\mathcal{E}))$ such that

$$X_t = X_0 + \int_0^t f(\tau)d\Psi_\tau, \quad t \geq 0$$

as required.

QED.

§3. Quasi-free fermion processes [3]

We consider the gauge-invariant quasi-free state ω on the CAR C^*-algebra $\mathfrak{A}(\mathfrak{h})$, $\mathfrak{h} = L^2(\mathbb{R}^+)$, given by

$$\omega(b^*(u)b(v)) = \int_0^\infty u(s)\overline{v(s)}\rho(s)ds$$

where $0 \leq \rho(s) \leq 1$ a.e.

Let $\mathfrak{H}$ be the GNS representation Hilbert space associated with ω on $\mathfrak{A}$ and let Ω be the associated cyclic vector in $\mathfrak{H}$. Let b^* and b denote the creation and annihilation operators in this quasi-free representation: i.e. we shall identify $\mathfrak{A}$ with its representation on $\mathfrak{H}$. (This is permissible since any representation of $\mathfrak{A}(\mathfrak{h})$ is faithful.)

Let $\mathfrak{A}_t$ be the C^*-algebra generated by the $b^*(u)$ with $\operatorname{supp} u \subset [0,t]$, and denote by $\mathfrak{H}_t$ the closed subspace of $\mathfrak{H}$ generated by $\mathfrak{A}_t\Omega$, $t \geq 0$. $\{\mathfrak{H}_t: t \geq 0\}$ is a filtration of Hilbert spaces.

Set $b_t^* = b^*(\chi_{[0,t]})$, $b_t = b(\chi_{[0,t]})$, $t \geq 0$. $\{b_t^*: t \geq 0\}$ is the creation process, $\{b_t: t \geq 0\}$ is the annihilation process in the representation given by ω.

For any finite interval $I \subset \mathbb{R}^+$ and $x \in \mathfrak{A}_{\inf I}$ we define the stochastic integrals

$$\int_0^\infty db_t^* x\chi_I(t)\Omega = b^*(I)x\Omega \in \mathfrak{H}$$

and $$\int_0^\infty db_t x\chi_I(t)\Omega = b(I)x\Omega \in \mathfrak{H}$$

where $b^*(I) = b^*(\chi_I) = b(I)^*$.

The stochatic integrals for $\mathfrak{A}\Omega$-valued simple adapted integrals are defined by linearity.

<u>Theorem</u>. (Isometry property) [3]

For any simple $\mathfrak{A}\Omega$-valued adapted processes f,g we have

$$\|\int_0^\infty db_t^* f(t)\|^2 = \int_0^\infty (1-\rho(t))\|f(t)\|^2 dt$$

and

$$\|\int_0^\infty db_t g(t)\|^2 = \int_0^\infty \rho(t)\|g(t)\|^2 dt.$$

For the proofs and further details see [3].

As before, these relations allow us to define $\int_0^\infty db_t^* f(t)$ for any $f \in \mathcal{K}_1$, the Hilbert space completion of the $\mathfrak{A}\Omega$-valued adapted maps with respect to the semi-norm $f \mapsto (\int_0^\infty (1-\rho(t))\|f(t)\|^2 dt)^{\frac{1}{2}}$ and to define $\int_0^\infty db_t g(t)$ for any $g \in \mathcal{K}_2$, the corresponding completion with respect to the semi-norm $g \mapsto (\int_0^\infty \rho(t)\|g(t)\|^2 dt)^{\frac{1}{2}}$.

The isometry property persists, by construction, in these extended cases.

We note that, by a limiting argument, one sees that for $f \in \mathcal{K}_1$, $g \in \mathcal{K}_2$, Ω, $\int_0^\infty db^* f$ and $\int_0^\infty dbg$ are pairwise orthogonal vectors in $\mathfrak{H}$. Furthermore, if $P_t: \mathfrak{H} \to \mathfrak{H}_t$ is the orthogonal projection, then $P_t \int_0^\infty db^* f = \int_0^t db^* f$ and $P_t \int_0^\infty dbg = \int_0^t dbg$, $t > 0$.

<u>Definition</u>. A family $\{\xi_t : t > 0\}$ in $\mathfrak{H}$ is a martingale iff $P_s \xi_t = \xi_s$ $\forall 0 < s < t$.

<u>Theorem</u>. Let $\xi \in \mathfrak{H}$. Then there exists a unique $\alpha \in \mathbb{C}$ and unique elements $f \in \mathcal{K}_1$ and $g \in \mathcal{K}_2$ such that

$$\xi = \alpha\Omega + \int_0^\omega db^* f + \int_0^\omega dbg .$$

<u>Proof</u>. The uniqueness follows from the isometry properties and the orthogonality of three terms.

For the existence, we look as before, at an appropriate total set. Indeed, let $\mathcal{V}$ denote the set of vectors of the form Ω, or $b^\#(I_1)\ldots b^\#(I_n)\Omega$, some n. ($b^\#$ denotes b^* or b). Then $\mathcal{V}$ is total in $\mathfrak{H}$. We may assume that the intervals I_i are either equal or disjoint. Furthermore, we may suppose that the latest interval is to the left of all other intervals (-using the CAR). But $b^*(I)^2 = b(I)^2 = 0$, so we have four types of vector in $\mathcal{V}$ to consider. Any element of $\mathcal{V}$ can be expressed as a linear combination of vectors of the form

(i) Ω ;

(ii) $b(I)x\Omega$, $x = b^\#(I_1)\ldots b^\#(I_k)$ some k and I is later than each I_i, $1 < i < k$;

(iii) $b^*(I')y\Omega$, $y = b^\#(I_1)\ldots b^\#(I_m)$ some m and I' is later than each I_i, $1 < i < m$;

(iv) $b^*(I'')b(I'')z\Omega$, $z = b^\#(I_1)\ldots b^\#(I_j)$ some j and I'' is later than each I_i, $1 < i < j$.

Now, $b(I)x\Omega = \int_0^\infty db_s x\chi_I(s)\Omega$, and

$$b^*(I')y\Omega = \int_0^\infty db_s^* y\chi_{I'}(s)\Omega$$

and so vectors of the form (ii) or (iii) are given as stochastic integrals.

To deal with case (iv), we subdivide the interval I" into n subintervals $J_1,\dots,J_n$ of equal length. Then

$$\begin{aligned} b^*(I'')b(I'')z &= \sum_{i,j} b^*(J_i)b(J_j)z \\ &= \Big(\sum_{i<j} + \sum_{i>j} + \sum_{i=j}\Big) b^*(J_i)b(J_j)z \\ &= -\sum_j b(J_j)b^*(J_j')z + \sum_i b^*(J_i)b(J_i')z \\ &\quad + \sum_i b^*(J_i)b(J_i)z \end{aligned}$$

where $J_j' = \bigcup_{k=1}^{j-1} J_k$, $J_1' = \emptyset$.

We have $\sum_j b(J_j)b^*(J_j')z\Omega \to \int_{I''} db_s b^*(s)z\Omega$ and $\sum_i b^*(J_i)b(J_i')z\Omega \to \int_{I''} db_s^* b(s)z\Omega$ in $\mathcal{H}$ as $n \to \infty$. Furthermore, using the definition of ω, it is a straightforward matter to verify that if we denote $\sum_i b^*(J_i)b(J_i) - \int_{I''} \rho(s)ds$ by δ_n, then we obtain

$$\omega(\delta_n^*\delta_n) = \sum_i \int_{J_i} \rho(s)ds \int_{J_i} (1-\rho(s))ds < \frac{|I''|^2}{n} .$$

Hence, using $\delta_n z = z\delta_n$ (by the CAR), we get

$$\begin{aligned} \omega(z^*\delta_n^*\delta_n z) &= \omega(\delta_n^* z^* z \delta_n) \\ &< \|z\|^2 \omega(\delta_n^*\delta_n) \\ &\to 0 \quad \text{as} \quad n \to \infty. \end{aligned}$$

Thus, we may write $b^*(I'')b(I'')z\Omega$ as

$$\int_0^\infty \rho(s)\chi_{I''}(s)dsz\Omega + \int_0^\infty db_s^* b(s)\chi_{I''}(s)z\Omega - \int_0^\infty db_s b^*(s)\chi_{I''}(s)z\Omega .$$

(See also [12].)

Now, $z\Omega \in \mathcal{V}$ and z has 'degree' 2 lower than that of $b^*(I'')b(I'')z$ and so by repeating the whole argument sufficiently many times we deduce that any vector in the linear span of $\mathcal{V}$ has the required form.

The result for arbitrary $\xi \in \mathcal{H}$ now follows as before from the density of the span of $\mathcal{V}$ in $\mathcal{H}$ together with the orthogonality and isometry properties.

QED.

Remark. The formula

$$b^*(I)b(I)\Omega = \int_I db^*b\Omega - \int_I dbb^*\Omega + \int_I \rho ds\Omega$$

can be viewed either as a Doob-Meyer decomposition [4,5] or as an Itô product formula [12].

Corollary. Let $\{\xi_t: t \geq 0\}$ be an $\mathfrak{H}$-valued martingale. Then there exist unique $\alpha \in \mathbb{C}$, $f \in \mathcal{K}_{1,loc}$, $g \in \mathcal{K}_{2,loc}$ such that

$$\xi_t = \alpha\Omega + \int_0^t db^*f + \int_0^t dbg, \qquad t \geq 0,$$

where $\mathcal{K}_{1,loc}$ and $\mathcal{K}_{2,loc}$ are the a.e. adapted elements of $L^2_{loc}(\mathbb{R}^+,(1-\rho(s))ds;\mathfrak{H})$ and $L^2_{loc}(\mathbb{R}^+,\rho(s)ds;\mathfrak{H})$, respectively [3].

Proof. This follows from the theorem exactly as for the Itô-Clifford case simply by "projecting down to time t."

QED.

§4. Stochastic integrals as operators

The stochastic integrals have been defined as vectors in some Hilbert space, so it is natural to ask if one can recover operators from these vectors; e.g. for simple integrands it is clear that one can define the various stochastic integrals as operators. For the Itô-Clifford case this poses no problem by virtue of the general theory of non-commutative integration [15,18]; elements of $L^2(\mathcal{C})$ can be canonically identified with (possibly unbounded) closed operators on Fock space. In the quasi-free theory, the situation is not so straightforward. Indeed, if $\rho = 0$, then whilst $b(I) = \int_I db_s$ makes good sense, we have $\int_I db_s\Omega = 0$. $b(I)$ cannot be recovered from 0.

Fortunately, the situation is better when $0 < \rho < 1$ a.e., i.e. ω has no Fock part. In this case, Ω is a cyclic and separating vector for $\mathfrak{A}''$ and one can use the Tomita-Takesaki modular theory to advantage.

Suppose then that $0 < \rho < 1$ a.e. Let S be the conjugation operator: S is the closure of the conjugate linear map S_0: $x\Omega \mapsto x^*\Omega$, $x \in \mathfrak{A}$. (It is usual to allow $x \in \mathfrak{A}''$ to define S_0, but by Kaplansky's density theorem we may restrict to $x \in \mathfrak{A}$.)

Let $\xi \in \mathfrak{H}$, and define L_ξ: $\mathfrak{A}'\Omega \to \mathfrak{H}$ by L_ξ: $y'\Omega \mapsto y'\xi$. Then one sees that L_ξ is (a possibly unbounded operator) affiliated to $\mathfrak{A}''$. If $\xi \in D(S)$, then one can show that L_ξ is closeable. (See [3].) Thus, if ξ is a quasi-free stochastic integral, we will obtain ξ as a closeable operator (namely $\bar{L}_\xi$) whenever we know that $\xi \in D(S)$.

Now, any stochastic integral $\xi = \int db^*f$, say, is the limit, in $\mathfrak{H}$, of a sequence of stochastic integrals $\xi_n = \int db^*f_n$ where the f_n are simple adapted $\mathfrak{A}\Omega$-valued processes; say, $f_n = h_n\Omega$, where h_n is $\mathfrak{A}$-valued. Then $\zeta_n \in D(S_0)$ and $S_0\zeta_n = \int h_n^* db\Omega$

$= \int db\, \beta(h_n^*)\Omega$, where β is the parity operator; using the CAR.

We will have $\zeta \in D(S)$ whenever both ζ_n and $S_0\zeta_n$ converge in $\mathcal{H}$; i.e. whenever ζ_n converges in $\mathcal{H}_{+1} = D(S)$ with the norm $\|.\|_{+1}$ given by $\|\xi\|_{+1}^2 = \|\xi\|^2 + \|S\xi\|^2$.

The isometry property in $\mathcal{H}_{+1}$ is easily seen to be $\|\int db^* h\Omega\|_{+1}^2 = \int(1-\rho)\|h\Omega\|^2 ds + \int\rho\|Sh\Omega\|^2 ds$ for any $\mathfrak{A}\Omega$-valued simple adapted process $h\Omega$.

If one completes the set of $\mathfrak{A}\Omega$-valued processes with respect to the norm determined by the right hand side of this isometry relation, then the resulting stochastic integral vector will, by construction, belong to $\mathcal{H}_{+1} = D(S)$ and thus will define a closeable operator as outlined above. Of course, one treats the stochastic integrals with respect to b similarly.

A similar analysis can be carried out for quasi-free boson stochastic integrals. The details are contained in [3].

§5. Multiparameter theory

We shall just make a few remarks here concerning multiparameter quantum stochastic integration. For concreteness, let us consider only the Clifford process over $L^2(\mathbb{R}_+^2)$.

For $z \in \mathbb{R}_+^2$, let $\mathcal{C}_z$ denote the von Neumann algebra $\{\Psi(u): \text{supp}\, u \subset R_z\}''$ where R_z is the rectangle whose diagonal has end-points $\{0,z\}$ and whose sides are parallel to the coordinates axes.

If R is any rectangle in $\mathbb{R}_+^2$ with sides parallel to the axes and whose lower left-hand corner has coordinates $z_0 = (x_0,y_0)$, and if $a \in \mathcal{C}_{z_0}$, we define the stochastic integral of $a\chi_R$ to be

$$\iint_{R_{z'}} a\chi_R(z)d\Psi_z = a\Psi(R \cap R_{z'}).$$

This is extended by linearity and one obtains an isometry relation just as for the 2-parameter Brownian motion theory (-Brownian sheet). The resulting integral is a martingale. (One can develop a multiparameter theory of belated integrals extending [6]. The above stochastic integrals are then such belated integrals.)

One can define non-commutative Wong-Zakai stochastic integrals of the second kind [7,16,17] and again one has appropriate isometry properties. These stochastic integrals are martingales and are orthogonal to those described at the beginning of this section. There is also a non-commutative stochastic Fubini's theorem for such integrals.

The details will appear elsewhere.

It is natural to conjecture that there is a martingale representation theorem for 2-parameter quantum martingales, extending the classical theorem of Wong and Zakai [17].

References

1. Applebaum, D.B., Hudson, R.L.: Fermion Itô's formula and stochastic evolutions. Commun. Math. Phys. 96, 473-496 (1984).

2. Barnett, C., Streater, R.F., Wilde, I.F.: The Itô-Clifford integral. J. Funct. Anal. 48, 172-212 (1982).

3. Barnett, C., Streater, R.F., Wilde, I.F.: Quasi-free quantum stochastic integrals for the CAR and CCR. J. Funct. Anal. 52, 19-47 (1983).

4. Barnett, C., Wilde, I.F.: Natural processes and Doob-Meyer decompositions over a probability gage space. J. Funct. Anal. 58, 320-334 (1984).

5. Barnett, C., Wilde, I.F.: The Doob-Meyer decomposition for the square of Ito-Clifford L^2-martingales. In Quantum probability and applications II, Lecture Notes in Mathematics 1136, Eds. Accardi, L., von Waldenfels, W., Springer-Verlag 1985.

6. Barnett, C., Wilde, I.F.: Belated integrals. J. Funct. Anal. 66, 283-307 (1986).

7. Cairoli, R., Walsh, J.B.: Stochastic integrals in the plane. Acta Math. 134, 111-183 (1975).

8. Hudson, R.L., Lindsay, J.M.: A non-commutative martingale representation theorem for non-Fock quantum Brownian motion. J. Funct. Anal. 61, 202-221 (1985).

9. Hudson, R.L., Lindsay, J.M., Parthasarathy, K.R.: Stochastic integral representation of some quantum martingales in Fock space. In Proc. of Symp. on Stoch. D.E. and Appl. (Warwick 1985), Ed. Elworthy, K.D. Pitman Lecture Note Series.

10. Hudson, R.L., Parthasarathy, K.R.: Quantum Itô's formula and stochastic evolutions. Commun. Math. Phys. 93, 301-323 (1984).

11. Journé, J-L., Meyer, P.A.: In Séminaire Probabilité XX, Lecture Notes in Mathematics 1204, Eds. Aźema, J., Yor, M., Springer-Verlag 1986.

12. Lindsay, J.M.: Fermion martingales. Probab. Th. Rel. Fields 71, 307-320 (1986).

13. Lindsay, J.M., Wilde, I.F.: On non-Fock boson stochastic integrals. J. Funct. Anal. 65, 76-82 (1986).

14. Parthasarathy, K.R., Sinha, K.B.: Stochastic integral representation of bounded quantum martingales in Fock spaces. J. Funct. Anal. 67, 126-151 (1986).

15. Segal, I.E.: A non-commutative extension of abstract integration. Ann. of Math. 57, 401-457 (1953).

16. Walsh, J.B.: Martingales with a multidimensional parameter and stochastic integrals in the plane. In Lectures in probability and statistics, Lecture Notes in Mathematics 1215, Eds. del Pino, G., Rebolledo, R., Springer-Verlag 1986.

17. Wong, E., Zakai, M.: Martingales and stochastic integrals for processes with a multi-dimensional parameter. Z. Warsch. verw. Gebiete 29, 109-122 (1974).

18. Yeadon, F.J.: Non-commutative L_p-spaces. Math. Proc. Camb. Phil. Soc. 77, 91-102 (1975).

MIX
Papier aus verantwortungsvollen Quellen
Paper from responsible sources
FSC® C105338

If you have any concerns about our products,
you can contact us on
ProductSafety@springernature.com

In case Publisher is established outside the EU,
the EU authorized representative is:
Springer Nature Customer Service Center GmbH
Europaplatz 3, 69115 Heidelberg, Germany

Printed by Libri Plureos GmbH
in Hamburg, Germany